Hilbert's Tenth Problem

NEW MATH MONOGRAPHS

For information about Cambridge University Press mathematics publications visit
http://www.cambridge.org/mathematics

CAMBRIDGE UNIVERSITY PRESS
Cambridge, New York, Melbourne, Madrid, Cape Town, Singapore, São Paulo

Cambridge University Press
The Edinburgh Building, Cambridge CB2 2RU, UK

Published in the United States of America by Cambridge University Press, New York

www.cambridge.org
Information on this title: www.cambridge.org/9780521833608

First published 2007

Printed in the United Kingdom at the University Press, Cambridge

A catalog record for this publication is available from the British Library

ISBN-13 978-0-521-83360-8 hardback
ISBN-10 0-521-83360-4 hardback

Hilbert's Tenth Problem

Diophantine Classes and Extensions to Global Fields

ALEXANDRA SHLAPENTOKH

Department of Mathematics
East Carolina University

To my thesis adviser Harold N. Shapiro, who taught me not to be scared.

Contents

Acknowledgements

The author would like to thank the following people for helpful suggestions and corrections: Gunther Cornelissen, Martin Davis, Jospeh Flenner, Dino Lorenzini, Laurent Moret-Bailly, and Bjorn Poonen.

The author also acknowledges support from the National Science Foundation (DMS-0354907).

1
Introduction

1.1 In the beginning

The subject of this book dates back to the beginning of the twentieth century. In 1900, at the International Congress of mathematicians, David Hilbert presented a list of problems, which exerted great influence on the development of mathematics in the twentieth century. The tenth problem on the list had to do with solving Diophantine equations. Hilbert was interested in the construction of an algorithm which could determine whether an arbitrary polynomial equation in several variables had solutions in the integers. If we translate Hilbert's question into modern terms, we can say that he wanted a program taking coefficients of a polynomial equation as input and producing a "yes" or "no" answer to the question "Are there integer solutions?" This problem became known as Hilbert's Tenth Problem (HTP).

It took some time to prove that the algorithm requested by Hilbert did not exist. At the end of the sixties, building on the work of Martin Davis, Hilary Putnam, and Julia Robinson, Yuri Matiyasevich proved that Diophantine sets over $\mathbb{Z}$ were the same as recursively enumerable sets and, thus, that Hilbert's Tenth Problem was unsolvable. The original proof and its immediate implications have been described in detail. The reader is referred to, for example, a book by Matiyasevich (see [52] – the original Russian edition – or [53], an English translation), an article by Davis (see [12]) or an article by Davis, Matiyasevich, and Robinson (see [14]). The solution of the original Hilbert's Tenth Problem gave rise to a whole new class of problems, some of which are the subject of this book.

The question posed by Hilbert can of course be asked of any recursive ring. In other words, given a recursive ring R, we can ask whether there exists an algorithm capable of determining when an arbitrary polynomial equation over R has solutions in R. Since the time when the solution of Hilbert's Tenth Problem was obtained, this question has been answered for many rings. In this book we will describe the developments in the subject pertaining to subrings of global fields: number fields and algebraic function fields over finite fields of constants. While there has been significant progress in the subject, many interesting questions are still unanswered. Chief among them are the questions of solvability of the analog of Hilbert's Tenth Problem over $\mathbb{Q}$ and the ring of algebraic integers of an arbitrary number field. Recent results of Poonen brought us "arbitrarily close" to solving the problem for $\mathbb{Q}$ but formidable obstacles still remain.

A question which is closely related to the analogs of Hilbert's Tenth Problem over number fields is the question of the Diophantine definability of $\mathbb{Z}$. As we will see in this book, the Diophantine definability of $\mathbb{Z}$ over a ring of characteristic zero contained in a field which is not algebraically closed implies the unsolvability of Hilbert's Tenth Problem for this ring. In general, questions of Diophantine definability are of independent number-theoretic and model theoretic interest. In particular, the question of the Diophantine definability of $\mathbb{Z}$ over $\mathbb{Q}$ has generated a lot of interest. Barry Mazur has made several conjectures which imply that such a definition does not exist. In this book we will discuss some of these conjectures and their consequences for generalizations of Hilbert's Tenth Problem to other domains.

For various technical reasons, which we will endeavor to make clear in this book, greater progress has been made for the function fields over finite fields of constants. In particular, we do know that the analog of Hilbert's Tenth Problem is unsolvable over all global function fields of positive characteristic over finite fields of constants. The main unanswered questions here have to do with Diophantine definability. In particular, we still do not know whether $\mathcal{S}$-integers have a Diophantine definition over a function field though in some senses we have come "arbitrarily close" to such a definition.

I would also like to address the main motivation in writing this book. What was wanted was a single coherent account of various methods employed so far in generalizing Hilbert's Tenth Problem to domains other than $\mathbb{Z}$ that are contained in global fields. In particular, I wanted to highlight the expected similarities and differences in the way various problems were solved over number fields and function fields of positive characteristic. In my opinion the relative comparison of these two cases brings to light the nature of the difficulty encountered over the number fields: the existence of archimedean valuations.

The material contained in the book will require some familiarity on the part of the reader with number theory and recursion (computability) theory. The required background information is collected with references in Appendix A (recursion theory) and Appendix B (number theory). As a general reference for recursion theory we suggest *Theory of Recursive Functions and Effective*

Computability by H. Rogers, McGraw-Hill, 1967. Unfortunately, there is no single reference for the number-theoretic material used in the book. However, the reader can find most of the necessary material in *Field Arithmetic* by M. Jarden and M. Fried, second edition, Springer Verlag, 2005 (this book also contains material pertaining to recursion theory), *Algebraic Number Fields* by J. Janusz, Academic Press, 1973, *Introduction to Theory of Algebraic Functions of One Variable* by C. Chevalley, Mathematical Surveys, volume 6, AMS, Providence, 1951, and *An Invitation to Arithmetic Geometry*, by D. Lorenzini, Graduate Studies in Mathematics, volume 9, AMS, 1997. Understanding Poonen's results in Chapter 12 will require some familiarity with elliptic curves. For this material the reader can consult *The Arithmetic of Elliptic Curves* by Joseph Silverman, Springer Verlag, 1986.

Before proceeding further we should also settle on the future use of some terms. Given a ring R, we will call the analog of Hilbert's Tenth Problem over R the "Diophantine problem of R". The expression "Diophantine (un)solvability of a ring R" will refer to the (un)solvability of the Diophantine problem of R. All the rings in the book will be assumed to be integral domains with the identity. We will also settle on a fixed algebraic closure of $\mathbb{Q}$ contained in the field of complex numbers and assume that all the number fields occurring in the book are subfields of this algebraic closure. Similarly, for each prime p, we will fix an algebraic closure of a rational function field over a p-element field of constants and assume that any global function field of characteristic p occurring in this book is a subfield of this algebraic closure. On occasion we will talk about the compositum of abstract fields. For these cases we will also maintain an implicit assumption throughout the book that all the fields in question are subfields of the same algebraically closed field.

Finally, a few words about the structure of this book and its possible uses as a text for a class. Chapters 1–3 contain the introductory material necessary to familiarize the reader with the terminology and to establish a connection between the algebraic and logical concepts presented in this book. Chapters 4–12 are the technical core of the book: Chapter 4 discusses the definability of order at a prime over global fields; Chapters 5–7 cover Diophantine classes of number fields; Chapters 8–10 go over the analogous material for function fields; Chapter 11 addresses Mazur's conjectures and their relation to the issues of Diophantine definability; Chapter 12 describes Poonen's results on undecidability and Mazur's conjectures for "large" subrings of $\mathbb{Q}$. The ideas described in Chapters 4–10 are essentially number-theoretic in nature, while Chapters 11 and 12 add geometric flavor to the mix. Finally, Chapter 13 briefly surveys some issues related to the problems discussed in the book but not covered by the book.

An experienced reader can probably skip most of Chapters 1–3 except for the definition of Diophantine generation (Definition 2.1.5) and the relation between Diophantine generation and HTP (Section 3.4). The chapters on the definition of order at a prime in number fields and function fields and on Mazur's conjectures are fairly self-contained and can be read independently. Understanding Poonen's results does require knowing the statement of the modified Mazur's conjectures (Section 11.2) and the material on Diophantine models in Section 3.4.

Parts of the book could be used as a text for an undergraduate course. For an algebra course, one could cover the following chapters and sections: Chapters 1–3 and Sections 6.3 and 7.1–7.3. Such a course would thus include some general ideas on Diophantine definability and would discuss in detail HTP over the rings of integers of number fields.

There are several options for a semester-long graduate course which would assume some background in algebraic number theory. One option would be to cover the Diophantine classes of number fields; using Chapters 1–3, Sections 4.1 and 4.2, and Chapters 5–7. Another option would be to cover the analogous material for function fields, using Chapters 1–3, Sections 4.1 and 4.3, and Chapters 8–10. A third possibility would be to cover Mazur's conjectures and Poonen's results, using Chapters 1–3, Sections 11.2 and 11.4, and Chapter 12. Such a course would also require a background in elliptic curves. The appendices should be used as needed for all the course versions.

The key to the whole subject lies in the notions of Diophantine definition and Diophantine sets, which we describe and discuss in the next section.

1.2 Diophantine definitions and Diophantine sets

Definition 1.2.1. Let R be an integral domain. Let m, n be positive integers and let $\mathcal{A} \subset R^n$. Then we will say that $\mathcal{A}$ has a Diophantine definition over R if there exists a polynomial

$$f(y_1, \ldots, y_n, x_1, \ldots, x_m) \in R[y_1, \ldots, y_n, x_1, \ldots, x_m]$$

such that for all $(t_1, \ldots, t_n) \in R^n$,

$$(t_1, \ldots, t_n) \in \mathcal{A} \Leftrightarrow \exists x_1, \ldots, x_m, f(t_1, \ldots, t_n, x_1, \ldots, x_m) = 0.$$

The set $\mathcal{A}$ is called *Diophantine over R*.

We can now state the precise result obtained by Matiyasevich.

Theorem 1.2.2. *The Diophantine sets over $\mathbb{Z}$ coincide with the recursively (computably) enumerable sets.*

The negative answer to Hilbert's problem is an immediate corollary of this theorem since not all the recursively enumerable (r.e.) sets are recursive. (For the definitions of recursive (computable) and recursively enumerable sets and their relationship to each other, see Definitions A.1.2, A.1.3, and A.2.1, Lemma A.2.2, and Proposition A.2.3 in Appendix A.)

Indeed, suppose that we had an algorithm taking the coefficients of a polynomial equation as inputs and determining whether the polynomial equation has a solution. Let $A \subset \mathbb{N}$ be a recursively enumerable but not recursive set. By the theorem above, there would exist a polynomial $f(y, x_1, \ldots, x_m)$ with integer coefficients such that $f(t, x_1, \ldots, x_m) = 0$ has integer solutions if and only if $t \in A$. Given a specific $t \in \mathbb{N}$, we could use t and other coefficients of f as the required input for our algorithm and determine whether $f(t, x_1, \ldots, x_m) = 0$ has solutions $(x_1, \ldots, x_m)$ in $\mathbb{Z}$. But this would also determine whether $t \in A$. Since, by assumption, there is no algorithm to determine membership in A, we must conclude that Hilbert's Tenth Problem is unsolvable.

Having seen how Matiyasevich's theorem implies the unsolvability of Hilbert's Tenth Problem via its characterization of the Diophantine sets, we would like to consider some alternative descriptions of Diophantine sets which will shed some light on the nature of our subject. The definition of Diophantine sets used above naturally identifies these sets as number-theoretic objects. Matiyasevich's theorem tells us that these sets also belong in recursion theory. However, as we will see from the lemma below, one could also consider Diophantine sets as sets definable in the language of rings by positive existential formulas and thus a subject of model theory. Finally, Diophantine sets are also projections of algebraic sets and consequently belong in algebraic geometry. Thus, the reader can imagine that the flavor of the discussion can vary widely depending on how one views Diophantine sets. In this book we display a pronounced bias towards the number-theoretic view of the matter at hand, though we will make some forays into geometry in our discussion of Mazur's conjectures and Poonen's results.

As we have mentioned in the previous section, Diophantine definitions can be used to establish the unsolvability of the Diophantine problem for other rings. Before we can explain in more detail how this is done, we have to make the following observation.

Lemma 1.2.3. *Let R be a ring whose quotient field K is not algebraically closed.* (Here we remind the reader that by assumption all the rings in this book are integral domains.) *Let*

$$\{f_i(x_1, \ldots, x_r), i = 1, \ldots, m\}$$

be a finite collection of polynomials over R. Then there exists a polynomial $H(x_1, \ldots, x_r) \in R[x_1, \ldots, x_r]$ *such that the system*

$$\begin{cases} f_1(x_1, \ldots, x_r) = 0; \\ \vdots \\ f_m(x_1, \ldots, x_r) = 0. \end{cases}$$

has solutions in R if and only if $H(x_1, \ldots, x_r) = 0$ *has solutions in R.*

Proof. It is enough to prove the lemma for the case $m = 2$. Let $h(x)$ be a polynomial with no roots in K. Assume that $h(x) = a_0 + a_1 x + \cdots + a_n x^n$, where $a_0, \ldots, a_n \in R$ and $a_n \neq 0$. Further, note that

$$g(x) = x^n h\left(\frac{1}{x}\right) = a_0 x^n + a_1 x^{n-1} + \cdots + a_n$$

is also a polynomial without roots in K. Indeed, if for some $b \neq 0$, we have that $g(b) = 0$, then $b^n h(1/b) = 0$ and consequently $h(1/b) = 0$. Finally, since $a_n \neq 0$, we know that $g(0) \neq 0$. Next consider

$$H(x_1, \ldots, x_r) = \sum_{i=0}^{n} a_i f_1^{n-i}(x_1, \ldots, x_r) f_2^i(x_1, \ldots, x_r).$$

It is clear that if for some r-tuple $(b_1, \ldots, b_r) \in R^r$

$$f_1(b_1, \ldots, b_r) = f_2(b_1, \ldots, b_n) = 0,$$

then $H(b_1, \ldots, b_r) = 0$. Conversely, suppose for some r-tuple $(b_1, \ldots, b_r) \in R^r$ we have that $H(b_1, \ldots, b_r) = 0$ and $f_1(b_1, \ldots, b_r) \neq 0$. Then

$$h\left(\frac{f_2(b_1, \ldots, b_r)}{f_1(b_1, \ldots, b_r)}\right) = 0.$$

However, if $f_2(b_1, \ldots, b_r) \neq 0$ then

$$g\left(\frac{f_1(b_1, \ldots, b_r)}{f_2(b_1, \ldots, b_r)}\right) = 0.$$

□

We can derive two consequences from this lemma. First, we note that, over fields which are not algebraically closed, having an algorithm for solving an arbitrary single polynomial equation is equivalent to having an algorithm for solving a finite system of polynomial equations. Second, we note that we can allow a Diophantine definition to consist of several polynomial equations without changing the nature of the relation.

We should also note here that for some algebraic geometers the restriction of Diophantine definitions to exactly one polynomial, as opposed to finitely many, might seem unnatural. In our defense we offer two arguments. As demonstrated by the lemma above, this distinction makes no difference for the global fields which are the main subjects of this book, and historically questions related to Hilbert's Tenth Problem have been phrased as questions about a single polynomial.

Equipped with the preceding lemma, we can now establish the following.

Proposition 1.2.4. *Let $R_1 \subset R_2$ be two recursive (i.e. computable) rings. Suppose that the fraction field of R_2 is not algebraically closed. Assume that the Diophantine problem of R_1 is undecidable and that R_1 has a Diophantine definition over R_2. Then the Diophantine problem of R_2 is also undecidable.*

Proof. Let $f(t, x_1, \ldots, x_r)$ be a Diophantine definition of R_1 over R_2. Let $g(t_1, \ldots, t_k)$ be a polynomial over R_1, and consider the following system:

$$\begin{cases} g(t_1, \ldots, t_k) \quad = 0; \\ f(t_1, x_1, \ldots, x_r) = 0; \\ \quad \vdots \\ f(t_k, x_1, \ldots, x_r) = 0. \end{cases} \tag{1.2.1}$$

Clearly the equation $g(t_1, \ldots, t_k) = 0$ will have solutions in R_1 if and only if the system in (1.2.1) above has solutions in R_2. Further, by the preceding lemma, since both rings are recursive, given coefficients of g there is an algorithm to construct a polynomial $T(g)(t_1, \ldots, t_k, x_1, \ldots, x_r) \in R_2[t_1, \ldots, t_k, x_1, \ldots, x_r]$ such that the corresponding polynomial equation $T(g)(t_1, \ldots, t_k, x_1, \ldots, x_r) = 0$ has solutions over R_2 if and only if (1.2.1) has solutions in R_2.

Suppose now that the Diophantine problem of R_2 is decidable. Then for each polynomial g over R_1 we can effectively decide whether $g(t_1, \ldots, t_r) = 0$ has solutions in R_1 by first algorithmically constructing $T(g)$ and then algorithmically determining whether $T(g) = 0$ has solutions in R_2. Thus the Diophantine problem of R_1 is decidable in contradiction of our assumption, and we must conclude that the Diophantine problem of R_2 is not decidable. □

Remark 1.2.5. In this proof we used the notions of "algorithm" and "recursive ring" rather informally. We will formalize this discussion in the chapter on weak presentations.

Almost all the known results (except for Poonen's theorem) concerning the unsolvability of the Diophantine problem of rings of algebraic numbers have been obtained by constructing a Diophantine definition of $\mathbb{Z}$ over these rings.

Before we present details of these and other constructions we would like to enlarge somewhat the context of our discussion by introducing the notions of Diophantine generation, Diophantine equivalence, and Diophantine classes. These concepts will serve several purposes. They will provide a uniform language for the discussion of "Diophantine relations" between rings with the same and different quotient fields. They will allow us to view the existing results within a unified framework. Finally these concepts will point to some natural directions for possible investigation of more general questions of Diophantine definability.

2

Diophantine classes: definitions and basic facts

In this chapter we will introduce the notion of Diophantine generation, which will eventually lead us to the notion of Diophantine classes. We will also obtain the first relatively easy results on Diophantine generation and develop some methods applicable to all global fields: number fields and function fields. Most of the material for this chapter has been derived from [94].

2.1 Diophantine generation

We will start with a first modification of the notion of Diophantine definition.

Definition 2.1.1. Let R be an integral domain with a quotient field F. Let k, m be positive integers and let $A \subset F^k$. Assume further that there exists a polynomial

$$f(a_1, \ldots, a_k, b, x_1, \ldots, x_m)$$

with coefficients in R such that

$$\begin{aligned} &\forall a_1, \ldots, a_k, b, x_1, \ldots, x_m \in R, \\ &f(a_1, \ldots, a_k, b, x_1, \ldots, x_m) = 0 \quad \Rightarrow \quad b \neq 0 \end{aligned} \tag{2.1.1}$$

and

$$\begin{aligned} A = \{(t_1, \ldots, t_k) \in F^k | \exists a_1, \ldots, a_k, b, x_1, \ldots, x_m \in R, \\ bt_1 = a_1, \ldots, bt_k = a_k, \ f(a_1, \ldots, a_k, b, x_1, \ldots, x_m) = 0\}. \end{aligned} \tag{2.1.2}$$

Then we will say that A is *field-Diophantine* over R and will call f a field-Diophantine definition of A over R.

Next we will see that the notion of field-Diophantine definition is a proper extension of the notion of Diophantine definition that we discussed in the introduction.

Lemma 2.1.2. *Suppose that R, A, F, k, m are as in Definition 2.1.1. Assume further that $A \subset R^k$ and that F is not algebraically closed. Then A has a Diophantine definition over R if and only if it has a field-Diophantine definition over R.*

Proof. First we assume that A has a field-Diophantine definition over R and show that A also has a Diophantine definition over R. Let

$$g = (a_1, \ldots, a_k, b, x_1, \ldots, x_m)$$

be a field-Diophantine definition of A over R. Then

$$f(t_1, \ldots, t_k, b, x_1, \ldots, x_m) = g(t_1 b, \ldots, t_k b, b, x_1, \ldots, x_m)$$

is a Diophantine definition of A over R in the sense that, for all $t_1, \ldots, t_k \in R$,

$$\exists b, x_1, \ldots, x_m \in R,\ f(t_1, \ldots, t_k, b, x_1, \ldots, x_m) = 0 \quad \Leftrightarrow \quad (t_1, \ldots, t_k) \in A.$$

Indeed, suppose that, for some $t_1, \ldots, t_k, b, x_1, \ldots, x_m \in R$,

$$f(t_1, \ldots, t_k, b, x_1, \ldots, x_m) = 0.$$

Then

$$g(t_1 b, \ldots, t_k b, b, x_1, \ldots, x_m) = 0$$

and consequently

$$b \neq 0,$$

while

$$(t_1 b/b, \ldots, t_k b/b) = (t_1, \ldots, t_k) \in A.$$

Conversely, suppose that $(t_1, \ldots, t_k) \in A$. Then by our assumption on g,

$$\exists x_1, \ldots, x_m, b \in R, \quad g(bt_1, \ldots, bt_k, b, x_1, \ldots, x_m) = 0.$$

Thus, there exist $x_1, \ldots, x_m, b \in R$ such that

$$f(t_1, \ldots, t_k, b, x_1, \ldots, x_m) = 0.$$

Suppose now that $f(t_1, \ldots, t_k, x_1, \ldots, x_m)$ is a Diophantine definition of A over R. Then consider the following system of equations:

$$\begin{cases} f(a_1, \ldots, a_k, x_1, \ldots, x_m) = 0; \\ \hfill b = 1. \end{cases} \tag{2.1.3}$$

Let $g(a_1, \ldots, a_k, b, x_1, \ldots, x_m)$ be a polynomial over R such that, for all

$$(a_1, \ldots, a_k, b, x_1, \ldots, x_m) \in F,$$
$$g(a_1, \ldots, a_k, b, x_1, \ldots, x_m) = 0 \tag{2.1.4}$$

is equivalent to $(a_1, \ldots, a_k, b, x_1, \ldots, x_m)$ being a solution to the system (2.1.3). Then g is a field-Diophantine definition of A. Indeed, suppose that equation (2.1.4) holds with all the variables taking values in R. Then $b = 1 \neq 0$ and

$$f(a_1, \ldots, a_k, x_1, \ldots, x_m) = 0.$$

This means, of course, that

$$(a_1, \ldots, a_k) = (a_1/b, \ldots, a_k/b) \in A.$$

Now suppose that $(a_1, \ldots, a_k) \in A$. Then there exists $(x_1, \ldots, x_k) \in R$ such that $g(a_1, \ldots, a_k, 1, x_1, \ldots, x_m) = 0$. Thus the lemma is true. □

We will next establish a couple of easy but useful properties of field-Diophantine definitions and sets.

Lemma 2.1.3. *Let R be a ring whose fraction field is not algebraically closed. Then the intersections, unions, and cartesian products of (field)-Diophantine sets of R are (field)-Diophantine over R.*

Proof. First we observe that we can obtain a (field-)Diophantine definition of the union by multiplying the (field)-Diophantine definitions of the constituent sets. Second, we can consider a cartesian product as an intersection. Indeed, let $A \subseteq R^m$, $B \subseteq R^n$ be two (field-)Diophantine sets. Then

$$A \times B = \{(\bar{x}, \bar{y}) \in R^{m+n} : \bar{x} \in A\} \cap \{(\bar{x}, \bar{y}) \in R^{m+n} : \bar{y} \in B\},$$

where both sets in the intersection are clearly (field)-Diophantine, assuming that A and B are. Finally to deal with the intersection we can use the same method as in Lemma 1.2.3. □

Lemma 2.1.4. *Let R be an integral domain with a quotient field F. Let $A \subset F^k$ for some positive integer k. Let m be a positive integer less than or equal to k. Assume that A has a field-Diophantine definition over R. Let*

$$B = \big\{(x_1, \ldots, x_r) \in F^r \mid x_i = P_i(y_1, \ldots, y_m), \\ (y_1, \ldots, y_m, H_{m+1}(y_1, \ldots, y_m), \ldots, H_k(y_1, \ldots, y_m)) \in A\big\},$$

where $P_1, \ldots, P_r, H_{m+1}, \ldots, H_k \in F[y_1, \ldots, y_m]$. Then B also has a field-Diophantine definition over R.

Proof. Let $f(u_1, \ldots, u_k, u, z_1, \ldots, z_s)$ be a field-Diophantine definition of A over R. Then

$$\begin{aligned} B = \big\{(x_1, \ldots, x_r) \in F^r \mid \exists u_1, \ldots, u_k, u, z_1, \ldots, z_s \in R, \\ x_i = P_i(u_1/u, \ldots, u_m/u), i = 1, \ldots, r, \\ uH_j(u_1/u, \ldots, u_m/u) = u_j, j = m+1, \ldots, k, \\ f(u_1, \ldots, u_k, u, z_1, \ldots, z_s) = 0\big\}. \end{aligned}$$

Let d_H be the maximum of the degrees of $H_{m+1}, \ldots, H_k$ and let D_H be a common denominator with respect to R of all the coefficients of $H_{m+1}, \ldots, H_k$. Let

$$\bar{H}_j(u_1, \ldots, u_m, u) = D_H u^{d_H} H_j(u_1/u, \ldots, u_m/u), \qquad j = m+1, \ldots, k.$$

Let d be the maximum of the degrees of $P_1, \ldots, P_r$, let D be a common denominator of the coefficients of $P_1, \ldots, P_r$ with respect to R, and let

$$\begin{aligned} \bar{P}_i(u_1, \ldots, u_m, u) &= u^d D P_i(u_1/u, \ldots, u_m/u) \\ &\in R[u_1, \ldots, u_m, u], \qquad i = 1, \ldots, r. \end{aligned}$$

Then

$$\begin{aligned} B &= \big\{(x_1, \ldots, x_r) \in F^r \mid \exists u_1, \ldots, u_k, u, z_1, \ldots, z_s \in R, \\ &\quad u\bar{H}_j(u_1, \ldots, u_m, u) = D_H u^{d_H} u_j, j = m+1, \ldots, k, \\ &\quad u^d D x_i = \bar{P}_i(u_1, \ldots, u_m, u), i = 1, \ldots, r, \\ &\quad f(u_1, \ldots, u_k, u, z_1, \ldots, z_s) = 0\big\} \\ &= \big\{(x_1, \ldots, x_r) \in F^r \mid \exists U, U_1, \ldots, U_r, u_1, \ldots, u_k, u, z_1, \ldots, z_s \in R, \\ &\quad Ux_i = U_i, U = Du^d, U_i = \bar{P}_i(u_1, \ldots, u_m, u), i = 1, \ldots, r, \\ &\quad \bar{H}_j(u_1, \ldots, u_m, u) = D_H u^{d_H - 1} u_j, j = m+1, \ldots, k, \\ &\quad f(u_1, \ldots, u_k, u, z_1, \ldots, z_s) = 0\big\}. \end{aligned} \tag{2.1.5}$$

Note that (2.1.5) implies that $u \neq 0$ and $U \neq 0$. □

Given a ring R, we can now consider constructing polynomial definitions over R for any subset of the quotient field of R. It will turn out that it is also useful to be able to do this not just for the subsets of the quotient field but also for the subsets of finite extensions of the quotient field. To accomplish this goal we extend the notion of Diophantine definition further. The new extended notion is called *Diophantine generation* and it will allow us to form Diophantine classes.

Definition 2.1.5. Let R_1, R_2 be two rings with quotient fields F_1 and F_2 respectively. Assume that neither F_1 nor F_2 is algebraically closed. Let F be a finite extension of F_1 such that $F_2 \subset F$. Further, assume that for some basis $\{\omega_1, \ldots, \omega_k\}$ of F over F_1 there exists a polynomial $f(a_1, \ldots, a_k, b, x_1, \ldots, x_m)$ with coefficients in R_1 such that

$$f(a_1, \ldots, a_k, b, x_1, \ldots, x_m) = 0 \quad \Rightarrow \quad b \neq 0 \tag{2.1.6}$$

and

$$R_2 = \left\{ \sum_1^k t_i \omega_i \mid \exists a_1, \ldots, a_k, b, x_1, \ldots, x_m \in R_1, \right.$$
$$\left. bt_1 = a_1, \ldots, bt_k = a_k, f(a_1, \ldots, a_k, b, x_1, \ldots, x_m) = 0 \right\}. \tag{2.1.7}$$

Then we will say that R_2 is *Dioph-generated* over R_1 and denote this fact by

$$R_2 \leq_{Dioph} R_1.$$

We will also call $f(a_1, \ldots, a_k, b, x_1, \ldots, x_m)$ a *defining polynomial* of R_2 over R_1, we will call $\Omega = \{\omega_1, \ldots, \omega_k\}$ a *Diophantine basis* of R_2 over R_1 and we will call F the *defining field* for the basis Ω.

Our next project is to answer some natural questions about Diophantine generation.

1. Is the notion of Diophantine generation a proper extension of the notion of Diophantine and field-Diophantine definitions? In other words we want to answer the following question. Suppose that $R_2 \subset F_1$. Then, is saying that $R_2 \leq_{Dioph} R_1$ equivalent to saying that R_1 is field-Diophantine over R_1? (Answer: Yes.)
2. Is Diophantine generation dependent on a particular Diophantine basis? In other words, what happens if we change the basis of the field F over F_1? Will the relationship be preserved? (Answer: Diophantine generation is not dependent on any specific basis of F over F_1. If one basis of F over F_1 is a Diophantine basis of R_2 over R_1 then any basis of F over F_1 is a Diophantine basis of R_2 over R_1.)
3. Is the defining field unique? Can we use a bigger field? Can we use a smaller field? (Answer: Defining fields are not unique. Any field containing F_1F_2 can be a defining field.)
4. Is this relationship transitive? In other words, is the use of the symbol "$\leq_{Dioph}$" justified? (Answer: Yes, the relationship is transitive.)

In the following sequence of lemmas we will tackle the questions of the basis and the field first. We start with an easy lemma which follows directly from the definition of Dioph-generation.

Lemma 2.1.6. *Let R_1, R_2 be integral domains with quotient fields F_1 and F_2 respectively. Let F be a finite extension of F_1 such that $F_2 \subset F$. Then, if there exists a basis $\Omega = \{\omega_1, \ldots, \omega_k\}$ of F over F_1 and a set $A_\Omega \subset F_1^k$ with a field-Diophantine definition over R_1 such that*

$$R_2 = \left\{ \sum_{i=1}^{k} z_i \omega_i | (z_1, \ldots, z_k) \in A_\Omega \right\}, \tag{2.1.8}$$

we can conclude that $R_2 \leq_{Dioph} R_1$. Conversely, if F and Ω are respectively a defining field and a corresponding Diophantine basis of R_2 over R_1 then R_2 has a representation of the form (2.1.8), where $A_\Omega \subset F_1^k$ is field-Diophantine over R_1.

Notation 2.1.7. A_Ω will be called a defining set for the basis Ω.

The next lemma will tell us that if $R_2 \leq_{Dioph} R_1$ then we can always use F_1F_2 as a defining field and any basis of F_1F_2 over F_1 as a Diophantine basis.

Lemma 2.1.8. *Let R_1, R_2 be integral domains with quotient fields F_1 and F_2 respectively. Assume that $R_2 \leq_{Dioph} R_1$.* (Here we should remind the reader that $R_2 \leq_{Dioph} R_1$ implies that F_1F_2 is of finite degree over F_1 and that F_1F_2 is not algebraically closed.) *Let $\Omega = \{\omega_1, \ldots, \omega_k\}$ be a Diophantine basis of R_2 over R_1, F_Ω being the corresponding defining field. Let $\Lambda = \{\lambda_1, \ldots, \lambda_m\}$ with $m \leq k$ be a basis of a field F_Λ over F_1, with $F_\Lambda \subseteq F_\Omega$. Then the ring $R_\Lambda = R_2 \cap F_\Lambda \leq_{Dioph} R_1$, and Λ is a Diophantine basis of R_Λ over R_1.*

Proof. Since $F_\Lambda \subseteq F_\Omega$, we have that for $i = 1, \ldots, m$

$$\lambda_i = \sum_{j=1}^{k} c_{i,j} \omega_j,$$

where, for all i, j, we have that $c_{i,j} \in F_1$. Further, if necessary we can reorder Ω in such a way that the matrix $(c_{i,j})$, $i, j = 1, \ldots, m$, is non-singular. Next note the following:

$$R_\Lambda = \left\{ \sum_{i=1}^{m} z_i \lambda_i \mid z_i \in F_1 \text{ and } \exists (t_1, \ldots, t_k) \in A_\Omega, \sum_{i=1}^{m} z_i \lambda_i = \sum_{j=1}^{k} t_j \omega_j \right\},$$

where A_Ω, the defining set for the basis Ω, is field-Diophantine over R_1. Given our ordering of Ω, by Lemma B.10.5 the equality $\sum_{i=1}^{m} z_i\lambda_i = \sum_{j=1}^{k} t_j\omega_j$ is equivalent to the system

$$z_i = P_i(t_1, \ldots, t_m), \qquad i = 1, \ldots, m,$$
$$t_{m+j} = T_j(t_1, \ldots, t_m), \qquad j = 1, \ldots, k - m,$$

where for each i and each j we have that P_i and T_j are fixed polynomials over F_1 depending on the choice and ordering of Ω and Λ only. Thus

$$R_\Lambda = \left\{ \sum_{i=1}^{m} z_i\lambda_i \mid z_i = P_i(t_1, \ldots, t_m), \right.$$
$$\left. (t_1, \ldots, t_m, T_1(t_1, \ldots, t_m), \ldots, T_{k-m}(t_1, \ldots, t_m)) \in A_\Omega \right\}.$$

Therefore, by Lemma 2.1.4,

$$R_\Lambda = \left\{ \sum_{i=1}^{m} z_i\lambda_i \mid (z_1, \ldots, z_m) \in A_\Lambda \right\},$$

where A_Λ is field-Diophantine over R_1. Hence $R_\Lambda \leq_{Dioph} R_1$ by Lemma 2.1.6. □

We are now able to prove that the defining fields have the desired property.

Corollary 2.1.9. *Let R_1, R_2 be integral domains with fraction fields F_1 and F_2 respectively. Assume that $R_2 \leq_{Dioph} R_1$. Then we can choose F_1F_2 as a defining field and any basis of F_1F_2 over F_1 as a Diophantine basis of R_2 over R_1.*

Proof. Let F be a defining field from the Definition 2.1.5 of Diophantine generation. Then, since $F_1F_2 \subseteq F$, from Lemma 2.1.8 it follows that $R_2 \cap F_1F_2 = R_2 \leq_{Dioph} R_1$, F_1F_2 being a defining field and any basis of F_1F_2 a defining basis of R_2 over R_1. □

Lemma 2.1.8 has another consequence, answering one of the questions posed above.

Corollary 2.1.10. *Let R_1, R_2 be integral domains with the quotient fields F_1 and F_2 respectively. Assume that $F_2 \subseteq F_1$ and $R_2 \leq_{Dioph} R_1$. Then R_2 has a field-Diophantine definition over R_1.*

Proof. By Corollary 2.1.9 we can select $F_1F_2 = F_1$ as our defining field and let the basis of F_1 over F_1 consist of $\{1\}$. By Lemma 2.1.6 we have that R_2 satisfies the equation in (2.1.8) with $k = 1$ and $\omega_1 = 1$. Thus we have

$$R_2 = \{z_1 \cdot 1 \mid z_1 \in A_\Omega\},$$

where A_Ω has a field-Diophantine definition over R_1. Therefore $R_2 = A_\Omega$ and R_2 has a field-Diophantine definition over R_1. □

The next lemma will demonstrate that we can always make a defining field larger. This fact will be important for proving the transitivity of Dioph generation.

Lemma 2.1.11. *Let R_1, R_2 be integral domains with quotient fields F_1, F_2 respectively. Assume that $R_2 \leq_{Dioph} R_1$. Let F be any finite extension of F_1 containing F_1F_2, and let $\Gamma = \{\gamma_1, \ldots, \gamma_m\}$ be any basis of F over F_1. Then F is a defining field and Γ is a Diophantine basis of R_2 over R_1.*

Proof. By Lemma 2.1.8, we know that F_1F_2 is a defining field and that any basis $\Omega = \{\omega_1, \ldots, \omega_k\}$, $k \leq m$, of F_1F_2 over F_1 is a Diophantine basis of R_2 over R_1. Thus, for some polynomial $f(u_1, \ldots, u_k, u, x_1, \ldots, x_r)$ over R_1, we have that

$$R_2 = \left\{ \sum_{i=1}^{k} t_i\omega_i \mid \exists u, u_1, \ldots, u_k, x_1, \ldots, x_r, \right.$$
$$\left. ut_1 = u_1, \ldots, ut_k = u_k, f(u_1, \ldots, u_k, u, x_1, \ldots, x_r) = 0 \right\}$$

and $f(u_1, \ldots, u_k, u, x_1, \ldots, x_r) = 0 \Rightarrow u \neq 0$. Since $F_1F_2 \subset F$, for $i = 1, \ldots, k$

$$\omega_i = \sum_{j=1}^{m} \frac{y_{i,j}}{w}\gamma_j,$$

where for all i, j we have that $y_{i,j}, w \in R_1$, and $w \neq 0$. Thus

$$R_2 = \left\{ \sum_{i=1}^{k} \sum_{j=1}^{m} \frac{y_{i,j}}{w}\gamma_j t_i \mid \exists u, u_1, \ldots, u_k, x_1, \ldots, x_r, \right.$$
$$\left. ut_1 = u_1, \ldots, ut_k = u_k, f(u_1, \ldots, u_k, u, x_1, \ldots, x_r) = 0 \right\},$$

so that

$$R_2 = \left\{ \sum_{j=1}^{m} \left(\sum_{i=1}^{k} \frac{y_{i,j}}{w} t_i \right) \gamma_j \mid \exists u, u_1, \ldots, u_k, x_1, \ldots, x_r, \right.$$
$$\left. ut_1 = u_1, \ldots, ut_k = u_k, f(u_1, \ldots, u_k, u, x_1, \ldots, x_r) = 0 \right\}.$$

Let

$$T_j = \sum_{i=1}^{k} \frac{y_{i,j}}{w} t_i;$$

then

$$wuT_j = \sum_{i=1}^{k} y_{i,j} u_i \in R_1.$$

Let $U_j = \sum_{i=1}^{k} y_{i,j} u_i$ and let $U = wu$. Then

$$R_2 = \left\{ \sum_{j=1}^{m} T_j \gamma_j \mid \exists U_1, \ldots, U_m, U, u_1, \ldots, u_k, u, x_1, \ldots, x_r \in R_1, \right.$$
$$UT_1 = U_1, \ldots, UT_m = U_m, U = wu,$$
$$U_j = \sum_{i=1}^{k} y_{i,j} u_i, \; j = 1, \ldots, m,$$
$$\left. f(u_1, \ldots, u_k, u, x_1, \ldots, x_r) = 0 \right\}.$$

Note that $f(u_1, \ldots, u_k, u, x_1, \ldots, x_r) = 0$ guarantees that $U \neq 0$. □

The next lemma considers a property of field-Diophantine definitions.

Lemma 2.1.12. *Let R be an integral domain with a fraction field F that is not algebraically closed. Let $m, n \in \mathbb{Z}_{>0}$. Let $A, B \subset F^n$, $C \subset F^m$, where B, C are field-Diophantine over R. Suppose that for some $f(\bar{X}, \bar{V}) \in R[\bar{X}, \bar{V}]$, where $\bar{X} = (X_1, \ldots, X_n)$ and $\bar{V} = (V_1, \ldots, V_m)$, we have that $A = \{(a_1, \ldots, a_n) \in B | \exists c_1, \ldots, c_m \in C : f(\bar{a}, \bar{c}) = 0\}$. Then A is field-Diophantine over R.*

Proof. Let $f_B(\bar{X}, Y, \bar{Z}) \in R[\bar{X}, Y, \bar{Z}]$, where $\bar{X} = (X_1, \ldots, X_n)$ and $\bar{Z} = (Z_1, \ldots, Z_r)$, be a field-Diophantine definition of B over R. Let $f_C(\bar{U}, W, \bar{V}) \in R[\bar{U}, W, \bar{V}]$, where $\bar{U} = (U_1, \ldots, U_m)$ and $\bar{V} = (V_1, \ldots, V_l)$, be a field-Diophantine definition of C over R. Next let

$$g(\bar{X}, Y, \bar{U}, W) = (YW)^{\deg(f)} f(X_1/Y, \ldots, X_n/Y, U_1/W, \ldots, U_m/W).$$

Let $D \subset F^n$ be such that

$$D = \big\{\bar{a} \in F^n : \exists \bar{x} \in R^n, \bar{v} \in R^l, w, y \in R, \bar{z} \in R^r, \bar{u} \in R^m : \\ y\bar{a} = \bar{x} \wedge g(\bar{x}, y, \bar{u}, w) = 0 \wedge f_B(\bar{x}, y, \bar{z}) = 0 \wedge f_C(\bar{u}, w, \bar{v}) = 0\big\}.$$

Then $D = A$. Indeed, suppose that $\bar{a} \in D$. Then $f_B(\bar{x}, y, \bar{z}) = 0 \wedge f_C(\bar{u}, w, \bar{v}) = 0$ implies that $y \neq 0$ and $w \neq 0$. Thus $g(\bar{x}, y, \bar{u}, w) = 0$ implies that

$$f(x_1/y, \ldots, x_n/y, u_1/w, \ldots, u_k/w) = 0.$$

Since $f_B(\bar{x}, y, \bar{z}) = 0 \wedge f_C(\bar{u}, w, \bar{v}) = 0$, we also have that $\bar{a} = \bar{x}/y \in B, \bar{c} = \bar{u}/w \in C$, and $f(\bar{a}, \bar{c}) = 0$. Thus $\bar{a} \in A$.

Suppose now that $\bar{a} \in A$. Then for some $\bar{c} \in C$ we have that $f(\bar{a}, \bar{c}) = 0$. Since $A \subset B$, for some $\bar{x} \in R^n$, $y \in R \setminus \{0\}$, and $\bar{z} \in R^r$ we have that $y\bar{a} = \bar{x}$ and $f_B(\bar{x}, y, \bar{z}) = 0$. Similarly, since $\bar{c} \in C$, for some $\bar{u} \in R^m$, $w \in R \setminus \{0\}$, and $\bar{v} \in R^l$ we have that $w\bar{c} = \bar{u}$ and $f_C(\bar{u}, w, \bar{v}) = 0$. Finally, since $f(\bar{a}, \bar{c}) = 0$, $w\bar{c} = \bar{u}$, and $y\bar{a} = \bar{x}$, $y \neq 0$, $w \neq 0$, we have that $g(\bar{x}, y, \bar{u}, w) = 0$. Thus $\bar{a} \in D$.

Since D is clearly field-Diophantine over F, the lemma holds. □

Our next step is to introduce a new notation designed to simplify discussion of the transitivity of Dioph-generation.

Notation 2.1.13. Let G/F be a finite field extension. Let $\Omega = \{\omega_1, \ldots, \omega_k\}$ be a basis of G over F. Let $B \subset G^n$, for some positive integer n. Then define $B^\Omega \subset F^{kn}$ to be the set such that

$$(a_{1,1}, \ldots, a_{k,n}) \in B^\Omega \quad \Leftrightarrow \quad \left(\sum_{i=1}^{k} a_{i,1}\omega_i, \ldots, \sum_{i=1}^{k} a_{i,n}\omega_n\right) \in B$$

Using this notational scheme for rings R_1, R_2 such that $R_2 \leq_{Dioph} R_1$ with a Diophantine basis Ω as above, we can now conclude by Lemma 2.1.6 that $R_2^\Omega \subset F_1^n$ is field Diophantine over R_1, where F_1 is the fraction field of R_1.

The following proposition is a generalization of Lemma 2.1.6.

Proposition 2.1.14. *Let R_1, R_2 be integral domains with quotient fields F_1 and F_2 respectively, such that $R_2 \leq_{Dioph} R_1$. Let F be a defining field and let $\Omega = \{\omega_1, \ldots, \omega_k\}$ be a Diophantine basis of R_2 over R_1. Let $B \subset F_2^n$ have a* *field-Diophantine definition over R_2. Then B^{Ω} has a field-Diophantine definition over R_1.*

Proof. Let $f(a_1, \ldots, a_n, a, x_1, \ldots, x_r)$ be a field-Diophantine definition of B over R_2. Then

$$B^{\Omega} = \left\{ (c_{1,1}, \ldots, c_{k,n}) \in \left(R_2^{\Omega}\right)^n \,\middle|\, \left(\sum_{i=1}^{k} c_{i,1}\omega_i, \ldots, \sum_{i=1}^{k} c_{i,n}\omega_i\right) \in B \right\}$$

$$= \left\{ (c_{1,1}, \ldots, c_{k,n}) \in \left(R_2^{\Omega}\right)^n \,\middle|\, \exists Y, Z_1, \ldots, Z_n, X_1, \ldots, X_r \in R_2, \right.$$

$$Y \sum_{i=1}^{k} c_{i,1}\omega_i = Z_1, \quad \ldots, \quad Y \sum_{i=1}^{k} c_{i,n}\omega_i = Z_n,$$

$$\left. f(Z_1, \ldots, Z_n, Y, X_1, \ldots, X_r) = 0 \right\}. \tag{2.1.9}$$

(Here we remind the reader that the last equation in (2.1.9) implies $Y \neq 0$). Using our new notational scheme and remembering that $R_2^{\Omega} \subset F^k$, we now have

$$B^{\Omega} = \left\{ (c_{1,1}, \ldots, c_{k,n}) \in \left(R_2^{\Omega}\right)^n \,\middle|\, \exists \bar{y}, \bar{z}_1, \ldots, \bar{z}_n, \bar{x}_1, \ldots, \bar{x}_r \in R_2^{\Omega}, \right.$$

$$\sum_{i=1}^{k} y_i\omega_i \left(\sum_{i=1}^{k} c_{i,j}\omega_i\right) = \sum_{i=1}^{k} z_{i,j}\omega_i, \; j = 1, \ldots, n,$$

$$f\left(\sum_{i=1}^{k} z_{i,1}\omega_i, \ldots, \sum_{i=1}^{k} z_{i,n}\omega_i, \sum_{i=1}^{k} y_i\omega_i, \right.$$

$$\left.\left. \sum_{i=1}^{k} x_{i,1}\omega_i, \ldots, \sum_{i=1}^{k} x_{i,r}\omega_i \right) = 0 \right\}. \tag{2.1.10}$$

Let $h(U, V, W) = UV - W$. Then using coordinate polynomials with respect to the basis Ω (see appendix section B.7), we have that

$$B^{\Omega} = \{(c_{1,1} \ldots, c_{k,n}) \in \left(R_2^{\Omega}\right)^n \,| \exists \bar{y}, \bar{z}_1, \ldots, \bar{z}_n, \bar{x}_1, \ldots, \bar{x}_r \in R_2^{\Omega} :$$
$$h_i^{\Omega}(\bar{y}, \bar{c}_j, \bar{z}_j) = 0, i = 1, \ldots, k, \; j = 1, \ldots, n,$$
$$f_i^{\Omega}(\bar{z}_1, \ldots, \bar{z}_n, \bar{y}, \bar{x}_1, \ldots, \bar{x}_r) = 0, i = 1, \ldots, n\}.$$

Now the lemma follows by Lemmas 2.1.3 and 2.1.12. □

We are now ready to prove the main theorem of this section: the transitivity of Dioph-generation.

Theorem 2.1.15. *Let R_1, R_2, R_3 be integral domains with quotient fields F_1, F_2, F_3 respectively. Assume that F_1, F_2, F_3 are all subfields of a field F which is not algebraically closed. Assume also that all the extensions $F/F_i, i = 1, 2, 3$, are finite. Finally, assume that $R_2 \leq_{Dioph} R_1$ and $R_3 \leq_{Dioph} R_2$. Then $R_3 \leq_{Dioph} R_1$.*

Proof. From the previous discussion we know that we can select F as a defining field for both the pairs (R_1, R_2) and (R_2, R_3). Let $\Omega = \{\omega_1, \ldots, \omega_k\}$ be a Diophantine basis for R_2 over R_1 such that F is the corresponding defining field, and likewise let $\Lambda = \{\lambda_1, \ldots, \lambda_n\}$ be a Diophantine basis of R_3 over R_2 with corresponding defining field F. Further, by Lemma 2.1.6 we can write

$$R_3 = \left\{ \sum_{i=1}^{n} z_i \lambda_i | (z_1, \ldots, z_n) \in A_\Lambda \subseteq F_2^n \right\},$$

where A_Λ has a field-Diophantine definition over R_2. By Proposition 2.1.14, A_Λ^Ω has a field-Diophantine definition over R_1. Thus

$$R_3 = \left\{ \sum_{i=1}^{n} \sum_{j=1}^{k} y_{i,j} \lambda_i \omega_j | (y_{1,1}, \ldots, y_{n,k}) \in A_\Lambda^\Omega \subseteq F_1^{nk} \right\}$$
$$= \left\{ \sum_{s=1}^{k} \left(\sum_{i,j} y_{i,j} A_{i,j,s} \omega_s \right) | (y_{1,1}, \ldots, y_{n,k}) \in A_\Lambda^\Omega \right\},$$

where

$$z_i = \sum_{j=1}^{k} y_{i,j} \omega_j, \qquad \sum_{s=1}^{k} A_{i,j,s} \omega_s = \lambda_i \omega_j, \quad A_{i,j,s} \in F_1.$$

Let

$$B_\Omega = \left\{ (t_1, \ldots, t_k) \in F_1^k \mid t_s = \sum_{i,j} y_{i,j} A_{i,j,s}, (y_{1,1}, \ldots, y_{n,k}) \in A_\Lambda^\Omega \right\}.$$

Then by Lemma 2.1.4 B_Ω has a field-Diophantine definition over R_1. Next we note that

$$R_3 = \left\{ \sum_{s=1}^{k} t_s \omega_s \mid (t_1, \ldots, t_k) \in B_\Omega \right\}.$$

Finally we conclude that $R_3 \leq_{Dioph} R_1$ by Lemma 2.1.6. □

We will now exploit the transitivity of Dioph-generation to obtain more general properties of this relation. We will start with a very common application of the transitivity of Diophantine generation.

Lemma 2.1.16. Going all the way down *Suppose that* $R_3 \subset R_2 \subset R_1$ *are integral domains whose fraction fields are not algebraically closed. Assume further that* $R_2 \leq_{Dioph} R_1$ *and* $R_3 \leq_{Dioph} R_2$. *Then* $R_3 \leq_{Dioph} R_1$. (See the following figure.)

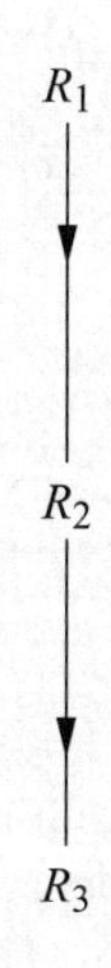

Lemma 2.1.17. Going up and then down *Let* $R_2 \subseteq R_1$ *be integral domains having quotient fields* F_1, F_2 *respectively. Assume that* F_1/F_2 *is a finite extension and that* F_1 *is not algebraically closed. Assume further that* $R_1 \leq_{Dioph} R_2$. *Then, for any* $A \subset R_1$ *that is Diophantine over* R_1, $A \cap R_2$ *is Diophantine over* R_2. (See the following figure.)

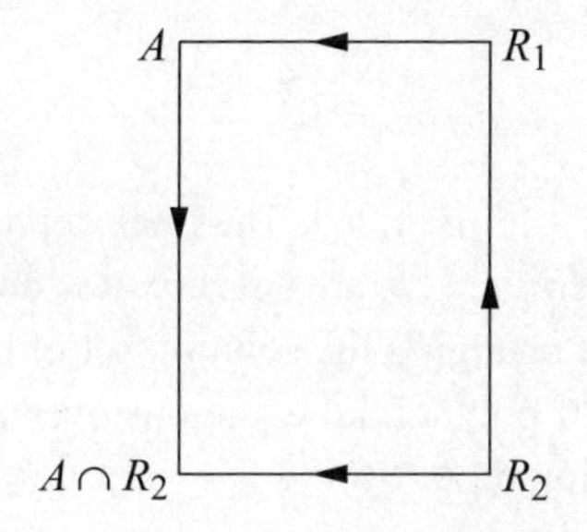

Proof. Let $f(t, \bar{y}) = f(t, y_1, \ldots, y_l)$ be a Diophantine definition of A over R_1 and let $\Omega = \{\omega_1, \ldots, \omega_n\}$ be a Diophantine basis of R_1 over R_2 with

$$P(Z_1, \ldots, Z_n, B, X_1, \ldots, X_m) = P(\bar{Z}, B, \bar{X})$$

being the corresponding defining polynomial. Then $t \in A \cap R_2$ if and only if $t \in R_2$ and

$$\exists y_1, \ldots, y_l \in R_1, \qquad f(t, y_1, \ldots, y_l) = 0. \tag{2.1.11}$$

However, (2.1.11) is true if and only if

$$\begin{cases} \exists \bar{a}_1, \ldots, \bar{a}_l \in R_2^n, \quad b_1, \ldots, b_l \in R_2, \quad \exists \bar{x}_1, \ldots, \bar{x}_m \in R_2^n, \\ P(\bar{a}_1, b_1, \bar{x}_1) = 0, \\ \vdots \\ P(\bar{a}_l, b_l, \bar{x}_l) = 0, \\ f\left(t, \sum_{j=1}^{n} \frac{a_{1,j}}{b_1}\omega_j, \quad \ldots, \quad \sum_{j=1}^{n} \frac{a_{l,j}}{b_l}\omega_j\right) = 0, \end{cases} \tag{2.1.12}$$

where $\bar{a}_i = (a_{i,1}, \ldots, a_{i,n})$, $\bar{x}_i = (x_{i,1}, \ldots, x_{i,m})$.

Now let $f_i^{\Omega}(u, \bar{v}_1, \ldots, \bar{v}_l)$, $i = 1, \ldots, n$, where $\bar{v}_j = (v_{j,1}, \ldots, v_{j,n})$, be the coordinate polynomials of $f(t, \bar{y})$ with respect to the basis Ω and the variables $y_1, \ldots, y_l$. (We remind the reader again to see appendix section B.7 for a discussion of coordinate polynomials.) Then we can rewrite the above system as

$$\begin{cases} \exists \bar{a}_1, \ldots, \bar{a}_l \in R_2^n, \quad b_1, \ldots, b_l \in R_2, \quad \exists \bar{x}_1, \ldots, \bar{x}_m \in R_2^n, \\ P(\bar{a}_1, b_1, \bar{x}_1) = 0, \\ \vdots \\ P(\bar{a}_l, b_l, \bar{x}_l) = 0, \\ f_1^{\Omega}(t, \bar{a}_1/b_1, \ldots, \bar{a}_l/b_l) = 0, \\ \vdots \\ f_n^{\Omega}(t, \bar{a}_1/b_1, \ldots, \bar{a}_l/b_l) = 0, \end{cases} \tag{2.1.13}$$

where $\bar{a}_j/b_j = (a_{j,1}/b_j, \ldots, a_{j,n}/b_1)$. The final step is to note that since "P"-equations guarantee that $b_1, \ldots, b_l$ are not zero, we can multiply "f"-equations by $\prod_{j=1}^{l} b^{\deg(f)}$ without changing the solution set of the system. The last step will produce a system of polynomial equations over R_2 which will constitute the Diophantine definition of $A \cap R_2$. □

Using transitivity again can also produce a "going down and then up" method.

Lemma 2.1.18. Going down and then up Let $R_1 \subset R_2 \subset R_3$ be integral domains with quotient fields F_1, F_2, F_3 respectively. Assume that F_3 is not algebraically closed and that F_3/F_1 is a finite extension. Suppose also that

$R_1 \leq_{Dioph} R_3$ and $R_2 \leq_{Dioph} R_1$. Then $R_2 \leq_{Dioph} R_3$.

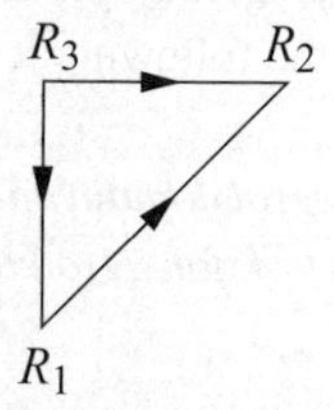

The next property is a restatement of Lemma 2.1.3 using new terminology.

2.1.19. The finite intersection property *Let $R_i \subset R$, $i = 1, \ldots, m$, be rings such that the quotient field of R is not algebraically closed and for all $i = 1, \ldots, m$ we have that $R_i \leq_{Dioph} R$. Then $\bigcap_{i=1}^{m} R_i \leq_{Dioph} R$.*

Proof. Since $R_i \subset R$ and $R_i \leq_{Dioph} R$, we can conclude that R_i has a Diophantine definition $f_i(t, x_1, \ldots, x_{n_i})$ over R. Then for all $x \in R$ we have that there exist $x_{1,1}, \ldots, x_{m,n_m} \in R$ with $f_i(x, x_{i,1}, \ldots, x_{i,n_i}) = 0$, $i = 1, \ldots, m$, if and only if $x \in \bigcap_{i=1}^{m} R_i$. □

2.2 Diophantine generation of integral closure and Dioph-regularity

In this section we will discuss two important properties of Diophantine generation over integrally closed subrings of global fields: the Dioph-generation of integral closure and the fraction field of a ring. We will start with integral closure for rings of $\mathcal{W}$-integers. The description of these rings can be found in appendix section B.1.

Proposition 2.2.1. *Let K be a global field, let R be an integrally closed ring with a quotient field K, let G be a finite extension of K, and let R_G be the integral closure of R in G. Then $R_G \leq_{Dioph} R$.*

Proof. Let $n = [G : K]$, and let $\Omega = \{\omega_1, \ldots, \omega_n\} \subset O_K$ be a basis of G over K. Then by Lemma B.1.27 and Lemma B.4.12, every element of R_G can be written as $\sum_{i=1}^{n} (a_i/D)\omega_i$, where $a_i \in R$ and D is the discriminant of the basis – a fixed (non-zero) constant of R. Further, every element of R_G has to satisfy a monic irreducible polynomial of degree at most n over R. Hence $R_G \leq_{Dioph} R$. □

We will now discuss the following question. When can the quotient field of an integral domain be Dioph-generated over an integral domain? The answer to this question is contained in the following lemma.

Lemma 2.2.2. *Let R be an integral domain and let F be its quotient field. Then $F \leq_{Dioph} R$ if and only if the set of non-zero elements of R has a Diophantine definition over R.*

Proof. First assume that the set of non-zero elements of R has a Diophantine definition over R. Let $f(y, x_1, \ldots, x_n)$ be a Diophantine definition of the set of non-zero elements of R. Then clearly

$$F = \{t | \exists b, a, x_1, \ldots, x_n \in R, bt = a, f(b, x_1, \ldots, x_n) = 0\}.$$

Conversely, suppose that $F \leq_{Dioph} R$ and let $g(z, y, x_1, \ldots, x_n)$ be a defining polynomial, i.e.

$$F = \{w | \exists y, z, x_1, \ldots, x_n : yw = z \wedge g(z, y, x_1, \ldots, x_n) = 0\}.$$

Then $g(1, y, x_1, \ldots, x_n)$ will be a Diophantine definition of the set of the non-zero elements of R. □

We now give the property discussed above a name.

Definition 2.2.3. Let R be an integral domain such that the set of its non-zero elements has a Diophantine definition over the ring. Then R will be called *Dioph-regular*.

Our next task is to show that all the rings that we shall consider in this book are Dioph-regular. The proof below is a generalization of the proof by Denef [18] for rings of algebraic integers.

Proposition 2.2.4. *Let K be a global field, and let $\mathcal{W}$ be any collection of non-archimedean primes of K. Then the ring $O_{K,\mathcal{W}}$ is Dioph-regular.*

Proof. First, assume that the complement of $\mathcal{W}$ contains at least two primes, $\mathfrak{p}_1$ and $\mathfrak{p}_2$. Let $a_i \equiv 0 \mod \mathfrak{p}_i$, $a_i \in O_K$, for $i = 1, 2$ and $(a_1, a_2) = 1$. Such $a_1, a_2 \in O_K$ exist by the Strong Approximation Theorem (Theorem B.2.1). Let $x \in O_{K,\mathcal{W}_K}$. Then $x \neq 0$ if and only if the following equation has solutions in $O_{K,\mathcal{W}}$:

$$xw = (u_1 a_1 - 1)(u_2 a_2 - 1).$$

Indeed, suppose that $x = 0$; then either a_1 or a_2 is invertible in $O_{K,\mathcal{W}}$. This is not true by the choice of $\mathfrak{p}_1$ or $\mathfrak{p}_2$. Suppose now that $x \neq 0$. Then let $\mathfrak{A}_1\mathfrak{A}_2/\mathfrak{B}$ be the divisor of x, where $\mathfrak{A}_1, \mathfrak{A}_2, \mathfrak{B}$ are integral divisors and $\mathfrak{A}_i$ and the divisor of a_i is relatively prime. (See Definition B.1.23 for the definition of the divisor of an element.) By the Strong Approximation Theorem, there exists $u_i \in O_K$ such that $u_i \equiv a_i^{-1} \bmod \mathfrak{A}_i$. Thus, the divisor of $(u_1a_1 - 1)(u_2a_2 - 1)$ is of the form $\mathfrak{A}_1\mathfrak{A}_2\mathfrak{C}$, where $\mathfrak{C}$ is an integral divisor. Hence, the divisor of w is of the form $\mathfrak{B}\mathfrak{C}$ and consequently $w \in O_K \subset O_{K,\mathcal{W}}$.

We now remove the assumption that the complement of $\mathcal{W}$ has at least two primes. First of all, if $\mathcal{W}$ contains all the primes then $O_{K,\mathcal{W}} = K$ and the proposition is trivially true. Second, if the complement of $\mathcal{W}$ contains only one prime $\mathfrak{p}$ then let M be a finite extension of K in which $\mathfrak{p}$ splits into distinct factors. (Such an extension exists by Lemma B.4.14.) Let $O_{M,\mathcal{W}_M}$ be the integral closure of $O_{K,\mathcal{W}}$ in M. By Proposition B.1.22, we have that $O_{M,\mathcal{W}_M}$ is a ring of $\mathcal{W}_M$-integers and $\mathcal{W}_M$ contains at least two primes of M. By the argument above, the set of non-zero elements of $O_{M,\mathcal{W}_M}$ has a Diophantine definition over $O_{M,\mathcal{W}_M}$. At the same time by Lemma 2.2.2, it is the case that $O_{M,\mathcal{W}_M} \leq_{Dioph} O_{K,\mathcal{W}}$. Thus we can use the "going up and then down" method (see Subsection 2.1.17) to complete the proof. □

Note 2.2.5. The importance of Dioph-regularity is obvious. If a ring R is Dioph-regular then any set which can be Dioph-generated over its quotient field can be Dioph-generated over the ring, by the transitivity of Dioph-generation.

2.3 Big picture: Diophantine family of a ring

The properties of Diophantine generation discussed in the preceding sections allow us to define an equivalence relationship based on Diophantine generation.

Definition 2.3.1. Let R_1, R_2 be rings with quotient fields F_1 and F_2 respectively. Assume that F_1 and F_2 are contained in some field F, not algebraically closed, and that F_1F_2/F_1 and F_1F_2/F_2 are finite extensions. Then call R_1 and R_2 *Dioph-equivalent* if $R_1 \leq_{Dioph} R_2$ and $R_2 \leq_{Dioph} R_1$. We will denote this relation by $R_2 \equiv_{Dioph} R_1$.

Clearly, $\equiv_{Dioph}$ is an equivalence relation. We will call the resulting equivalence classes the *Diophantine classes*. It is also clear that $\leq_{Dioph}$ is a relation on Diophantine classes. Given an integrally closed ring R, we will consider all the integrally closed rings whose quotient fields are either finite extensions or finite subextensions of the quotient field of R. Call the set of all these rings a

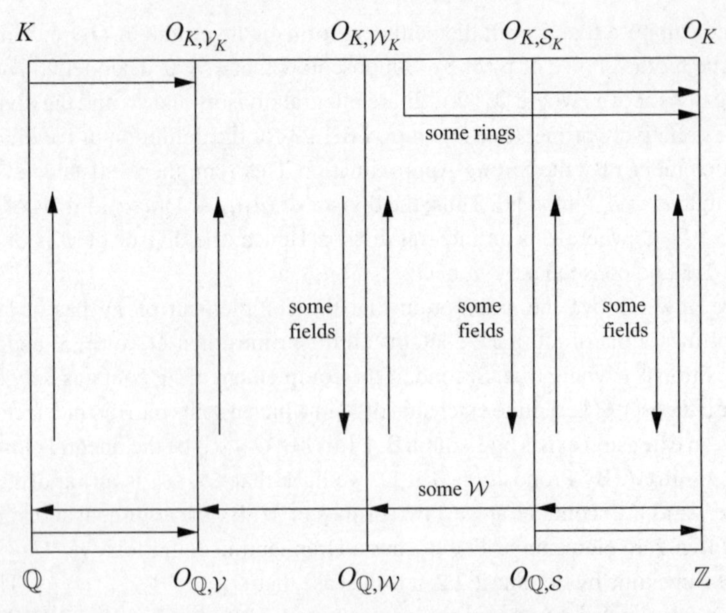

Figure 2.1 Horizontal and vertical problems for the Diophantine family of $\mathbb{Z}$. For any set of rational primes $\mathcal{U}$, $O_{\mathbb{Q},\mathcal{U}}$ is a ring of rational $\mathcal{U}$-integers.

Diophantine family of R. We can now rephrase some of the the main problems discussed in this book.

1. What is the structure of the Diophantine family of $\mathbb{Z}$? In other words, given an integrally closed ring R contained in a number field we want to know whether $R \leq_{Dioph} \mathbb{Z}$ and whether $\mathbb{Z} \leq_{Dioph} R$.
2. Let $\mathbb{F}_p$ be a finite field and let t be transcendental over $\mathbb{F}_p$. Then what is the structure of the Diophantine family of $\mathbb{F}_p[t]$? As above we can rephrase the questions as follows. Let K be a finite extension of $\mathbb{F}_p(t)$ and let R be an integrally closed ring whose quotient field is K. Then when is $R \leq_{Dioph} \mathbb{F}_p[t]$ and when is $\mathbb{F}_p[t] \leq_{Dioph} R$?

These questions clearly comprise many problems of varying difficulty. Our next task is to classify them and determine which problems can be dealt with quickly. We also remind the reader that any integrally closed subring of a global field is a ring of $\mathcal{W}$-integers by Proposition B.1.27. So our intention is to study the Diophantine classes of $\mathcal{W}$-integers.

2.3.2. Horizontal and vertical problems for Diophantine family of $\mathbb{Z}$ We will start with Figure 2.1,fig 1 where $\mathcal{S}$ is a finite set of rational primes, $\mathcal{V}$ is a set of rational primes whose complement is finite, and $\mathcal{W}$ is an infinite set of primes whose complement is also infinite. Further, K is a number field and $\mathcal{S}_K, \mathcal{V}_K, \mathcal{W}_K$ are the sets of K-primes lying above $\mathcal{S}$, $\mathcal{V}$, and $\mathcal{W}$ respectively. The arrows represent the direction of Dioph-generation, for pairs of rings where we know the answer, at least in some cases. More precisely, if we have $R_1 \to R_2$ in Figure 2.1 then $R_2 \leq_{Dioph} R_1$. We divide the problems into two main groups: horizontal and vertical. Horizontal problems concern two rings with the same quotient field. Vertical problems concern pairs of rings whose quotient fields are different: one is a proper subfield of the other. Using results from Section 2.2 we can solve one horizontal problem and a family of vertical problems. The Dioph-regularity of $\mathbb{Z}$ tells us that $\mathbb{Q} \leq_{Dioph} \mathbb{Z}$. Further, since we know that integral closures of the rings of $\mathcal{W}$-integers in the extensions are Dioph-generated, we can solve all the upward-going vertical problems.

We can also determine when $O_{\mathbb{Q},\mathcal{U}} \leq_{Dioph} \mathbb{Z}$ for any set of rational primes $\mathcal{U}$. The answer is a direct consequence of Matiyasevich's result and will be discussed in detail in the following chapter.

We know how to solve the downward vertical problem for some subrings of totally real fields, for their extensions of degree 2, for fields with exactly one pair of non-real embeddings, and for some other number fields. These problems will be discussed in Chapter 7. In all the number fields we can solve the "short" horizontal problem in any subring (i.e. define integrality at finitely many primes). We will discuss this problem in Chapter 4. In some subrings of totally real fields and their totally complex extensions of degree 2 we aim to solve "longer" horizontal problems, i.e. to define integrality at infinitely many primes. The partial solution of this problem will be discussed in Chapter 7.

2.3.3. Horizontal and vertical problems for Diophantine family of $\mathbb{F}_p[t]$

We will now consider the problems associated with the Diophantine family of a polynomial ring over a finite field of constants. Consider Figure 2.2.fig 2 Here we use notation analogous to that used in Figure 2.1. We also assume that $\mathcal{S}$, $\mathcal{W}$, $\mathcal{V}$ contain the valuation which is the pole of t. Further, K as before is a finite extension of the ground field $\mathbb{F}_p(t)$, and O_K, $O_{K,\mathcal{S}_K}$, $O_{K,\mathcal{W}_K}$, $O_{K,\mathcal{V}_K}$ are again integral closures of subrings of the ground field, in this case $\mathbb{F}_p[t]$, $O_{\mathbb{F}_p(t),\mathcal{S}}$, $O_{\mathbb{F}_p(t),\mathcal{W}}$, and $O_{\mathbb{F}_p(t),\mathcal{V}}$ respectively in K. For the same reason as in the case of number fields, the upward vertical problem follows from the fact that integral closure is Dioph-generated for the rings under consideration. The "short" horizontal problem for function fields (i.e. the definability of integrality at finitely many primes) will be discussed in Chapter 4. The downward vertical problem for function fields, which has been solved to some extent for all global function fields, will be discussed in Chapter 10. Finally, the "long" horizontal problem

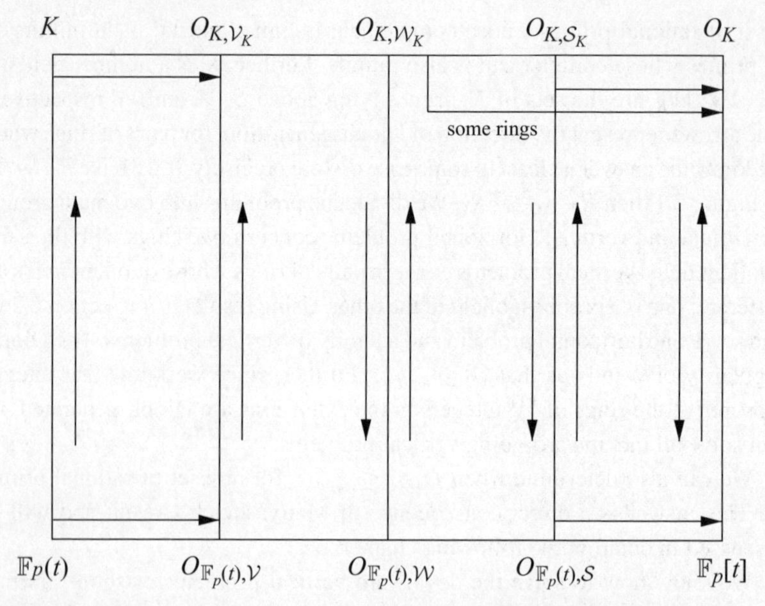

Figure 2.2 Horizontal and vertical problems for the Diophantine family of $\mathbb{F}_p[t]$.

(i.e. the definability of integrality at infinitely many primes) for function fields has also been partially solved for all function fields and is also described in Chapter 10.

3

Diophantine equivalence and Diophantine decidability

In this chapter we will take a closer look at what Diophantine generation and Diophantine equivalence tell us about Diophantine decidability and definability over countable rings. We have already touched on these questions in our introduction. There we talked about the relationship between Diophantine definitions and Diophantine undecidability. To make this discussion more precise over rings other than the ring of rational integers, we will need to determine what the analog of a recursive function (or, more informally, an algorithm) is over these rings. To formalize the notion of an algorithm over countable structures, one uses presentations. If it exists, a recursive presentation of a given field F is a homomorphism from F into a field whose elements are natural numbers. Under this homomorphism all the field operations of F are interpreted by restrictions of recursive functions and the image of F is a recursive set. (Here we remind the reader that Appendix A contains definitions of recursive functions and recursive sets, as well as a list of references.) Not all fields and rings have such presentations. A field or ring which has such a presentation is called recursive. However, as we will see below, this notion of a presentation is too "strong" for our purposes. Presentations which are more suitable for a discussion of Diophantine questions are called "weak presentations." We describe these presentations in the following section. Finally, we note that most of this chapter is based on [95].

3.1 Weak presentations

We start with a couple of definitions.

Definition 3.1.1. Let R be a countable ring such that there exists an injective map $j : R \to \mathbb{N}$ with the following properties. There exist recursive (computable) functions $P_+, P_-, P_\times : \mathbb{N}^2 \to \mathbb{N}$ such that for all $x, y \in F$ we have that

$$P_+(j(x), j(y)) = j(x+y),$$
$$P_-(j(x), j(y)) = j(x-y),$$
$$P_\times(j(x), j(y)) = j(xy).$$

Then j is called a weak presentation of R as a ring.

Next let F be a countable field. Let $j : F \to \mathbb{N}$ be a weak presentation of F as a ring and assume additionally that there exists a recursive function $P_/ : \mathbb{N}^2 \to \mathbb{N}$ such that for all non-zero $y \in F$ we have that

$$P_/(j(x), j(y)) = j(x/y).$$

Then j is called a weak presentation of F as a field.

Definition 3.1.2. Let $R \subseteq R'$ be countable rings and let $j : R \to \mathbb{N}$, $j' : R' \to \mathbb{N}$ be weak presentations of R and R' as rings respectively. Then we say that j' is an extension of j if $j'_{|R} = \psi \circ j$, where ψ is a recursive injective function with a recursive range or in other words the inverse of ψ is also recursive.

Note that not every weak presentation of a ring R is extendable to a weak presentation of its quotient field F as a field. (For example one can construct weak presentations of $\mathbb{Q}$ as a ring which are not weak presentations of $\mathbb{Q}$ as a field. In other words, division is not translated by a restriction of a computable function. For more details about such constructions see [97].)

While not all countable fields or integral domains are recursive, they all have weak presentations.

Proposition 3.1.3. *Let R be a countable integral domain. Then R has a weak presentation as a ring and its quotient field has a weak presentation as a field.*

Proof. By Proposition A.7.13 we have that $R \subseteq F \subseteq K$, where K is a computable field. Let $j : K \to \mathbb{N}$ be a computable presentation of K. Then j restricted to R and F is a weak presentation of R and F respectively. □

3.2 Some properties of weak presentations

To connect the weak presentations to the subject of Diophantine definability and decidability we need to discuss some of the properties of weak presentations. First we need to introduce additional notation.

Notation 3.2.1. Let K be a countable field and let $j : K \to \mathbb{N}$ be a weak presentation of K. Let $f : K^n \to K$ be a function defined on $D \subset K^n$. Then $j(f)$ will denote a function from $j(D)$ to $j(K)$ defined by the following. For all $(a_1, \ldots, a_n) \in D$ we have that $j(f)(j(a_1), \ldots, j(a_n)) = j(f(a_1, \ldots, a_n))$.

Next we note an obvious but important corollary of the definition of a weak presentation.

Lemma 3.2.2. *Let R be a countable ring and let $j : R \to \mathbb{N}$ be a weak presentation of R. Let $P(X_1, \ldots, X_n)$ be a polynomial over R. Then $j(P) : j(R) \to j(R)$ is a restriction of a recursive function.*

Proof. A polynomial function is a composition of finitely many binary additions, subtractions, and multiplications. Therefore one can proceed by induction on the number n of times that addition, subtraction, or multiplication is used in the construction of P. The argument will require the use of Definition A.1.2 concerning recursive (computable) functions and Lemma A.1.12. The details are left to the reader. □

If we are considering a weak presentation of a field, then it is just as easy to see that rational functions are also translated by restrictions of recursive functions.

Corollary 3.2.3. *Let K be a countable field. Let $j : K \to \mathbb{N}$ be a weak presentation of K as a field. Let $W \in K(X_1, \ldots, X_m)$ be a rational function over K. Then $j(W)$ is a restriction of a recursive function.*

Proof. Let $W(X_1, \ldots, X_m) = Q_1(X_1, \ldots, X_m)/Q_2(X_1, \ldots, X_m)$ where Q_1, Q_2 are polynomial functions over K without common factors in $K[X_1, \ldots, X_m]$. Clearly for all $X_1, \ldots, X_m \in K$ with $Q(X_1, \ldots, X_m) \neq 0$,

$$j(W)(j(X_1), \ldots, j(X_m)) = P_/\big(j(Q_1)(j(X_1), \ldots, j(X_m)), j(Q_2)(j(X_1), \ldots, j(X_m))\big).$$

Since $j(Q_1), j(Q_2), P_/$ are all restrictions of recursive functions, by the definition of a recursive function so is $j(W)$. □

Our next task is to show that weak presentations are extendable under finite algebraic extensions.

Proposition 3.2.4. *Let F be a countable field and let $j : F \to \mathbb{N}$ be a weak presentation of F. Let K be a finite extension of F of degree n. Let $\Omega = \{\omega_1 = 1, \ldots, \omega_n\}$ be a basis of K over F. Then there exists a weak presentation $J : K \to \mathbb{N}$ such that J restricted to K is equal to $\psi \circ j$, where $\psi : \mathbb{N} \to \mathbb{N}$ is recursive with recursive inverse, and there exist recursive coordinate functions $C_1, \ldots, C_n : \mathbb{N} \to \mathbb{N}$ such that for any element $x \in K$ we have that*

$$x = \sum_{i=1}^{n} \left(J^{-1} \circ C_i \circ J(x)\right) \omega_i \tag{3.2.1}$$

and

$$C_i \circ J(x) \in J(F) \tag{3.2.2}$$

for all $i = 1, \ldots, n$.

Proof. Let $G_n : \mathbb{N}^n \to \mathbb{N}$ be the recursive (computable) function defined in Lemma A.1.14. We will construct $J : K \to \mathbb{N}$ by setting $J(\sum_{i=1}^{n} a_i \omega_i) = G_n(j(a_1), \ldots, j(a_n)) = \prod_{i=1}^{n} p_i^{j(a_i)}$. First we should observe that for $x \in F$ we have that $J(x) = 2^{j(x)}$, and therefore by Proposition A.1.11 the first requirement of the lemma is satisfied. Second, for $m \in \mathbb{Z}_{>0}$ and for $i = 1, \ldots, n$, let $C_i(m) = \text{ord}_{p_i} m$, where p_i is the ith prime in the ascending listing of all rational primes. It is clear that C_i satisfies (3.2.1) and (3.2.2) and, by Proposition A.1.11, for all $i = 1, \ldots, n$ we know C_i to be recursive. Finally, to make J into a weak presentation we need to show that the images of field operations are extendable to total recursive functions. Let $J(\text{op})$ denote the translations of the field operations under J and let P_{op} denote, as before, the translation of the field operations under j. Then define

$$J(\pm)(m_1, m_2) = \prod_{i=1}^{n} p_i^{P_{\pm}(C_i(m_1), C_i(m_2))}.$$

For multiplication and division the definition is a bit more complicated. For all $i, j = 1, \ldots, k$, let

$$B_{i,j,1}, \ldots, B_{i,j,k} \in F$$

be such that

$$\omega_i \omega_j = \sum_{r=1}^{k} B_{ijr} \omega_r.$$

Then for $a_1, \ldots, a_n, b_1, \ldots, b_n \in F$,

$$\sum_{i=1}^{n} a_i \omega_i \sum_{j=1}^{n} b_j \omega_j = \sum_{i,j} a_i b_j \omega_i \omega_j = \sum_{i,j,k} a_i b_j B_{i,j,k} \omega_k = \sum_{k=1}^{n} \left(\sum_{i,j} B_{i,j,k} a_i b_j \right) \omega_k.$$

Let $H_k(a_1, \ldots, b_n) = \sum_{i,j} B_{i,j,k} a_i b_j$, and note that $H_k(T_1, \ldots, T_{2n})$ depends on the basis Ω only. Then

$$\sum_{i=1}^{n} a_i \omega_i \sum_{j=1}^{n} b_j \omega_j = \sum_{k=1}^{n} H_k(a_1, \ldots, a_n, b_1, \ldots, b_n)\omega_k.$$

By Lemma 3.2.2, for all $k = 1, \ldots, n$ we know that $j(H_k)$ is extendable to a recursive function. Thus we can define

$$J(\times)(m_1, m_2) = \prod_{i=1}^{n} p_i^{j(H_i)(C_1(m_1),\ldots,C_n(m_1),C_1(m_2),\ldots,C_n(m_2))}.$$

Next we move to the translation of division. Since we have shown that the J-translation of multiplication in K is a restriction of a recursive function, it is enough to show that the J-translation of finding the multiplicative inverse is a restriction of a recursive function. By Lemma B.10.6, there exist

$$T_1, \ldots, T_n, Q \in F[x_1, \ldots, x_n],$$

depending on F, G, and Ω only, such that $\sum_{i=1}^{n} a_i \omega_i \neq 0$ if and only if

$$Q(a_1, \ldots, a_n) \neq 0$$

and

$$\left(\sum_{i=1}^{k} a_i \omega_i \right)^{-1} = \sum_{i=1}^{n} \frac{T_i(a_1, \ldots, a_n)}{Q(a_1, \ldots, a_n)} \omega_i.$$

Thus, we can define

$$J(^{-1})(m) = \prod_{i=1}^{n} p_i^{P_/(j(T_i)(C_1(m),\ldots,C_n(m)),\, j(Q)(C_1(m),\ldots,C_n(m)))}.$$

□

We are now ready to start establishing connections between Diophantine definability and decidability. Our first goal is to show that Diophantine definability assures relative enumerability under any weak presentation.

Proposition 3.2.5. *Let R be a countable integral domain. Let $j : R \to \mathbb{N}$ be a weak presentation of R as ring. Let A be a Diophantine subset of R^l.* Then *$j(A) \leq_e j(R)$.*

Proof. First note that if A is finite (and this includes the case $A = \emptyset$) then A is recursive (computable) by Lemma A.1.10 and recursively enumerable (r.e.) by Lemma A.2.2. So without loss of generality we can assume that A is infinite.

Let

$$\psi : \mathbb{N} \to j(R)$$

be any function enumerating $j(R)$. Let $P(t_1, \dots, t_l, x_1, \dots, x_n)$ be a Diophantine definition of A over R. Let

$$F(k, n+l) = (a_1, \dots, a_{n+l}),$$

where $a_i = \text{ord}_{p_i} k$, be a function defined by Lemma A.1.14. Then by this lemma $F(k, n+l)$ is recursive. Let $\pi_i : \mathbb{N}^{n+1} \to \mathbb{N}$ be the projection on the ith coordinate for $1 \leq i \leq n+1$. Our next function definition will use a minimization operator μ from Definition A.1.1. Define

$$\bar{\phi} : \mathbb{N} \to j(A) \subset j(R^l)$$

in the following fashion. First define a function $\nu : \mathbb{N} \to \mathbb{N}$ by

$$\nu(0) = (\mu t)\big[j(P)\big(\psi(\pi_1(F(t, n+l))), \dots, \psi(\pi_{n+l}(F(t, n+l)))\big) = j(0)\big],$$

$$\nu(m) = (\mu t)\big[t > \nu(m-1) \wedge j(P)\big(\psi(\pi_1(F(t, n+l))), \dots, \psi(\pi_{n+l}(F(t, n+l)))\big) = j(0)\big].$$

Now define

$$\bar{\phi}(m) = \big(\psi(\pi_1(F(n+l, \nu(m)))), \dots, \psi(\pi_l(F(n+l, \nu(m))))\big).$$

Note that the relation inside the μ-operator is recursive by Lemmas 3.2.2 and A.1.4 and Corollary A.1.7. Thus ϕ is constructed from basic functions (as in Definition A.1.2) and ψ using finitely many applications of composition, minimization, and recursion. Therefore, $j(A) \leq_e j(R)$ by Definition A.3.2.

Informally, the relative enumerability of A follows from the following argument. To compute $\bar{\phi}(m)$ perform the following steps. Consider the list of $(n+l)$-tuples of elements of $j(R)$ generated by ψ. Plug each $(n+l)$-tuple into $j(P)$ to see whether the result is $j(0)$, starting with the first $(n+l)$-tuple not processed in the calculation of $\bar{\phi}(0), \dots, \bar{\phi}(m-1)$. Record the first l elements of the first $(n+l)$-tuple satisfying the condition above as the value of $\bar{\phi}(m)$. □

We now prove a more general version of Proposition 3.2.5.

Proposition 3.2.6. *Let R_1, R_2 be two integral domains such that $R_2 \leq_{Dioph} R_1$. Let F_1, F_2 be the fraction fields of R_1 and R_2 respectively, and let $j : F_1 \to \mathbb{N}$ be a weak presentation of F_1. Then there exists a weak presentation $J : F_1F_2 \to \mathbb{N}$ such that J is an extension of j in the sense of Definition 3.1.2, and for any such extension J we have that $J(R_2) \leq_e J(R_1)$.*

Proof. By the definition of Diophantine generation 2.1.5, there exists $f(a_1, \ldots, a_k, b, x_1, \ldots, x_m)$ with coefficients in R_1 such that

$$f(a_1, \ldots, a_k, b, x_1, \ldots, x_m) = 0 \quad \Rightarrow \quad b \neq 0, \tag{3.2.3}$$

and

$$R_2 = \left\{ \sum_{i=1}^{k} t_i \omega_i \mid \exists a_1, \ldots, a_k, b, x_1, \ldots, x_m \in R_1, \right.$$
$$\left. bt_1 = a_1, \ldots, bt_k = a_k, f(a_1, \ldots, a_k, b, x_1, \ldots, x_m) = 0 \right\}.$$

In other words,

$$R_2 = \left\{ \sum_{i=1}^{k} \frac{a_i}{b} \omega_i \mid (a_1, \ldots, a_k, b) \in A \subset R_1 \right\},$$

where A is a Diophantine subset of R_2^{k+1}. By Proposition 3.2.4 there exists a weak presentation

$$J : F_1 F_2 \to \mathbb{N}$$

such that J is an extension of j. By Proposition 3.2.5, $J(A) \leq_e J(R_1)$. Let

$$U(X_1, \ldots, X_k, Y) = \sum_{i=1}^{k} \frac{X_i}{Y} \omega_i$$

be a rational function over $F_1 F_2$. Then $J(U)$ is a restriction of a recursive function by Corollary 3.2.3. Now we can proceed in a manner analogous to that used in the proof of Proposition 3.2.5. Let $\bar{\phi} : \mathbb{N} \to \bar{J}(A)$ be any enumeration of $\bar{J}(A)$. Then define

$$\xi(m) = J(U)\big(\pi_1(\bar{\phi}(m)), \ldots, \pi_{k+1}(\bar{\phi}(m))\big).$$

Since $\xi(m)$ is a recursive function enumerating R_2, we conclude that $J(R_2) \leq_e \bar{J}(A)$ and therefore, by the transitivity of relative enumerability (see the remarks following Definition A.3.2), $J(R_2) \leq_e J(R_1)$. □

Remark 3.2.7. Proposition A.6.5 shows that any two recursive presentations of a finitely generated ring or field are related by a recursive function, i.e. one presentation is a composition of the other and a recursive function. It also shows that Turing degree and enumeration degree structures of the finitely generated rings and fields are invariant with respect to recursive presentations. Actually, the proof of the proposition shows a bit more. It shows that weak presentations preserve the enumeration degree structure for a finitely generated ring or field. (Weak presentations do not necessarily preserve the Turing degree structure of finitely generated objects. See for example [40].) Thus, under any weak presentation the class of r.e. sets is the same class of subsets for a finitely generated ring or field. Therefore Matiyasevich's theorem (Theorem 1.2.2), holds independently of the weak presentation chosen for $\mathbb{Z}$.

Before we leave this section, we note that Proposition 3.2.6 tells us that the notion of weak presentation is the right notion to use in the study of Diophantine and existential definability. First, both weak presentations and existentially definable sets exist over any ring. Second, Diophantine classes fit completely inside relative enumerability classes invariant under weak presentations of finitely generated objects. We explore the last fact further by using it to show that there are infinitely many Diophantine classes.

3.3 How many Diophantine classes are there?

As a consequence of the fact that, over $\mathbb{Z}$, recursive enumerability and Diophantine definability are the same we have the following result.

Lemma 3.3.1. Let $\mathcal{A} = \{p_n\}$ *be a set of rational primes. Then* $O_{\mathbb{Q},\mathcal{A}} \leq_{Dioph} \mathbb{Z}$ *if and only if* $\mathcal{A}$ *is recursively enumerable.*

Proof. Suppose that $\mathcal{A}$ is an r.e. (c.e.) set of primes. Then U, the set of all natural numbers that can be written as products of primes from $\mathcal{A}$, is also r.e. by Proposition A.4.2. By Matiyasevich's theorem, 1.2.2, U is Diophantine over $\mathbb{Z}$. Let $f_U(t, x_1, \ldots, x_m)$ be the Diophantine definition of U over $\mathbb{Z}$. Hence we can set

$$O_{\mathbb{Q},\mathcal{A}} = \left\{\frac{x}{y} \mid x \in \mathbb{Z}, y \in \mathbb{Z}, \exists x_1, \ldots, x_m \in \mathbb{Z}, f_U(y, x_1, \ldots, x_m\} = 0\right\}$$

and thus $O_{\mathbb{Q},\mathcal{A}} \leq_{Dioph} \mathbb{Z}$.

Suppose now that $O_{\mathbb{Q},\mathcal{A}} \leq_{Dioph} \mathbb{Z}$. Then

$$O_{\mathbb{Q},\mathcal{A}} = \left\{\frac{x}{y} \mid x \in \mathbb{Z}, y \in \mathbb{Z}, \exists x_1, \ldots, x_n \in \mathbb{Z}, g(x, y, x_1, \ldots, x_m) = 0\right\}$$

for some polynomial $g(x, y, x_1, \ldots, x_m) \in \mathbb{Z}[x, y, x_1, \ldots, x_m]$. Let

$$U_{\mathbb{Z}} = \{y \in \mathbb{Z} \mid \exists x_1, \ldots, x_n \in \mathbb{Z}, g(1, y, x_1, \ldots, x_m) = 0\}.$$

Then $U_{\mathbb{Z}}$ is a set that is Diophantine over $\mathbb{Z}$ and, by Proposition 3.2.5 and Remark 3.2.7, for any weak presentation j we have that $j(U_{\mathbb{Z}}) \leq_e j(\mathbb{Z})$, where

$j(\mathbb{Z})$ is r.e. Consequently, $j(U_{\mathbb{Z}})$ is also an r.e. set. Let

$$U = U_{\mathbb{Z}} \cap \mathbb{N} = \{|m| \, | m \in U_{\mathbb{Z}}\}.$$

By Proposition A.6.1 and, again, Remark 3.2.7, for any weak presentation j of $\mathbb{Z}$ we now have that $U \equiv_e j(U) \equiv_e j(U_{\mathbb{Z}})$ and therefore that U is r.e. Finally, by Proposition A.4.2 we have that $U \equiv_e \mathcal{A}$. Hence $\mathcal{A}$ is r.e. □

Unfortunately, all aspects of the relationship between Diophantine definability and enumerability for rings other than $\mathbb{Z}$ are far from clear. However from Proposition 3.2.5 we do get the following fact.

Proposition 3.3.2. *There are infinitely many Diophantine classes.*

Proof. We will show that the subrings of $\mathbb{Q}$ are partitioned into infinitely many Diophantine classes. On the one hand, from Proposition A.6.4 and Remark 3.2.7, given a set of rational primes

$$\mathcal{A} = \{p_i, i \in I\},$$

where $p_1 \leq p_2 \cdots$ is a listing of all primes in ascending order and I is any subset of the natural numbers, we have $J(O_{\mathbb{Q},\mathcal{A}}) \equiv_e J(I) \equiv_e I$ under any weak presentation J of $\mathbb{Q}$ as a field. On the other hand, given two sets of rational primes $\mathcal{A}_1, \mathcal{A}_2$ then by Proposition 3.2.5

$$O_{\mathbb{Q},\mathcal{A}_1} \equiv_{Dioph} O_{\mathbb{Q},\mathcal{A}_2} \Rightarrow J(O_{\mathbb{Q},\mathcal{A}_1}) \equiv_e J(O_{\mathbb{Q},\mathcal{A}_2})$$

under any weak presentation of $\mathbb{Q}$ (as a field or ring). Since there are infinitely many enumeration classes (see Proposition A.3.3), we must conclude that there are infinitely many Diophantine classes. □

3.4 Diophantine generation and Hilbert's Tenth Problem

In this section we want to investigate the very close connection between Diophantine generation and Diophantine undecidability. First we need to formalize exactly what we mean by the unsolvability (or solvability) of HTP for an arbitrary countable ring.

Definition 3.4.1. Let R be a countable ring. We will say that HTP is decidable (undecidable) over R if there exists (does not exist) a presentation $J : R \to \mathbb{N}$

and a recursive function f from the space of finite sequences of natural numbers into the set $\{0, 1\}$ such that

$$f(d, m, k, A_{i_1,\dots,i_m,j_1,\dots,j_k}, 0 \le i_k, j_l \le d)(n_1, \dots, n_m) = 1$$

if and only if $(n_1, \dots, n_m) = (J(x_1), \dots, J(x_m))$, where $x_1, \dots, x_m$ are elements of the Diophantine set whose definition over R is the polynomial

$$H(x_1, \dots, x_m, y_1, \dots, y_k) = \sum_{0 \le i_k, j_l \le d} a_{i_1,\dots,i_m,j_1,\dots,j_k} x_1^{i_1} \cdots x_m^{i_m} y_1^{j_1} \cdots y_k^{j_k},$$

and

$$J(a_{i_1,\dots,i_m,j_1,\dots,j_k}) = A_{i_1,\dots,i_m,j_1,\dots,j_k}.$$

Note that an immediate corollary of this definition is the fact that the solvability of HTP under J requires that every Diophantine set is recursive under J.

Remark 3.4.2. Looking at the formalization above one can conclude immediately that, despite the fact that we have not put any assumptions on J, without loss of generality we can assume that J is a strongly recursive presentation of R. Indeed, first suppose that J is not recursive. Then either $J(R)$ is not recursive or one or more of the graphs of addition, subtraction, and multiplication is or are not recursive. Since R is Diophantine over R and the graphs of the ring operations are also Diophantine over R, we conclude that under J there are non-recursive Diophantine sets and that in this case HTP is not solvable under J. Finally, assume that J is a recursive but not strongly recursive presentation of R. (See Definition A.7.3 for the definition of a strongly recursive presentation.) Then under J the image of the set $D = \{(x, y) : \exists z \in R, x = yz\}$ is not recursive. So in this case R again has a Diophantine set which is undecidable under J.

We also should note that by Matiyasevich's theorem and Corollary A.6.6, $\mathbb{Z}$ has undecidable Diophantine subsets under *any* presentation, and so we can continue to say that HTP is unsolvable over $\mathbb{Z}$.

Next we introduce a new notion which will play an important role in this and later sections in this book – the notion of a Diophantine model.

Definition 3.4.3. Let R_1, R_2 be two rings such that the following statements are true.

1. There exists a map

$$\bar{\phi} = (\phi_1, \dots, \phi_k) : R_1 \to R_2^k$$

such that for any Diophantine set $D_1 \subseteq R_1^l$ there exists a Diophantine set $D_2 \subset R_2^{lk}$ with the following properties:

$$D = \{(\bar{\phi}(a_1), \ldots, \bar{\phi}(a_l)) : (a_1, \ldots, a_l) \in D_1\} \subseteq D_2,$$

and for

$$\bar{D} = \{(\bar{\phi}(a_1), \ldots, \bar{\phi}(a_l)) : (a_1, \ldots, a_l) \notin D_1\}$$

we have that

$$\bar{D} \cap D_2 = \emptyset.$$

2. Under some recursive presentations J_1 and J_2, of R_1 and R_2 respectively, the map $J_2 \circ \phi_i \circ J_1^{-1}$ is a restriction of a recursive function for all $i = 1, \ldots, k$.

Then we will say that R_2 has a Diophantine model of R_1.

The raison d'être of the Diophantine models is made clear by the following proposition whose proof we leave to the reader.

Proposition 3.4.4. *Let R_1, R_2 be two rings such that R_2 has a Diophantine model of R_1 under some presentations J_1 and J_2 of R_1 and R_2 respectively. Assume that one of the following statements is true.*

- *HTP is undecidable over R_1 under J_1, and there exists an effective procedure such that, given the J_1-codes of the coefficients of a Diophantine definition over R_1 of any Diophantine subset D_1 of R_1 as its input, this procedure will produce the J_2-codes of the coefficients for a Diophantine definition over R_2 of D_2, the Diophantine subset of R_2 containing the image of D_1 in the Diophantine model.*
- *There exists an undecidable (under J_1) Diophantine set over R_1 (and therefore HTP is undecidable over R_1 under J_1).*

Then HTP is undecidable over R_2 under J_2.

The proposition below provides the connection between Diophantine models and Diophantine generation.

Proposition 3.4.5. *Let $R_1 \leq_{Dioph} R_2$ be two rings with residue fields F_1 and F_2 respectively. Let F be a field such that $F_1F_2 \subset F$, F/F_2 is a finite extension of degree n, and F is not algebraically closed. Assume that R_2 is strongly recursive. Then the following statements are true.*

1. *There exists a recursive presentation J of F such that the following conditions are satisfied:*
    - *R_2 is recursive under J;*
    - *the coordinate functions from F to F_2 (as defined in Proposition 3.2.3) are recursive under J with respect to some basis of F over F_2;*
    - *there exists a recursive mapping $\bar{\Lambda} = (\Lambda_1, \Lambda_2) : J(F_2) \to J(R_2)^2$ defined by $\bar{\Lambda}(J(x)) = (J(x_1), J(x_2))$, where $x_1, x_2 \in R_2, x_2 \neq 0$ and $x = x_1/x_2$.*
2. *If $J(R_1)$ is recursive then*
    - *R_2 has a Diophantine model of R_1,*
    - *for any set $D_1 \subseteq R_1$, Diophantine over R_1, a Diophantine definition of D_2 – the set containing the image of D_1 but no element of its complement – can be effectively (under J) and uniformly in D_1 constructed over R_2 from a Diophantine definition of D_1.*

Proof. The first assertion of the proposition is satisfied by Propositions A.7.4 and A.7.7. Next assume that $J(R_2)$ is recursive and let $\Omega = \{\omega_1, \ldots, \omega_n\}$ be a basis of F over F_2 such that under J the coordinate functions $(C_1, \ldots, C_n)$ with respect to Ω are recursive. Let

$$(c_1, \ldots, c_n) : F \to F_2$$

be such that $c_i = J^{-1} \circ C_i \circ J$. Also let $\lambda_i = J^{-1} \circ \Lambda_i \circ J$.

Since $R_1 \leq_{Dioph} R_2$, for some

$$f(A_1, \ldots, A_n, B, Z_1, \ldots, Z_m) \in R_2[A_1, \ldots, A_n, B, Z_1, \ldots, Z_m],$$

we have

$$R_1 = \left\{ \sum_{i=1}^{n} \frac{a_i}{b} \omega_i : \exists z_1, \ldots, z_m \in R_2, f(a_1, \ldots, a_n, b, z_1, \ldots, z_m) = 0 \right\}, \tag{3.4.1}$$

where $f(a_1, \ldots, a_n, b, z_1, \ldots, z_m) = 0 \Rightarrow b \neq 0$. So given $x \in R_1$, define for $i = 1, \ldots, n$

$$\phi_{2i-1}(x) = \lambda_1(c_i(x)), \qquad \phi_{2i}(x) = \lambda_2(c_i(x)),$$

and for $X \in \mathbb{N}$ let

$$\Phi_{2i-j}(X) = \begin{cases} J \circ \phi_{2i-j} \circ J^{-1}(X) = \Lambda_k(C_i(X)) & \text{if } X \in J(R_1) \\ 0 & \text{if } X \notin J(R_1). \end{cases}$$

Here either $j = 1$ and $k = 1$ or $j = 0$ and $k = 2$. Observe that, given our assumptions, $\bar{\Phi} = (\Phi_1, \ldots, \Phi_n)$ is recursive. Next let D_1 be a Diophantine subset of R_1^l with Diophantine definition $g(X_1, \ldots, X_l, T_1, \ldots, T_r)$. Let D_2 consist of all $(y_1, \ldots, y_{2n}) \in R_2^{2n}$ such that

$$\begin{aligned} \exists \bar{a}_1, \ldots, \bar{a}_l, \bar{u}_1, \ldots, \bar{u}_r &\in R_2^n, \\ b_1, \ldots, b_l, v_1, \ldots, v_r, x_1, \ldots, x_l, t_1, \ldots, t_r &\in R_2, \\ \bar{z}_1, \ldots, \bar{z}_l, \bar{w}_1, \ldots, \bar{w}_r &\in R_2^m, \end{aligned}$$

satisfying the following system of equations:

$$\begin{cases} y_{2i-1,j} b_j = y_{2i,j} a_{i,j}, & j = 1, \ldots, l, \quad i = 1, \ldots, n, \\ b_j x_j = \sum_{i=1}^n a_{i,j} \omega_i, \\ f(\bar{a}_j, b_j, \bar{z}_j) = 0, & j = 1, \ldots, l, \\ g(x_1, \ldots, x_l, t_1, \ldots, t_r) = 0, \\ v_j t_j = \sum_{i=1}^n u_{i,j} \omega_i, & j = 1, \ldots, r \\ f(\bar{u}_j, v_j, \bar{w}_j) = 0, & j = 1, \ldots, r. \end{cases} \tag{3.4.2}$$

In other words $(y_1, \ldots, y_{2n})$ is in D_2 if and only if for all $j = 1, \ldots, l$ either for some $i = 1, \ldots, n$ we have that $y_{2i,j} = 0$ or $x_j = \sum_{i=1}^n (y_{2i-1,j}/y_{2i,j})\omega_i \in D_1$. Since

$$x_j = \sum_{i=1}^n \frac{\lambda_1(c_i(x_j))}{\lambda_2(c_i(x_j))} \omega_i$$

and for all $j = 1, \ldots, l$ and $i = 1, \ldots, n$ we have that $\lambda_2(c_i(x_j)) \neq 0$, we conclude that $(\bar{\phi}(x_1), \ldots, \bar{\phi}(x_l)) \in D_2$ if and only if $(x_1, \ldots, x_l) \in D_1$. Thus the proposition holds. □

It is clear from Proposition 3.4.5 that in the case of Diophantine generation the first clause of Proposition 3.4.4 applies, and we have the following corollary.

Corollary 3.4.6. *Let R_1 and R_2 be any two rings with $R_1 \leq_{Dioph} R_2$ and HTP unsolvable over R_1. Then HTP is unsolvable over R_2.*

Proof. By Remark 3.4.2 we can assume that R_2 is strongly recursive. Let F be a defining field and let Ω be a Diophantine basis for R_1 over R_2. Then by Proposition 3.4.5 there exists a recursive presentation J of F such that

- $J(R_2)$ is strongly recursive,
- the coordinate functions with respect to basis Ω are recursive,
- the function $\bar{\Lambda}$, as defined in Proposition 3.4.5, is recursive.

Suppose now that $J(R_1)$ is not recursive. As before, $R_1 \leq_{Dioph} R_2$ implies that (3.4.1) holds for some

$$f(x_1, \ldots, x_n, y, z_1, \ldots, z_m) \in R_2[x_1, \ldots, x_n, y, z_1, \ldots, z_m].$$

Let D_{R_1} consist of all the $2n$-tuples $(y_1, \ldots, y_{2n})$ such that there exist $a_1, \ldots, a_n$, $b \in R_2$ satisfying the following system of equations:

$$\begin{cases} y_{2i-1}b = y_{2i}a_i, & i = 1, \ldots, n, \\ f(a_1, \ldots, a_n, b, z_1, \ldots, z_m) = 0. \end{cases} \tag{3.4.3}$$

Then $(y_1, \ldots, y_{2n}) \in D_{R_1}$ if and only if for some $i = 1, \ldots, n$ we have that $y_{2i} = 0$ or $\sum_{i=1}(y_{2i-1}/y_{2i})\omega_i \in R_1$. Thus for any positive integer $l \in J(F)$ we have that $l \in J(R_1)$ if and only if

$$(\Lambda_1(C_1(l)), \Lambda_2(C_1(l)), \quad \ldots, \quad \Lambda_1(C_n(l)), \Lambda_2(C_n(l))) \quad \in \quad \bar{J}(D_{R_1}).$$

Therefore, if $\bar{J}(D_{R_1})$ is recursive then so is $J(R_1)$. Consequently, if $J(R_1)$ is not recursive then R_2 has an undecidable Diophantine set under J and HTP is undecidable under this presentation.

Finally, suppose that $J(R_1)$ is recursive. Then the conditions of Proposition 3.4.5 are satisfied and, by the first clause of Proposition 3.4.4, HTP is undecidable over R_2 under J in this case also. Hence HTP is undecidable under any presentation of R_2. □

We should also like to note here that in practice one can almost always make use of the second clause of Proposition 3.4.4, because there are undecidable Diophantine sets over $\mathbb{Z}$. In particular the following proposition is true.

Proposition 3.4.7. *Let R be a ring with a recursive presentation such that for some positive integer k there exists a map*

$$\bar{\tau} = (\tau_1, \ldots, \tau_k) : \mathbb{Z} \to R^k$$

for which the images of the graphs of addition and multiplication are Diophantine over R. Then R has a Diophantine model of $\mathbb{Z}$ and HTP is undecidable over R.

Proof. First of all, using an inductive argument on the number of operations, similar to the argument used to prove Lemma 3.2.2, one can show that τ^l : $\mathbb{Z}^l \to R^{kl}$ will map any Diophantine subset of $\mathbb{Z}^l$ into a Diophantine subset of R^{kl}. So it remains to be shown that under any recursive presentation of R the map $\bar{\tau}$ will be recursive. Let $J : R \to \mathbb{N}$ be a recursive presentation of R.

By assumption, the set $(\bar{\tau}(m), \bar{\tau}(n), \bar{\tau}(m+n))$ is a Diophantine subset of R^{3k}. Therefore, by Proposition 3.2.5, given that $J(R)$ is recursive, the set

$$S = \{(J^k(\bar{\tau}(m)), J^k(\bar{\tau}(n)), J^k(\bar{\tau}(m+n))), \quad m \in \mathbb{N}\}$$

is recursively enumerable. Let f be a recursive function enumerating S. Define $g : \mathbb{N} \to R^k$ in the following manner. Let $g(0) = J^k(\tau(0))$, $g(3) = J^k(\tau(1))$, $g(6) = J^k(\tau(-1))$. Let m be a positive integer and assume that $g(3^{m-1})$ has been computed. Then define

$$g(3^m) = (\pi_{2k+1}(f(l)), \ldots, \pi_{3k}(f(l))),$$

where l is the smallest positive integer such that for some $\bar{u} \in J^k(R)$ we have that

$$f(l) = (g(3), g(3^{m-1}), \bar{u}).$$

Similarly, for any positive m, assuming that $g(2 \cdot 3^{m-1})$ has been defined already, define

$$g(2 \cdot 3^m) = (\pi_{2k+1}(f(l)), \ldots, \pi_{3k}(f(l))),$$

where l is the smallest positive integer such that for some $\bar{u} \in J^k(R)$ we have

$$f(l) = (g(6), g(2 \cdot 3^{m-1}), \bar{u}).$$

If $t \in \mathbb{N}$ is not of the form $2 \cdot 3^m$ or 3^m for some positive integer m then define $g(t) = (0, \ldots, 0)$ and observe that by the definition of recursive (computable) functions (using the encoding of k-tuples described in Lemma A.1.14), given our adopted presentation $J_{\mathbb{Z}}$ of $\mathbb{Z}$ (see Proposition A.6.1), $g(m)$ is a recursive function with $g(J_{\mathbb{Z}}(z)) = \tau(z)$ for all $z \in \mathbb{Z}$. □

Remark 3.4.8. We finish this chapter with a historical note. The term "Diophantine model" belongs to Gunther Cornelissen who introduced the terminology and the notion, in a slightly different form, in [5].

4

Integrality at finitely many primes and divisibility of order at infinitely many primes

In this chapter we will continue with the task of describing the known Diophantine classes of the rings of $\mathcal{W}$-integers of global fields. We will start with horizontal problems. The question which we will partially answer here is the following. Does the Diophantine class of a ring of $\mathcal{W}$-integers change if we add to or remove from $\mathcal{W}$ finitely many primes? As we will see below, we are able to show in many cases that the class does not change. We conjecture that this is true for all rings of $\mathcal{W}$-integers but are unable to prove this at the present time.

The main tool used so far to prove results of the type described in this section is the strong Hasse norm principle (see Theorem 32.9 of [79]). The ideas behind the construction of a Diophantine definition of integrality at finitely many primes presented below go back to the work of Ershov and of Penzin (see [65], [26]) and to the work of Julia Robinson on the arithmetic definability of rational integers in algebraic number fields (see [81] and [82]). Robinson used quadratic forms to carry out her construction. Later, Rumely generalized Robinson's methods in his paper on arithmetic definability over global fields (see [85]). In his paper Rumely used norm equations and the strong Hasse norm principle. Kim and Roush were the first to use this methodology for the purposes of showing the undecidability of some Diophantine problems over function fields (see [42]). They also were the first to use quadratic forms to show the existential definability of order at a prime over $\mathbb{Q}$, in [43]. The present author has also used various versions of the norm method to resolve some issues of existential definability, in [98], [93], and [102]. Finally, we note that somewhat different ideas were used by Eisenträger in [24] to address integrality at a prime over some global fields.

Before proceeding with the technical material, we note that in this chapter we will use Definition B.1.23 and Notation B.8.1 from the number theory appendix. We start with an exposition of the main technical methods to be used in this chapter.

4.1 The main ideas

In this section we survey the main ideas which are used in the following sections, omitting some technical details and the proofs of some facts in order to simplify the presentation. The omitted details and proofs will, however, appear in the later sections.

4.1.1. Norm equations are polynomial equations Let M/L be a finite extension of global fields. Then the equation $N_{M/L}z = h$ can be rewritten as a polynomial equation with variables and coefficients in L. It suffices to select a basis $\Omega = \{\omega_1, \ldots, \omega_n\}$ of M over L and write $z = \sum_{i=1}^{n} a_i\omega_i$. Then $N_{M/L}z$ becomes a polynomial in $a_1, \ldots, a_n$ with coefficients invariant under the action of any element of $\text{Gal}(M_G/L)$, where M_G is the Galois closure of M over L.

4.1.2. The connection between a norm and divisibility of order Assume that the extension M/L is cyclic and that $\mathfrak{p}$ is a prime of L not splitting in the extension. Then if $h \in L$ is the norm of an element of M, we have that $\text{ord}_{\mathfrak{p}}h \equiv 0 \mod [M : L]$. Thus if $\text{ord}_{\mathfrak{p}}h \not\equiv 0 \mod [M : L]$, the equation $N_{M/L}z = h$ does not have solutions in M.

4.1.3. How having a pole can produce a wrong order at a prime Let $u \in L$ be such that $\text{ord}_{\mathfrak{p}}u = -1$. Let $y \in L$. Let n be any natural number. Then y is integral at $\mathfrak{p}$ if and only if $n \mid \text{ord}_{\mathfrak{p}}(uy^n + u^n)$. From this it is easy to see that we have the following implication. Let $n = [M : L]$, where M/L is a cyclic extension. Let $\mathfrak{p}$ be a prime not splitting in this extension. Then if y has a pole at $\mathfrak{p}$, the equation $N_{M/L}(z) = uy^n + u^n$ has no solution in M. Unfortunately, this will go only halfway in producing the existential definability of order at a prime. We need to make sure that we have solutions to the norm equation when y has no pole at $\mathfrak{p}$. This is where the strong Hasse norm principle comes in.

4.1.4. The role of the strong Hasse norm principle The strong Hasse norm principle asserts that if an equation is cyclic then an element of the field below is a norm if and only if the element is a norm locally at every prime. In other words, if M/L is a cyclic extension as above and $h \in L$ then $N_{M/L}z = h$ has solutions in M if and only if for any $\mathfrak{q}$, a prime of L, and for any $\mathfrak{Q}$, a prime of M above it, $N_{M_{\mathfrak{Q}}/L_{\mathfrak{q}}}v = h$ has a solution $v = v(\mathfrak{Q}) \in M_{\mathfrak{Q}}$. Thus to ensure that our norm equation has a solution when h does not have a pole at $\mathfrak{p}$, we can work locally.

4.1.5. The making of a local norm To begin with, the problem will be immediately solved for all but finitely many primes. Indeed, if $\mathfrak{q}$ is a prime of L and h

is a unit at $\mathfrak{q}$, while $\mathfrak{q}$ is unramified in the local (and therefore global) extension, h is automatically a local norm at $\mathfrak{q}$. Thus, we just have to worry about the primes which are ramified and those which occur in the divisor of h.

Next we observe how to take care of the unramified primes occurring in the divisor of h. If for some unramified prime $\mathfrak{q}$ of L we have that $\text{ord}_{\mathfrak{q}} h \equiv 0 \bmod n = [M : L]$, then $h = \pi_{\mathfrak{q}}^n \varepsilon_{\mathfrak{q}}$, where $\pi_{\mathfrak{q}} \in L$ is of order 1 at $\mathfrak{q}$ and $\varepsilon_{\mathfrak{q}} \in L$ is a unit at $\mathfrak{q}$. Since M/L is Galois, the relative degree of any prime will be a divisor of n. (The relative degree of a prime and the divisibility requirement are explained in Proposition B.1.11 and Lemma B.4.1.) Thus we can write $n = f(\mathfrak{Q}/\mathfrak{q})n_1 = [M_{\mathfrak{Q}} : L_{\mathfrak{q}}]n_1$, where $\mathfrak{Q}$ lies above $\mathfrak{q}$ in M, and observe that $N_{M_{\mathfrak{Q}}/L_{\mathfrak{q}}} \pi_{\mathfrak{q}}^{n_1} = \pi_{\mathfrak{q}}^n$. Therefore h is a norm at $\mathfrak{q}$ if and only if $\varepsilon_{\mathfrak{q}}$ is a norm. But the last assertion is true if $\mathfrak{q}$ is not ramified by the argument above.

So how can we arrange for the order of h at all the "non-involved" primes occurring in its divisor to be divisible by n? We do this by introducing another extension G of L which will depend on h and be such that in the extension GL/L all the primes occurring in the divisor of h with orders not divisible by n ramify with ramification degree n. (Roughly speaking, we will be taking the nth root of h. However, we will have to be careful not to change the divisibility of order at $\mathfrak{p}$, and so some details will have to be adjusted.) Thus in the end we will be aiming to solve the norm equation $\mathbf{N}_{MG/LG} z = h$.

Finally we will have to deal with ramified primes, but only in the number field case. In the function field case, we can always use a constant field extension to avoid this issue. In the number field case we arrange for h to be an nth power in $L_{\mathfrak{q}}$ for every ramified prime $\mathfrak{q}$. To ensure this, it is enough, by Hensel's lemma, to show that $h\pi_{\mathfrak{q}}^{-\text{ord}_{\mathfrak{q}} h}$, where $\text{ord}_{\mathfrak{q}} \pi_{\mathfrak{q}} = 1$, is an nth power modulo a sufficiently high power of $\mathfrak{q}$.

4.2 Integrality at finitely many primes in number fields

In this section we will show that integrality at finitely many primes is existentially definable over a number field. We will start by describing the notation to be used in this section.

Notation 4.2.1.

- Let K denote a number field.
- Let $q > 2$ be a rational prime such that K contains a primitive qth root of unity ξ_q.

- Let $\mathfrak{P}_1, \ldots, \mathfrak{P}_l$ be primes of K not lying above q.
- Let $a \in O_K$ be such that a is not a qth power modulo $\mathfrak{P}_i$ for any $i = 1, \ldots, l$. (In particular, $\mathrm{ord}_{\mathfrak{P}_i} a = 0$ for all $i = 1, \ldots, l$.) Let $\prod_{i=1}^{l} \mathfrak{a}_i^{r_i}$ be the divisor of a. (Note that such an a always exists by the Strong Approximation Theorem; see Theorem B.2.1.)
- For each $i = 1, \ldots, l$, let A_i be the rational prime below $\mathfrak{a}_i$.
- Let $\mathfrak{C}$ be a prime of K distinct from the $\mathfrak{P}_i$ and relatively prime to aq.
- Let $\mathfrak{D} = \prod_j \mathfrak{d}_j^{d_j}$ an integral and possibly trivial divisor of K, be a product of K-primes $\mathfrak{d}_j$ that is distinct from all $\mathfrak{P}_i$, $\mathfrak{C}$, and prime factors of q and a.
- Let $g \in K$ satisfy the following conditions:
  (a) $g \equiv 1 \bmod aq^3$;
  (b) the divisor of g is of the form $\prod \mathfrak{P}_i^{s_i}/\mathfrak{C}\mathfrak{D}$, where $s_i \in \mathbb{N} \setminus \{0\}$, $s_i \not\equiv 0 \bmod q$. (Such a $g \in K$ exists for some $\mathfrak{D}$ as above, by Proposition B.2.2.)
- Let $r = q^{(3q[K:\mathbb{Q}])}(q^{(q[K:\mathbb{Q}])!} - 1)\left(\prod (A_i^{(q[K:\mathbb{Q}])!} - 1)\right)$.
- Let $s = \max(s_i, d_j, 3qr_i[K:\mathbb{Q}])$.
- Let $v \in O_K$ be such that $v \equiv 1 \bmod (q^3 a)^r$, $\mathrm{ord}_{\mathfrak{d}_j} v = d_j$, and $\mathrm{ord}_{\mathfrak{C}} v = 1$. (The existence of v follows from the Strong Approximation Theorem when it is applied in the same way as in the proof of Proposition B.2.2)

We start with a proposition containing the technical core of the section.

Proposition 4.2.2. *Let $x \in K$ and let*

$$h = (q^3 a)^r \left(g^{-1} x^{r(s+1)} + g^{-q}\right) + v^2.$$

Let $b \in O_K, b \not\equiv 0 \bmod \mathfrak{P}_i$ *for any* $i = 1, \ldots, l$. *Let* $c \in K, c \equiv gb^q \bmod \prod \mathfrak{P}_i$, $\mathrm{ord}_{\mathfrak{C}} c = -q - 1$, $\mathrm{ord}_{\mathfrak{d}_j} c = -qd_j - 1$ *and let c be integral at all the other primes.* (Such a c exists, again by the Strong Approximation Theorem.) *Let β_x be a root of*

$$T^q - (h^{-1} + c).$$

Let $\alpha \in \tilde{\mathbb{Q}}$ (the algebraic closure of $\mathbb{Q}$) be a root of the polynomial $X^q - a$. Then

1. $[K(\beta_x) : K] = q$;
2. *the equation*

$$\prod_{j=0}^{q-1} \left(a_0 + a_1 \xi_q^j \alpha + \cdots + a_{q-1} \xi_q^{j(q-1)} \alpha^{q-1}\right) = h \qquad (4.2.1)$$

has solutions $a_0, \ldots, a_{q-1} \in K(\beta_x)$ *if and only if* x *is integral at* $\mathfrak{P}_i, i = 1, \ldots, l$.

Proof. First of all we observe the following. If $\text{ord}_{\mathfrak{C}} x > 0$ then

$$\begin{aligned}\text{ord}_{\mathfrak{C}} h &= \text{ord}_{\mathfrak{C}}\big((q^3 a)^r \big(g^{-1} x^{r(s+1)} + g^{-q}\big) + v^2\big)\\ &= \min\big(\text{ord}_{\mathfrak{C}}\big(g^{-1} x^{r(s+1)} + g^{-q}\big), \text{ord}_{\mathfrak{C}} v^2\big)\\ &= \min(q, 2) = 2.\end{aligned}$$

If $\text{ord}_{\mathfrak{C}} x = 0$ then

$$\begin{aligned}\text{ord}_{\mathfrak{C}} h &= \text{ord}_{\mathfrak{C}}\big((q^3 a)^r \big(g^{-1} x^{r(s+1)} + g^{-q}\big) + v^2\big)\\ &= \min\big(\text{ord}_{\mathfrak{C}}\big(g^{-1} x^{r(s+1)} + g^{-q}\big), \text{ord}_{\mathfrak{C}} v^2\big)\\ &= \min(1, 2) = 1.\end{aligned}$$

If $\text{ord}_{\mathfrak{C}} x < 0$ then

$$\begin{aligned}\text{ord}_{\mathfrak{C}} h &= \text{ord}_{\mathfrak{C}}\big((q^3 a)^r \big(g^{-1} x^{r(s+1)} + g^{-q}\big) + v^2\big)\\ &= \min\big(\text{ord}_{\mathfrak{C}}\big(g^{-1} x^{r(s+1)} + g^{-q}\big), \text{ord}_{\mathfrak{C}} v^2\big)\\ &= \min(1 + r(s+1)\,\text{ord}_{\mathfrak{C}} x, q) < -2.\end{aligned}$$

Thus at $\mathfrak{C}$ either h has a pole of order greater than 2 or a zero of order 1 or 2. Therefore, at $\mathfrak{C}$ it is the case that $h^{-1} + c$ has a pole of order $q + 1$. Hence $h^{-1} + c$ is not a qth power in K and thus by Lemma B.4.11 we have that $[K(\beta_x) : K] = q$.

Before we proceed to the second assertion of the lemma we would like to note the following. By Lemma B.4.11, for any $x \in K$ it is the case that $\mathfrak{C}$ is completely ramified in the extension $K(\beta_x)/K$. Further, consider what happens to prime factors of $\mathfrak{D}$ under this extension: these primes behave like $\mathfrak{C}$. Indeed, if $\text{ord}_{\mathfrak{d}_j} x > 0$ then

$$\begin{aligned}\text{ord}_{\mathfrak{d}_j} h &= \text{ord}_{\mathfrak{d}_j}\big((q^3 a)^r \big(g^{-1} x^{r(s+1)} + g^{-q}\big) + v^2\big)\\ &= \min\big(\text{ord}_{\mathfrak{d}_j}\big(g^{-1} x^{r(s+1)} + g^{-q}\big), \text{ord}_{\mathfrak{d}_j} v^2\big)\\ &= \min(q d_j, 2d_j) = 2d_j.\end{aligned}$$

If $\text{ord}_{\mathfrak{d}_j} x = 0$ then

$$\begin{aligned}\text{ord}_{\mathfrak{d}_j} h &= \text{ord}_{\mathfrak{d}_j}\big((q^3 a)^r \big(g^{-1} x^{r(s+1)} + g^{-q}\big) + v^2\big)\\ &= \min\big(\text{ord}_{\mathfrak{d}_j}\big(g^{-1} x^{r(s+1)} + g^{-q}\big), \text{ord}_{\mathfrak{d}_j} v^2\big)\\ &= \min(d_j, 2d_j) = d_j.\end{aligned}$$

If $\text{ord}_{\mathfrak{d}_j} x < 0$ then

$$\begin{aligned}\text{ord}_{\mathfrak{d}_j} h &= \text{ord}_{\mathfrak{d}_j}\left((q^3 a)^r \left(g^{-1} x^{r(s+1)} + g^{-q}\right) + v^2\right)\\ &= \min\left(\text{ord}_{\mathfrak{d}_j}\left(g^{-1}x^{r(s+1)} + g^{-q}\right), \text{ord}_{\mathfrak{d}_j} v^2\right)\\ &= \min(d_j + r(s+1)\,\text{ord}_{\mathfrak{d}_j} x, qd_j) < -2d_j.\end{aligned}$$

Thus at $\mathfrak{d}_j$ either h has a pole of order greater than $2d_j$ or a zero of order d_j or $2d_j$. Therefore at $\mathfrak{d}_j$ we have that $h^{-1} + c$ has a pole of order $qd_j + 1$. So, just as for the case of $\mathfrak{C}$, by Lemma B.4.11, for all j, for any $x \in K$, we have that $\mathfrak{d}_j$ is completely ramified in the extension $K(\beta_x)/K$.

We now turn to the second assertion of the lemma. Note that if for some i we have that $\text{ord}_{\mathfrak{P}_i} x < 0$ then

$$\text{ord}_{\mathfrak{P}_i}\left(g^{-1}x^{r(s+1)} + g^{-q}\right) < 0$$

and

$$\begin{aligned}\text{ord}_{\mathfrak{P}_i}\left(g^{-1}x^{r(s+1)} + g^{-q}\right) &= \text{ord}_{\mathfrak{P}_i} g^{-1} x^{r(s+1)}\\ &= r(s+1)\text{ord}_{\mathfrak{P}_i} x - s_i \equiv -s_i \bmod q.\end{aligned}$$

Thus

$$\text{ord}_{\mathfrak{P}_i} h = \text{ord}_{\mathfrak{P}_i}\left((q^3a)^r\left(g^{-1}x^{r(s+1)} + g^{-q}\right) + v^2\right) \not\equiv 0 \bmod q.$$

However, if $\text{ord}_{\mathfrak{P}_i} x \geq 0$ then

$$\text{ord}_{\mathfrak{P}_i} h = \text{ord}_{\mathfrak{P}_i}\left((q^3a)^r\left(g^{-1}x^{r(s+1)} + g^{-q}\right) + v^2\right) = \text{ord}_{\mathfrak{P}_i} g^{-q} \equiv 0 \bmod q,$$

and

$$\text{ord}_{\mathfrak{P}_i} h = \text{ord}_{\mathfrak{P}_i}\left((q^3a)^r\left(g^{-1}x^{r(s+1)} + g^{-q}\right) + v^2\right) < 0.$$

Consequently, in this case $\text{ord}_{\mathfrak{P}_i} h \equiv 0 \bmod q$. (Compare with subsection 4.1.3.) Observe that in either case $h^{-1} \equiv 0 \bmod \mathfrak{P}_i$.

Next note that $h^{-1} + c \equiv b^q \not\equiv 0 \bmod \mathfrak{P}_i$. Thus, by Lemma B.4.11, in the extension $K(\beta_x)/K$ each $\mathfrak{P}_i$ is not ramified and splits completely. Therefore the residue fields of all the factors of $\mathfrak{P}_i$ in $K(\beta_x)$ are the same as the residue field of $\mathfrak{P}_i$ in K. Consequently, a is not a qth power modulo any factor of $\mathfrak{P}_i$ in $K(\beta_x)$ and each of these factors remains prime in the extension $K(\alpha, \beta_x)/K(\beta_x)$, again by Lemma B.4.11. This also means that $[K(\alpha, \beta_x) : K(\beta_x)] = q$ by Lemma B.4.11 yet again. Observe further that the equation (4.2.1) has solutions $a_0, \ldots, a_{q-1} \in K(\beta_x)$ if and only if

$$\mathbf{N}_{K(\alpha,\beta_x)/K(\beta_x)}(z) = h$$

has a solution $z \in K(\alpha, \beta_x)$.

Further, by Lemma B.4.11, that $\text{ord}_{\mathfrak{P}_i} h \not\equiv 0 \bmod q$ for some i implies that (4.2.1) does not have solutions in $K(\beta_x, \alpha)$. (Compare with Subsection 4.1.2.)

Next we show that if $\text{ord}_{\mathfrak{P}_i} h \equiv 0 \bmod q$ for all $i = 1, \ldots, l$ then (4.2.1) will have solutions in $K(\beta_x)$. By the strong Hasse norm principle, an element of $K(\beta_x)$ is a $K(\beta_x, \alpha)$-norm if and only if it is a norm locally at all the valuations. To determine which elements are local norms, as discussed in Subsection 4.1.5, we use two main devices. First we note that, locally, every unit is a norm in a non-ramified extension (see Lemma B.8.5). Second, using Lemmas B.8.3 and B.8.4 one can derive the following. Assume that γ, κ are units of a local field $K_{\mathfrak{q}}$ with $\text{ord}_{\mathfrak{q}}(\kappa - \gamma^q) > 2\,\text{ord}_{\mathfrak{q}}\, q + 1$. Then κ is a local qth power. Note further that in an extension of degree q every qth power of the field below is a norm of an element from the field above.

Suppose now that the order of h is divisible by q at all $\mathfrak{P}_i$. Let $\tilde{\mathfrak{P}}_i$ be a factor of $\mathfrak{P}_i$ in $K(\beta_x)$. Then $\text{ord}_{\tilde{\mathfrak{P}}_i} h$ is 0 mod q. Let $\pi_{\tilde{\mathfrak{P}}_i}$ be a local uniformizing parameter with respect to $\tilde{\mathfrak{P}}_i$ (in other words, $\text{ord}_{\mathfrak{P}_i} \pi_{\tilde{\mathfrak{P}}_i} = 1$). Then h is a norm locally at $\tilde{\mathfrak{P}}_i$ in the extension $K(\beta_x, \alpha)/K(\beta_x)$ if and only if

$$u = h\pi_{\tilde{\mathfrak{P}}_i}^{-\text{ord}_{\tilde{\mathfrak{P}}_i} h}$$

is a local norm at $\tilde{\mathfrak{P}}_i$. But u is unit at $\tilde{\mathfrak{P}}_i$ and $\tilde{\mathfrak{P}}_i$ is not ramified in the extension $K(\beta_x, \alpha)/K(\beta_x)$. Therefore, by the observation above, u is a norm locally with respect to $\tilde{\mathfrak{P}}_i$.

Next we note the following. On the one hand, if $\mathfrak{T}$ is a prime of K such that $\text{ord}_{\mathfrak{T}} h > 0$ then either $\text{ord}_{\mathfrak{T}} h \equiv 0 \bmod q$ or $\mathfrak{T}$ is ramified completely in the extension $K(\beta_x)/K$. Indeed, all the poles of c are poles of g, i.e. c has a pole at $\mathfrak{C}$ and at all the $\mathfrak{d}_j$, and we have already established that these primes are completely ramified in the extension in question. Thus, any zero of h which is not $\mathfrak{C}$ or a $\mathfrak{d}_j$ is also a pole of $h^{-1} + c$ and the assertion is true. On the other hand, by the construction of h, if $\mathfrak{T} \neq \mathfrak{P}_i$, $\mathfrak{T} \neq \mathfrak{C}$, $\mathfrak{T} \neq \mathfrak{d}_j$ is a prime of K such that $\text{ord}_{\mathfrak{T}} h < 0$ then $\text{ord}_{\mathfrak{T}} h \equiv 0 \bmod q$. Thus, if $\mathfrak{Z}$ is any prime of $K(\beta_x)$ which is not a factor of any $\mathfrak{P}_i$ and such that $\mathfrak{Z}$ occurs in the divisor of h then $\text{ord}_{\mathfrak{Z}} h \equiv 0 \bmod q$. Hence, if $\mathfrak{Z}$ is not a factor of any $\mathfrak{P}_i$ or of q and does not occur in the divisor of a then $\mathfrak{Z}$ is unramified in the extension $K(\beta_x, \alpha)/K(\beta_x)$ and h is a norm locally at $\mathfrak{Z}$.

Finally, we consider the primes $\mathfrak{Z}$ which are factors of q or occur in the divisor of a and thus may ramify in the extension $K(\beta_x, \alpha)/K$. First assume that x is integral at $\mathfrak{Z}$. Then

$$\text{ord}_{\mathfrak{Z}}(h - 1) = r\,\text{ord}_{\mathfrak{Z}}\, a \tag{4.2.2}$$

if $\mathfrak{Z}$ occurs in the divisor of a, and

$$\text{ord}_{\mathfrak{Z}}(h-1) = 3r\,\text{ord}_{\mathfrak{Z}}\,q \tag{4.2.3}$$

if $\mathfrak{Z}$ is a divisor of q.

If, however, x has a pole at $\mathfrak{Z}$ then by the definition of s we have that $(q^3a)^r x^{r(s+1)}$ has a pole at $\mathfrak{Z}$. Indeed, assume first that $\mathfrak{Z}$ is a factor of q. Then

$$\begin{aligned}\text{ord}_{\mathfrak{Z}}(q^3a)^r x^{r(s+1)} &= 3r\,\text{ord}_{\mathfrak{Z}}\,q + r(s+1)\,\text{ord}_{\mathfrak{Z}}\,x < 3r[K:\mathbb{Q}] - r(s+1)\\ &\leq 3r[K:\mathbb{Q}] - r(3[K:\mathbb{Q}]+1) < 0.\end{aligned}$$

Similarly, if $\mathfrak{Z}$ occurs in the divisor of a then

$$\begin{aligned}\text{ord}_{\mathfrak{Z}}(q^3a)^r x^{r(s+1)} &= r\,\text{ord}_{\mathfrak{Z}}\,a + r(s+1)\,\text{ord}_{\mathfrak{Z}}\,x\\ &< rr_i[K:\mathbb{Q}] - r(s+1)\\ &\leq rr_i[K:\mathbb{Q}] - r(3r_i[K:\mathbb{Q}]+1) < 0.\end{aligned}$$

Next we note that if $\mathfrak{Z}$ occurs in the divisor of a and $z \in \mathbb{N}$ is the size of the residue field of $\mathfrak{Z}$ then $r \equiv 0 \bmod (z-1)$. Indeed, the size of the residue field is A_i^f, where $1 \leq f \leq [K:\mathbb{Q}]$. Thus $f \mid ([K:\mathbb{Q}]q)!$ and $(A_i^f - 1) \mid (A_i^{([K:\mathbb{Q}]q)!} - 1)$. Finally, by construction $r \equiv 0 \bmod (A_i^{([K:\mathbb{Q}]q)!} - 1)$. Thus any $\mathfrak{Z}$-unit ε raised to the power r is equivalent to 1 modulo any prime $\mathfrak{Z}$ occurring in the divisor of a.

Suppose now that $\mathfrak{Z}$ is a factor of q. Let e be the ramification of this factor over $\mathbb{Q}$ and let f be its relative degree over $\mathbb{Q}$. Consider the size of the multiplicative group of the finite ring $O_K/\mathfrak{Z}^{3e}$. It is equal to $q^{3ef} - q^{f(3e-1)} = q^{3ef-f}(q^f-1)$. By construction,

$$r \equiv 0 \bmod q^{3ef} - q^{f(3e-1)} = q^{3ef-f}(q^f-1).$$

Thus, any element of K prime to a factor $\mathfrak{Z}$ of q in K and raised to the power r will be equivalent to 1 modulo $\mathfrak{Z}^{3e(\mathfrak{Z}/q)}$.

Next let $\mathfrak{Z}$ either be a factor of q or occur in the divisor of a. Let $\Pi = \Pi_{\mathfrak{Z}}$ be a local uniformizing parameter. Then for some non-zero $u \in \mathbb{N}$ we have that $(\Pi^u q^3 a x^{s+1})^r$ is a unit at $\mathfrak{Z}$. Further, from the discussion above it follows that if $\mathfrak{Z}$ occurs in the divisor of a then $(\Pi^u q^3 a x^{s+1})^r \equiv 1 \bmod \mathfrak{Z}$, and if $\mathfrak{Z}$ is a factor of q in K then $(\Pi^u q^3 a x^{s+1})^r \equiv 1 \bmod \mathfrak{Z}^{3e}$, where $e = e(\mathfrak{Z}/q)$. Next, consider $h(\Pi^u)^r$. Note that, for any $K(\beta_x)$-factor $\mathfrak{Z}_x$ of any K-prime $\mathfrak{Z}$, we have that h is a local norm at $\mathfrak{Z}_x$ in the extension $K(\beta_x, \alpha)/K(\beta_x)$ if and only if $h(\Pi^u)^r$ is a local norm at $\mathfrak{Z}_x$, since the local degree is either 1 or q and $r \equiv 0 \bmod q$. At the same time, depending on whether $\mathfrak{Z}$ is a factor of q or occurs in the divisor of a,

$$\begin{aligned}h(\Pi^u)^r &= (\Pi^u)^r\big((q^3a)^r\big(g^{-1}x^{r(s+1)} + g^{-q}\big) + v^2\big)\\ &= (\Pi^u)^r(q^3a)^r x^{r(s+1)} g^{-1} + (\Pi^u)^r (q^3a)^r g^{-q} + (\Pi^u)^r v^2\\ &\equiv g^{-1} \bmod \mathfrak{Z}^v \equiv 1 \bmod \mathfrak{Z}^v,\end{aligned}$$

where v is 1 if $\mathfrak{Z}$ occurs in the divisor of a and $v = 3e(\mathfrak{Z}/q)$ if $\mathfrak{Z}$ is a factor of q. Thus, by Hensel's lemma, $h(\Pi^u)^r$ is a qth power in K completed at $\mathfrak{Z}$. Consequently $h(\Pi^u)^r$ is also a qth power locally at all the factors $\mathfrak{Z}_x$ of $\mathfrak{Z}$ in $K(\beta_x)$ and therefore it is a $K(\alpha, \beta_x)$-norm in $K(\beta_x)$.

Thus h is a local norm in $K(\beta_x)$ at all the non-archimedean valuations. □

Since K contains the qth roots of unity for $q > 2$, K does not have any real embeddings and therefore all the archimedean completions of K are equal to $\mathbb{C}$. Hence h is automatically a local norm under all the archimedean completions.

Corollary 4.2.3. *The set of elements of K integral at $\mathfrak{P}_i$ for $i = 1, \ldots, l$ is Diophantine over K.*

Proof. First consider the left-hand side of equation (4.2.1) as a polynomial in the variables $a_0, \ldots, a_{q-1}$. The coefficients of this equation are in $K(\alpha)$. Further observation indicates that these coefficients are not moved by any element of the Galois group of $K(\alpha)$ over K and thus must be in K. So let

$$N(a_0, \ldots, a_{q-1}) = \prod_{i=0}^{q-1} \left(a_0 + a_1 \xi_q^i \alpha + \cdots + a_{q-1} \xi_q^{i(q-1)} \alpha^{q-1}\right) \in K[a_0, \ldots, a_{q-1}].$$

The remaining task, of rewriting all the equations so that all the variables range over K, can be carried out using coordinate polynomials and pseudo-coordinate polynomials as in Lemma B.7.5. We leave the details to the reader. □

We will now state the main result of this section.

Theorem 4.2.4. *Let M be any number field. Let $\mathfrak{p}$ be any prime of M. Then the set of elements of M integral at $\mathfrak{p}$ is Diophantine over M.*

Proof. Let $K = M(\xi_q)$, where ξ_q is a primitive qth root of unity. Let $\mathfrak{p} = \prod_{i=1}^{l} \mathfrak{P}_i^{e_i}$ be the factorization of $\mathfrak{p}$ in K. Note that since $[K : M] < q$ we have that $e_i < q$ for all i. Then by Corollary 4.2.3, the set I_K of elements of K integral at $\mathfrak{P}_1, \ldots, \mathfrak{P}_l$ is Diophantine over K. Thus by the "going up and then down" method (see Subsection 2.1.17), we have that $I_K \cap M$ is Diophantine over M. But $I_K \cap M$ is precisely the set of elements of M integral at $\mathfrak{p}$. □

4.3 Integrality at finitely many primes over function fields

In this section we will show that integrality at finitely many primes is existentially definable over function fields over finite fields of constants. In addition to Definition B.1.23 and Notation B.8.1, in this section we will also use the following notation.

4.3.1. Notation and assumptions

- Let K denote a function field of characteristic $p > 0$ over a finite field of constants C_K.
- Let $\tilde{K}$ denote the algebraic closure of K.
- Let q denote a rational prime number distinct from p.
- Let K contain all the qth roots of unity.
- Let K contain an element f with a divisor of the form

$$\frac{\prod_{i=1}^{m} \mathfrak{b}_i^{s_i}}{\prod_{j=1}^{l} \mathfrak{a}_j^{qr_j}}, \tag{4.3.1}$$

  where for $i = 1, \ldots, m$ and $j = 1, \ldots, l$ we have that $\mathfrak{a}_j$ and $\mathfrak{b}_i$ are distinct prime divisors of degree 1 and where $(s_i, q) = 1$.
  (The existence of f will be demonstrated later.)
- Let $\mathfrak{B} = \prod_{i=1}^{m} \mathfrak{b}_i$.
- Let $a \in C_K$ be such that the equation

$$x^q - a = 0 \tag{4.3.2}$$

  has no solution in K.
- Let $z_1 \in K$ be such that it has a pole of order 1 at $\mathfrak{a}_1$ and $z_1 \equiv b^q \bmod \mathfrak{B}$ for some $b \in C_K \setminus \{0\}$. (The existence of such a z_1 can be deduced from the Strong Approximation Theorem; see Theorem B.2.1.)
- Let $k > 0, k \not\equiv 0 \bmod q$, be greater than the order of any pole of z_1 in K. Then let $z_2 \in K$ be such that, for some $a \in \mathbb{N}$ with $a > \log_q kr_1 + 2$, it is the case that z_2 has a pole of order q^a at $\mathfrak{a}_1$, is equivalent to 1 mod $\mathfrak{B}$, and is integral at all the other primes of K. (The existence of $z_2 \in K$ follows from Lemma B.2.3.)
- Let $z = z_1 z_2$ and observe that:
  (a) $\text{ord}_{\mathfrak{a}_1} z = -q^a - 1 < -q^2 r_1 k - 1$;
  (b) $z \equiv b^q \bmod \mathfrak{B}$;
  (c) if $\text{ord}_{\mathfrak{c}} z < 0$ then $\text{ord}_{\mathfrak{c}} z > -k$ for any $\mathfrak{c} \neq \mathfrak{a}_1$.

The following proposition constitutes the technical core of this section.

Proposition 4.3.2. *Let $w \in K$. Let h be defined by the equation*

$$h = f^{-1}w^{q(s+1)} + f^{-q}, \tag{4.3.3}$$

where $s = \max(s_1, \ldots, s_m, r_1, \ldots, r_l)$. Let $\beta_w \in \tilde{K}$ be a root of the equation

$$T^q - (h^{-k} + z) = 0. \tag{4.3.4}$$

Then, for all w, we have that β_w is of degree q over K, α is of degree q over $K(\beta_w)$, and the following equation has solutions $a_0, \ldots, a_{q-1} \in K(\beta_w)$ if and only if $\text{ord}_{\mathfrak{b}_i} w \geq 0$ for all $i = 1, \ldots, m$:

$$\prod_{i=0}^{q-1} \left(a_0 + a_1\xi_q^i\alpha + \cdots + a_{q-1}\xi_q^{i(q-1)}\alpha^{q-1}\right) = h. \tag{4.3.5}$$

Proof. First of all, we will show that for all $w \in K$ we have that

$$[K(\beta_w) : K] = [K(\beta_w, \alpha) : K(\beta_w)] = q. \tag{4.3.6}$$

In order to show that (4.3.6) holds, we will show that in the extension $K(\beta_w)/K$ at least one prime will have ramification degree q while the degree of the extension is at most q. Since this extension above is separable, the presence of a totally ramified prime implies that adjoining β_w to K does not result in a constant field extension, by Lemma B.4.17. Thus, since α is of degree q over C_K, it will remain of degree q over the constant field of $K(\beta_w)$.

Observe that in K it is the case that f has a pole of order qr_1 at $\mathfrak{a}_1$, so that f^{-1} and f^{-q} have zeros of order qr_1 and q^2r_1 respectively at $\mathfrak{a}_1$. Therefore, if w has a pole at $\mathfrak{a}_1$ then

$$\text{ord}_{\mathfrak{a}_1} h = \text{ord}_{\mathfrak{a}_1} f^{-1}w^{(s+1)q} + f^{-q} = q(s+1)\,\text{ord}_{\mathfrak{a}_1} w + qr_1 < 0.$$

If w is a unit at $\mathfrak{a}_1$ then

$$\text{ord}_{\mathfrak{a}_1} h = \text{ord}_{\mathfrak{a}_1} f^{-1}w^{q(s+1)} + f^{-q} = -\text{ord}_{\mathfrak{a}_1} f = qr_1.$$

If w has a zero at $\mathfrak{a}_1$ then

$$\text{ord}_{\mathfrak{a}_1} h = \text{ord}_{\mathfrak{a}_1} f^{-1}w^{q(s+1)} + f^{-q} = -q\,\text{ord}_{\mathfrak{a}_1} f = q^2r_1.$$

Thus, at $\mathfrak{a}_1$, it is the case that h either has a pole or a zero of degree at most q^2r_1. Now consider $h^{-k} + z$. Since at $\mathfrak{a}_1$ we have that z has a pole of order greater than q^2kr_1,

$$\text{ord}_{\mathfrak{a}_1}(h^{-k} + z) = \text{ord}_{\mathfrak{a}_1} z = -(q^a + 1).$$

Therefore, by Lemma B.4.11, we know that $\mathfrak{a}_1$ will ramify completely in the extension $K(\beta_w)/K$. Hence, this extension is of degree q as noted above.

Since at least one prime is ramified completely and the extension is separable, the constant field of $K(\beta_w)$ is the same as the constant field of K. Thus α is of degree q over $K(\beta_w)$, as promised.

For future use, also note that any valuation that is a zero of h is also a pole of $h^{-k}+z$. Further, the order of $h^{-k}+z$ at any such valuation except for $\mathfrak{a}_1$ is divisible by q if and only if the order of h at this valuation is divisible by q. Thus, if h has a zero at some prime $\mathfrak{t}$ and $\text{ord}_{\mathfrak{t}} h \not\equiv 0$ modulo q in K then $\mathfrak{t}$ ramifies completely in the extensions $K(\beta_w)/K$.

We will now proceed to the proof of the second statement of the lemma. Note that, as in Subsection 4.1.1 and the number field case, again the existence of solutions $a_0, \ldots, a_{p-1} \in K(\beta_w)$ to (4.3.5) is equivalent to the existence of a $u \in K(\alpha, \beta_w)$ such that

$$\mathbf{N}_{K(\alpha,\beta_w)/K(\beta_w)}(u) = h. \tag{4.3.7}$$

Suppose that, as in Subsection 4.1.3 and the number field case, $w \in K$ has a pole at some $\mathfrak{b}_i$. Then, in K, we have that

$$\text{ord}_{\mathfrak{b}_i} h = \text{ord}_{\mathfrak{b}_i}\left(f^{-1}w^{q(s+1)} + f^{-q}\right) = q(s+1)\,\text{ord}_{\mathfrak{b}_i} w - s_i \not\equiv 0 \bmod q.$$

Further,

$$\text{ord}_{\mathfrak{b}_i} h < 0.$$

Next observe the following: $h^{-k}+z \equiv b^q \not\equiv 0 \bmod \mathfrak{B}$. Thus $\mathfrak{b}_i$ does not divide the discriminant of the power basis of β_w and therefore it does not ramify in the extension $K(\beta_w)/K$, by Lemma B.4.11. Now, since the polynomial $T^q - (h^{-k}+z)$ splits completely modulo $\mathfrak{b}_i$, by Lemma B.4.11, we may conclude that $\mathfrak{b}_i$ splits completely in the extension $K(\beta_w)/K$ and that the order of h at any factor of $\mathfrak{b}_i$ is not divisible by q in $K(\beta_w)$. Further, since there is no constant field extension, and $\mathfrak{b}_i$ is of degree 1 in K, each factor of $\mathfrak{b}_i$ will be of degree 1 in $K(\beta_w)$. Because (4.3.2) still has no solution in $K(\beta_w)$, we conclude that (4.3.2) has no solution modulo any factor of $\mathfrak{b}_i$ in $K(\beta_w)$. Hence, again by Lemma B.4.11, every factor of $\mathfrak{b}_i$ in $K(\beta_w)$ remains prime in $K(\beta_w, \alpha)$ and (4.3.7) will have no solution in $K(\alpha, \beta_w)$. (See Subsection 4.1.2 again, as in the number field case.)

Suppose now that w does not have a pole at any $\mathfrak{b}_i$ for any $i = 1, \ldots, m$. We will show that in this case (4.3.7) will have a solution in $K(\alpha, \beta_w)$. By the strong Hasse norm principle applied to the function field case, it is enough to show that for all primes $\mathfrak{t}$ of K we have that h is a local norm. Note that no prime ramifies in the extension $K(\alpha, \beta_w)/K(\beta_w)$. Thus, as discussed in Subsection 4.1.5 and in the number field case, if h is a unit at some prime $\mathfrak{t}$ of K then it is automatically a local norm at $\mathfrak{t}$ by Lemma B.8.5. Suppose that on the one hand $\mathfrak{t}$ is a pole of h.

Then either it is a factor of $\mathfrak{B}$ or it is a pole of w. Since w has no pole at any factor of $\mathfrak{B}$, direct calculation assures us that h will have a pole at every factor of $\mathfrak{B}$ of order divisible by q. On the other hand, if $\mathfrak{t}$ is a pole of w then, again by direct calculation, one can see that h will also have a pole at $\mathfrak{t}$ of order divisible by q.

Assume now that $\mathfrak{t}$ is a zero of h. Then by the argument above it is a pole of $h^{-k} + z$. Further, since $k \not\equiv 0 \bmod q$, for $\mathfrak{t} \neq \mathfrak{a}_1$ we have $\text{ord}_{\mathfrak{t}} h \not\equiv 0 \bmod q$ if and only if $\text{ord}_{\mathfrak{t}} h^{-k} + z \not\equiv 0 \bmod q$. Thus, if $\text{ord}_{\mathfrak{t}} h \not\equiv 0 \bmod q$, $\mathfrak{t}$ is ramified completely in the extension $K(\beta_w)/K$. As discussed above, $\mathfrak{a}_1$ ramifies in the extension $K(\beta_w)/K$ under all circumstances. Hence, if $\mathfrak{t}_{K(\beta_w)}$ is a prime of $K(\beta_w)$ lying above $\mathfrak{t}$ then $\text{ord}_{\mathfrak{t}_{K(\beta_w)}} h \equiv 0 \bmod q$.

Summarizing the above discussion, we conclude that for every prime $\mathfrak{t}_{K(\beta_w)}$ of $K(\beta_w)$ we have that

$$h = \pi_{\mathfrak{t}_{K(\beta_w)}}^{bq} \varepsilon_{\mathfrak{t}_{K(\beta_w)}},$$

where

$$\text{ord}_{\mathfrak{t}_{K(\beta_w)}} \pi_{\mathfrak{t}_{K(\beta_w)}} = 1,$$

b is an integer, and

$$\text{ord}_{\mathfrak{t}_{K(\beta_w)}} \varepsilon_{\mathfrak{t}_{K(\beta_w)}} = 0.$$

For every prime $\mathfrak{t}_{K(\beta_w,\alpha)}$ of $K(\beta_w, \alpha)$, let $n_{\mathfrak{t}_{K(\beta_w,\alpha)}}$ be the local degree at the prime $\mathfrak{t}_{K(\beta_w)}$ below in $K(\beta_w)$, i.e.

$$n_{\mathfrak{t}_{K(\beta_w,\alpha)}} = \left[K(\beta_w, \alpha)_{\mathfrak{t}_{K(\beta_w,\alpha)}} : K(\beta_w)_{\mathfrak{t}_{K(\beta_w)}}\right].$$

Since K contains the qth roots of unity, by Lemma B.4.11 the local degree is either 1 or q. If the local degree is 1 then h is automatically a norm. However, if the local degree is q then $\pi_{\mathfrak{t}_{K(\beta_w)}}^{bq}$ is a norm of $\pi_{\mathfrak{t}_{K(\beta_w)}}^{b}$ and, as above, since the extension is not ramified $\varepsilon_{\mathfrak{t}_{K(\beta_w)}}$ is also a norm. Thus h is a $K(\beta_w)$-norm of an element from $K(\beta_w, \alpha)$. □

We now state a corollary whose proof is completely analogous to that of Corollary 4.2.3.

Corollary 4.3.3. *The set of elements of K integral at $\mathfrak{b}_i$ for $i = 1, \ldots, m$ is Diophantine over K.*

The corollary above paves the way for the main result of this section.

Theorem 4.3.4. *Let M be a function field of positive characteristic p over a finite field of constants. Let $\mathfrak{p}$ be any prime of M. Then the set of elements of M integral at $\mathfrak{p}$ is Diophantine over M.*

Proof. The proof of this theorem is quite similar overall to the proof of Theorem 4.2.4, in particular in its use of the "going up and then down" method, but some details are different. Let $\mathfrak{q} \neq \mathfrak{p}$ be another prime of M. Let $f \in K$ be an element with a divisor of the form $\mathfrak{p}^c/\mathfrak{q}^r$ for some positive integers c, r. (We can always take $c = h_M \deg \mathfrak{q}, r = h_M \deg \mathfrak{p}$.) Let q be a rational prime different from p and prime to rc. Let $M_1 = M(\xi_q)$, where ξ_q is a primitive qth root of unity. By Lemma B.4.17, no prime of M will ramify in the extension $M(\xi_q)/M$. Thus the divisor of f in this extension is of the form

$$\frac{\prod_{i=1}^n \mathfrak{p}_i^c}{\prod_{j=1}^m \mathfrak{q}_j^r},$$

where $\mathfrak{p}_1, \ldots, \mathfrak{p}_n$ are distinct factors of $\mathfrak{p}$ in $M(\xi_q)$ and $\mathfrak{q}_1, \ldots, \mathfrak{q}_m$ are distinct factors of $\mathfrak{q}$ in $M(\xi_q)$. Let $v \in M(\xi_q)$ be such that for all j we have that $\text{ord}_{\mathfrak{q}_j} v \not\equiv 0 \bmod q$ and for some $b \neq 0$, for all i, we have that $v \equiv b^q \bmod \mathfrak{p}_i$. Such a v exists by the strong approximation theorem. Let $M_1 = M(\xi_q, \delta)$, where δ is a root of the polynomial $T^q - v$. By Lemma B.4.11, for all i we have that $\mathfrak{p}_i$ splits completely into distinct factors and for all j we have that $\mathfrak{q}_j$ is completely ramified in the extension. Thus in M_1 the divisor of f is now of the form

$$\frac{\prod_{i=1}^n \prod_{u=1}^q \mathfrak{r}_{i,u}^c}{\prod_{j=1}^m \mathfrak{v}_j^{qr}},$$

where for all i we have that $\mathfrak{r}_{i,1}, \ldots, \mathfrak{r}_{i,q}$ are all the distinct factors of $\mathfrak{p}_i$ and for all j we have that $\mathfrak{v}_j$ is the only factor of $\mathfrak{q}_j$ in M_1. Finally, let $C_{i,u}$ be the residue field of $\mathfrak{r}_{i,u}$ and let C_j be the residue field of $\mathfrak{v}_j$. Let

$$K = MC_{1,1} \cdots C_{n,q} C_1 \cdots C_m.$$

Then in K each prime factor of the divisor of f in M_1 will split into distinct factors of relative degree 1 by Lemmas B.4.15 and B.4.16. Thus in K the divisor of f will be of the form required by Notation 4.3.1, while the field K will contain the required roots of unity. By Corollary 4.3.3 the set I_K of elements of K integral at all the factors of $\mathfrak{p}$ in K is Diophantine over K. Thus, by the "going up and then down" method (see Subsection 2.1.17), $I_K \cap M$ is Diophantine over M. But $I_K \cap M$ consists precisely of elements of M integral at $\mathfrak{p}$. □

4.4 Divisibility of order at infinitely many primes over number fields

In this section we take the first steps towards being able to say something about order at infinite sets of primes. What we mean by "the divisibility of order at infinitely many primes" is that, for some rational prime q, the order of an element at primes of the field contained in an infinite set is divisible by q or is non-negative. As above, the proofs in the number field case and the function field case are very similar but not identical, primarily because in the function field case we can arrange for an extension to be unramified while this is not always possible in the number field case. As usual, we start with a notation list.

Notation 4.4.1.

- Let K denote a number field, and let $\mathcal{P}(K)$ denote the set of non-archimedean primes of K.
- Let q denote a prime number.
- Let t denote a prime number equivalent to 1 modulo q^2 such that K and $\mathbb{Q}(\xi_t)$, where ξ_t is a primitive tth root of unity, are linearly disjoint over $\mathbb{Q}$. (Such a prime exists by Theorem 5.9, Chapter IV of [37] and by Corollary B.3.11.)
- Let $\mathcal{T}$ be the set of prime factors of the divisor of t in K. Let $\mathfrak{T} = \prod_{\mathfrak{t}\in\mathcal{T}} \mathfrak{t}$.
- Let $a_1, \ldots, a_k \in O_K$ be integers representing every equivalence class modulo $\mathfrak{T}^q$.
- Let $\mathcal{A}$ be the set of primes of K such that $\mathfrak{a} \in \mathcal{A}$ if and only if $\mathfrak{a} \notin \mathcal{T}$ and for some $i = 1, \ldots, k$, we have that $\text{ord}_{\mathfrak{a}} a_i \neq 0$.
- Let $\mathcal{Q}$ be the set consisting of all the factors of the divisor of q in K. Note that q splits completely in the extension $\mathbb{Q}(\xi_t)$ by Lemma B.4.13.
- Let $x \in K \setminus \{0\}$ be such that $\text{ord}_{\mathfrak{a}} x \geq 0$ for all $\mathfrak{a} \in \mathcal{A}$.
- For $i = 1, \ldots, k$ and x as above, let $\alpha_i(x) = \alpha_i \in \tilde{\mathbb{Q}}$ be such that $\alpha_i^q = 1 + 1/a_i x$. Further, if $1 + 1/a_i x$ is the qth power of an element in K then let $\alpha_i(x) \in K$. (Here $\tilde{\mathbb{Q}}$ is the algebraic closer of $\mathbb{Q}$.)
- Let $G_i = K(\alpha_i(x))$.

Our first job is to establish the existence of a cyclic extension of K to use with the strong Hasse norm principle.

Lemma 4.4.2. *There exists $\delta \in \tilde{\mathbb{Q}}$ such that $K(\delta)/K$ is a cyclic extension of degree q, δ and all its conjugates are totally real, and factors of the K-divisor of t are the only primes possibly ramifying in this extension. Further, all the factors of q split completely into distinct factors.*

Proof. By assumption we have that K and $\mathbb{Q}(\xi_t)$ are linearly disjoint over $\mathbb{Q}$. Note further that $\mathbb{Q}(\xi_t + \xi_t^{-1})$ is a totally real cyclic extension of $\mathbb{Q}$ of degree $(t-1)/2 \equiv 0 \bmod q$. Next, if σ is a generator of $\text{Gal}(\mathbb{Q}(\xi_t + \xi_t^{-1})/\mathbb{Q})$ then let L be the fixed field of σ^q. In this case $L/\mathbb{Q}$ is a cyclic extension of $\mathbb{Q}$ of degree q.

Let $\delta \in O_L$ be such that $L = \mathbb{Q}(\delta)$. Then $K(\delta)/K$ is a cyclic extension of degree q by Lemma B.3.5, and δ and all its conjugates are real. Since elements of $\mathcal{T}$ are the only primes dividing the discriminant of the power basis of ξ_t, these are the only primes possibly ramifying in the extension $K(\xi_t)/K$, by Proposition 8, Section 2, Chapter III of [46], and therefore the only primes possibly ramifying in the extension $K(\delta)/K$, by Proposition B.1.12. Finally the last assertion of the lemma follows by B.4.7. □

Notation 4.4.3. From now on, let δ be the generator of a cyclic extension of degree q over K and let $\mathcal{V}$ be the set of all primes of K not splitting either in the extension $K(\delta)/K$ or in $\mathcal{A}$. (Note that by the argument above $\mathcal{Q} \not\subset \mathcal{V}$.)

Our next task is to consider the behavior of various primes in the extensions G_i/K, as well as the degrees of these extensions.

Lemma 4.4.4. *The following statements are true.*

1. *Either* $[G_i : K] = q$ *or* $G_i = K$.
2. *Suppose that there exists* $\mathfrak{p} \in \mathcal{V}$ *such that* $\mathrm{ord}_{\mathfrak{p}} x < 0$. *Then:*
   (a) $\mathfrak{p}$ *will have a factor* $\mathfrak{P}$ *of relative degree 1 in the extension* G_i/K *for all* $i = 1, \ldots, k$;
   (b) $\mathfrak{P}$ *will not split in the extension* $G_i(\delta)/G_i$;
   (c) $[G_i(\delta) : G_i] = q$.
3. *Let* $\mathfrak{p}$ *be a prime of* K *such that* $\mathrm{ord}_{\mathfrak{p}} a_i x > 0$. *Then either* $\mathrm{ord}_{\mathfrak{p}} a_i x \equiv 0 \bmod q$ *or* $[G_i : K] = q$ *and* $\mathfrak{p}$ *is ramified completely in the extension. (Thus in any case if* $\mathfrak{P}$ *is a prime above* $\mathfrak{p}$ *in* G_i *then* $\mathrm{ord}_{\mathfrak{P}} a_i x \equiv 0 \bmod q$.)
4. *For all* $x \in K$ *and all* $i = 1, \ldots, k$, *if* $[G_i(\delta) : G_i] \neq q$ *then* $K(\delta) = G_i(\delta) = G_i$. *If* $[G_i(\delta) : G_i] = q$ *and* $K \neq G_i$ *then the fields* $K(\delta)$ *and* G_i *are linearly disjoint over* K.
5. *Let* $\mathfrak{q}$ *be a prime of* K *such that* $\mathfrak{q} \notin \mathcal{V} \cup \mathcal{A} \cup \mathcal{T}$. *Let* $\mathfrak{Q}$ *be a prime of* G_i *above* $\mathfrak{q}$. *Then* $\mathfrak{Q}$ *splits completely in the extension* $G_i(\delta)/G_i$.
6. *If* $\mathfrak{Q}$ *is a prime of* $G_i(\delta)$ *ramified over* G_i *then it lies above a prime of* $\mathcal{T}$.

Proof. Consider the extension G_i/K. Note that $\alpha_i(x)$ is a root of the polynomial

$$T^q - \left(1 + \frac{1}{a_i x}\right). \tag{4.4.1}$$

and, unless $1 + 1/a_i x$ is a qth power in K, we have that $\alpha_i(x)$ generates an extension of degree q over K, by Lemma B.4.11. Thus, given our assumptions either $G_i = K$ or $[G_i : K] = q$.

Now let $\mathfrak{p} \in \mathcal{V}$ be a prime such that $\text{ord}_{\mathfrak{p}} x < 0$. If $G_i = K$ then assertions (2a)–(2c) of the lemma are trivially satisfied. Suppose now that $[G_i : K] = q$, and observe that $1 + 1/a_i x \equiv 1 \bmod \mathfrak{p}$. Thus $\mathfrak{p}$ does not ramify in the extension G_i/K, by Lemma B.4.11, and by Lemma B.4.12 the power basis of $\alpha_i(x)$ is an integral basis with respect to $\mathfrak{p}$. Further, we note that the polynomial (4.4.1) has a root modulo $\mathfrak{p}$. Therefore, by Lemma B.4.13, $\mathfrak{p}$ has factor $\mathfrak{P}$ of relative degree 1 in G_i. Also, by Lemma B.4.35 there exists $\delta_{\mathfrak{p}} \in K(\delta)$ such that $K(\delta) = K(\delta_{\mathfrak{p}})$, the power basis of $\delta_{\mathfrak{p}}$ is an integral basis of $K(\delta_{\mathfrak{p}})$ over K with respect to $\mathfrak{p}$, and the monic irreducible polynomial of $\delta_{\mathfrak{p}}$ over K remains irreducible over the residue field of $\mathfrak{p}$. Thus the monic irreducible polynomial of $\delta_{\mathfrak{p}}$ over K remains irreducible over the residue field of $\mathfrak{P}$ (and consequently over G_i). Hence, for any i, by Lemma B.4.12 $\mathfrak{P}$ does not split in the extension $G_i(\delta_{\mathfrak{p}})/G_i$, and $[G_i(\delta) : G_i] = q$.

Next let $\mathfrak{p}$ be a prime of K such that $\text{ord}_{\mathfrak{p}}\, a_i x > 0$. Then we have two possibilities: either

$$\text{ord}_{\mathfrak{p}} a_i x \equiv 0 \bmod q$$

or $1 + 1/a_i x$ is not a qth power in K and $\mathfrak{p}$ is ramified completely in the nontrivial extension G_i/K. In any case, if $\mathfrak{P}$ is a factor of $\mathfrak{p}$ in G_i then $\text{ord}_{\mathfrak{P}} a_i x \equiv 0 \bmod q$.

Next note that if $[G_i(\delta) : G_i] \neq q$ then $K(\delta)$ and G_i are not linearly disjoint over K by Lemma B.3.1. Since $K(\delta)/K$ is a Galois extension, by Lemma B.3.3 the lack of linear disjointness implies that $K(\delta) \cap G_i$ is strictly bigger than K. As we are dealing with extensions of prime degree, we must conclude that in this case $G_i = K(\delta) = G_i(\delta)$. However, if $[G_i(\delta) : G_i] = q$ then by Lemma B.3.1 we conclude that G_i and $K(\delta)$ are linearly disjoint over K.

Now, let $\mathfrak{q} \notin \mathcal{V} \cup \mathcal{A} \cup \mathcal{T}$ and assume that $K \neq G_i$. Let $\mathfrak{Q}$ be a prime of G_i above $\mathfrak{q}$ and let $\tilde{\mathfrak{Q}}$ be the $G_i(\delta)$-prime above $\mathfrak{Q}$. By Lemma B.4.1, it is enough to show that the decomposition group of $\tilde{\mathfrak{Q}}$ over G_i is trivial. Let $\sigma \in \text{Gal}(G_i(\delta)/G_i) \setminus \{\text{id}\}$. Without loss of generality, we can assume that $G_i(\delta) \neq G_i$ and, therefore, that G_i and $K(\delta)$ are linearly disjoint over K. Then, by Lemma B.3.5, we have that σ restricted to K is not the identity element of $\text{Gal}(K(\delta)/K)$. Note that $\tilde{\mathfrak{Q}} \cap K(\delta)$ lies above $\mathfrak{q}$. Since $\mathfrak{q}$ splits completely in $K(\delta)$, the decomposition group of $\tilde{\mathfrak{Q}} \cap K(\delta)$ is trivial and therefore $\sigma \mid_{K(\delta)}(\tilde{\mathfrak{Q}} \cap K(\delta)) \neq \tilde{\mathfrak{Q}} \cap K(\delta)$. Hence $\sigma(\tilde{\mathfrak{Q}}) \neq \tilde{\mathfrak{Q}}$ and the decomposition group of $\tilde{\mathfrak{Q}}$ is trivial.

Next, suppose that $\tilde{\mathfrak{Q}}$ is a prime of $G_i(\delta)$ ramified in the extension $G_i(\delta)/G_i$. Assuming the extension is not trivial, we must conclude that $\tilde{\mathfrak{Q}}$ is completely ramified and that the inertia group of $\tilde{\mathfrak{Q}}$ over G_i is of size q and equal to $\text{Gal}(G_i(\delta)/G_i)$. In this case, however, the inertia group of $\tilde{\mathfrak{Q}} \cap K(\delta)$ is equal to

$\mathrm{Gal}(K(\delta)/K)$ and therefore $\tilde{\mathfrak{Q}} \cap K(\delta)$ is completely ramified in the extension $K(\delta)/K$. □

Now we prove the main technical result of this section.

Proposition 4.4.5. *The equation*

$$\mathbf{N}_{G_i(\delta)/G_i}(z) = a_i x \tag{4.4.2}$$

has a solution $z \in G_i(\delta)$ for some i if and only if the order of every pole of x at primes of $\mathcal{V}$ is divisible by q.

Proof. Suppose that for some $\mathfrak{p} \in \mathcal{V}$ we have that $\mathrm{ord}_{\mathfrak{p}} x < 0$. Then, given our assumptions on x, we know that $\mathfrak{p} \notin \mathcal{A}$. Further, by Lemmas B.4.3 and 4.4.4, for any $i = 1, \ldots, k$ we know that $a_i x$ cannot be a norm in the extension $G_i(\delta)/G_i$ unless $\mathrm{ord}_{\mathfrak{P}} x \equiv 0 \bmod q$, where $\mathfrak{P}$ is a factor of $\mathfrak{p}$ in G_i of relative degree 1 over $\mathfrak{p}$ and $\mathfrak{P}$ does not split in the extension $G_i(\delta)/G_i$. But since $\mathfrak{P}$ is not ramified over $\mathfrak{p}$, we have that

$$\mathrm{ord}_{\mathfrak{P}} x \equiv 0 \bmod q \quad \Leftrightarrow \quad \mathrm{ord}_{\mathfrak{p}} x \equiv 0 \bmod q.$$

Suppose now that for all $\mathfrak{p} \in \mathcal{V}$ we have that

$$\mathrm{ord}_{\mathfrak{p}} x < 0 \quad \Rightarrow \quad \mathrm{ord}_{\mathfrak{p}} x \equiv 0 \bmod q.$$

Let $\mathfrak{q}$ be a prime of G_i. Let $\mathfrak{Q}$ be a prime above $\mathfrak{q}$ in $G_i(\delta)$. We would like to determine the possible values of local degree $[(G_i(\delta))_{\mathfrak{Q}} : (G_i)_{\mathfrak{q}}]$, where $(G_i(\delta))_{\mathfrak{Q}}, (G_i)_{\mathfrak{q}}$ are completions of $G_i(\delta)$ and G_i under $\mathfrak{Q}$ and $\mathfrak{q}$ respectively. First of all, if the global degree is equal to 1 then the local degree is 1. Second, if the global degree is not 1 then by Lemma 4.4.4 the global degree is q and the global extension is cyclic. Thus, by Lemma B.4.1, for any prime $\mathfrak{q}$ of G_i, the relative degree and the ramification degree are 1, or the relative degree is q and the ramification degree is 1, or the ramification degree is q and the relative degree is 1. In the case where the local degree is 1, we know that $a_i x$ is automatically a local norm. In the case where the prime is unramified and the relative degree is q, it is enough to arrange, as in the preceding sections, that the order of $a_i x$ at the prime is divisible by q. Finally, in the case of ramified primes it is enough to demonstrate that $a_i x$ is a qth power locally. Keeping this plan in mind, let us examine $a_i x$. We will divide the primes of G_i into five categories:

1. those lying above the primes in $\mathcal{P}(K) \setminus \mathcal{T}$ and such that they do not occur in the divisor of $a_i x$;

2. those lying above the primes of $\mathcal{P}(K) \setminus \mathcal{T}$ and such that they are zeros of $a_i x$;
3. those lying above the primes in $\mathcal{V}$ and such that they are poles of $a_i x$;
4. those lying above the primes in $\mathcal{P}(K) \setminus (\mathcal{V} \cup \mathcal{A} \cup \mathcal{T})$ and such that they are poles of $a_i x$;
5. those lying above the primes of $\mathcal{T}$.

First of all, by Lemma 4.4.4 any prime of G_i ramifying in the extension $G_i(\delta)/G_i$ lies above a $\mathcal{T}$-prime of K. Also from Lemma 4.4.4 we know that for any G_i-prime $\mathfrak{P}$ from categories 1–3, for all $i = 1, \ldots, k$ we have that $\text{ord}_{\mathfrak{P}} a_i x \equiv 0 \mod q$ and thus $a_i x$ is a local norm at such a prime. If $\mathfrak{P}$ is a prime from category 4 then, by Lemma 4.4.4, we have that $\mathfrak{P}$ splits completely in the extension $G_i(\delta)/G_i$ and therefore $a_i x$ is a local norm at any such prime for any i as listed above.

Now, for some $i = 1, \ldots, k$ we have that $a_i x$ is a qth power in $K_{\mathfrak{t}}$ for $\mathfrak{t} \in \mathcal{T}$. Indeed, for every $\mathfrak{t} \in \mathcal{T}$ we can write $x = \varepsilon(\mathfrak{t})\pi_{\mathfrak{t}}^{lq+r}$, where $\varepsilon(\mathfrak{t})$ is a unit at $\mathfrak{t}$, $\text{ord}_{\mathfrak{t}}\pi_{\mathfrak{t}} = 1, l \in \mathbb{Z}$, and $0 \leq r < q$. So we can pick a_i in such a way that for all $\mathfrak{t} \in \mathcal{T}$ we have that $a_i \equiv \pi_{\mathfrak{t}}^{q-r}\mu_{\mathfrak{t}} \mod \mathfrak{t}^q$ and $\mu_{\mathfrak{t}} \equiv \varepsilon_{\mathfrak{t}}^{-1} \mod \mathfrak{t}$ if $r > 0$ and $a_i = \mu_{\mathfrak{t}} \mod \mathfrak{t}$ otherwise. Then for some $j \in \mathbb{Z}$ we have that $a_i \times \pi_{\mathfrak{t}}^{j_q} \equiv 1 \mod \mathfrak{t}$ and, by a version of Hensel's Lemma B.8.2, it is the case that $a_i x$ is a qth power in $K_{\mathfrak{t}}$ for all $\mathfrak{t} \in \mathcal{T}$. Therefore, by choosing a_i correctly we can make sure that $a_i x$ is a norm locally at any prime potentially ramified in the extension $G_i(\delta)/G_i$.

Finally, we need to account for infinite primes. If a completion of G_i is $\mathbb{C}$ then of course every element of the completion is a norm (since the local degree is 1). If a completion of G_i is $\mathbb{R}$ then the completion of $G_i(\delta)$ is also $\mathbb{R}$, since all the conjugates of δ are real. Thus the local degree is 1 again and every element is a norm. □

We are now ready to state the main theorem of this section.

Theorem 4.4.6. *Let $\mathcal{W} = P(K) \setminus \mathcal{A}$. Then the set*

$$A_{\mathcal{V}} = \{h \in O_{K,\mathcal{W}} | \forall \mathfrak{p} \in \mathcal{V}, \text{ord}_{\mathfrak{p}} h \geq 0 \vee \text{ord}_{\mathfrak{p}} h \equiv 0 \mod q\} \tag{4.4.3}$$

is Diophantine over K.

Proof. The proof of this proposition is similar to the proof of Corollary 4.2.3 but we have a complication: $[G_i(\beta) : G_i]$ and $[G_i : K]$ can be either 1 or q. We will start as before with the norm polynomial,

$$N(X_0, \ldots, X_{q-1}) = \prod_{j=1}^{q} \sum_{i=0}^{q-1} X_i \sigma_j(\delta)^i, \tag{4.4.4}$$

where $\sigma_1 = \text{id}, \ldots, \sigma_{q-1}$ are all the automorphisms of $\mathbb{Q}(\delta)$. Note that all the coefficients of $N(X_0, \ldots, X_{q-1})$ are in $K(\delta)$ but are invariant under the action of $\text{Gal}(K(\delta)/K)$ and therefore are actually in K. We are interested in solving

$$N(X_0, \ldots, X_{q-1}) = a_i x, \tag{4.4.5}$$

with $X_0, \ldots, X_{q-1} \in G_i$. It is clear that we need to consider two cases, $G_i(\delta) \neq G_i$ and $G_i(\delta) = G_i$. In the first case equation (4.4.5) has solutions $X_0, \ldots, X_{q-1} \in G_i$ if and only if equation (4.4.2) has a solution $z \in G_i(\delta)$. Suppose now that $G_i = G_i(\delta)$. Then $G_i = K(\delta)$ and G_i contains all the conjugates of δ over K. Keeping this in mind, consider the following linear system:

$$\begin{pmatrix} 1 & \sigma_1(\delta) & \cdots & \sigma_1(\delta)^{q-1} \\ 1 & \sigma_2(\delta) & \cdots & \sigma_2(\delta)^{q-1} \\ \vdots & \vdots & \vdots & \vdots \\ 1 & \sigma_q(\delta) & \cdots & \sigma_q(\delta)^{q-1} \end{pmatrix} \begin{pmatrix} X_0 \\ X_1 \\ \vdots \\ X_{q-1} \end{pmatrix} = \begin{pmatrix} a_i x \\ 1 \\ \vdots \\ 1 \end{pmatrix} \tag{4.4.6}$$

Note that the determinant of the system is non-zero, because its square is the discriminant of the power basis of δ over K. Therefore this system can be solved in G_i and the solution will satisfy (4.4.5).

Summarizing the discussion, we can now assert that (4.4.5) has solutions $X_0, \ldots, X_{q-1}$ in G_i if and only if (4.4.2) has a solution $z \in G_i(\delta)$, which happens if and only if $x \in A_{\mathcal{V}}$. Now the assertion of the theorem follows from Lemma B.7.5. □

We will now address the issue of making the set $\mathcal{V}$ bigger.

Theorem 4.4.7. *Let K be a number field. Let q be a prime. Then for any $\varepsilon > 0$ there exists a set $\bar{\mathcal{V}}$ of K-primes of Dirichlet density greater than $1 - \varepsilon$ and a finite set $\bar{\mathcal{A}}$ such that the set*

$$A_{q,\bar{\mathcal{V}}} = \{h \in O_{K,\bar{\mathcal{W}}} \mid \forall \mathfrak{p} \in \bar{\mathcal{V}}, \text{ord}_{\mathfrak{p}} h \geq 0 \vee \text{ord}_{\mathfrak{p}} h \equiv 0 \bmod q\}, \tag{4.4.7}$$

where $\bar{\mathcal{W}} = \mathcal{P}(K) \setminus \bar{\mathcal{A}}$, is Diophantine over K. (For a definition of the Dirichlet density the reader is referred to appendix section B.5.)

Proof. Let $t_1, \ldots, t_n$ be prime numbers such that $t = t_i$ satisfies the conditions of Notation 4.4.1. For each $i = 1, \ldots, n$ define $\mathcal{A}_i$ and $\mathcal{V}_i$ as $\mathcal{A}$ and $\mathcal{V}$ in Notations 4.4.1 and 4.4.3 respectively. Let $\bar{\mathcal{V}} = \cup_{i=1}^n \mathcal{V}_i$, let $\bar{\mathcal{A}} = \cup_{i=1}^n \mathcal{A}_i$, and let $\bar{\mathcal{W}} = \mathcal{P}(K) \setminus \bar{\mathcal{A}}$. Then by Theorem 4.4.6 and from the fact that an intersection of Diophantine sets is Diophantine, the set

$$A_q = \{x \in O_{K,\bar{\mathcal{W}}} | \forall \mathfrak{p} \in \bar{\mathcal{V}}, \text{ord}_{\mathfrak{p}} x \geq 0 \vee \text{ord}_{\mathfrak{p}} x \equiv 0 \bmod q\}$$

is Diophantine over K. To finish the proof of the theorem, it is enough to show that the natural density of $\bar{\mathcal{V}}$ is greater than $1 - \varepsilon$ for sufficiently large n, since $K(\xi_{t_i})$ and $K(\xi_{t_j}, j = 1, \ldots, n, j \neq i)$ are linearly disjoint over K. (Linear disjointness is the consequence of an argument similar to that used to prove Lemma B.3.10.) This is done in the proof of Lemma B.5.7. □

We next consider the function field version of the above results.

4.5 Divisibility of order at infinitely many primes over function fields

Proposition 4.5.1. *Let q be a rational prime and let K be a function field over a finite field of constants of characteristic $p > 0$, $p \neq q$. Let $h \in K$, let α in the algebraic closure of K be the root of the equation $T^q - (1 + h^{-1})$, let β be an element of the algebraic closure of C_K, the constant field of K, of degree q over C_K, let $G = K(\alpha)$, let $\mathcal{W}$ be the set of all K-primes not splitting in the extension $K(\beta)/K$, and consider the following norm equation:*

$$\mathbf{N}_{G(\beta)/G} z = h. \tag{4.5.1}$$

Then this equation has a solution $z \in G(\beta)$ if and only if for all primes $\mathfrak{p} \in \mathcal{W}$, either h is integral at $\mathfrak{p}$ or $\text{ord}_{\mathfrak{p}} h \equiv 0 \bmod q$.

Proof. The proof of this proposition is very similar to the proof of Proposition 4.4.5, but the situation is actually simpler because, by Lemma B.4.17, there are no ramified primes in the extension $G(\beta)/G$. We leave the details to the reader. □

We now state the function field analog of Theorem 4.4.6.

Theorem 4.5.2. *Let q be a rational prime and let K be a function field over a finite field of constants of characteristic $p > 0$, $p \neq q$. Let β be an element of the algebraic closure of C_K, the constant field of K, of degree q over C_K. Let $\mathcal{W}$ be the set of all primes of K not splitting in the extension $K(\beta)/K$. Then the set*

$$A_{\mathcal{W}} = \{h \in K | \forall \mathfrak{p} \in \mathcal{W}, \text{ord}_{\mathfrak{p}} h \geq 0 \vee \text{ord}_{\mathfrak{p}} h \equiv 0 \bmod q\} \tag{4.5.2}$$

is Diophantine over K.

The proof of this theorem is almost identical to the proof of Theorem 4.4.6.

Finally, we show that the Dirichlet density of the set of "covered" primes is arbitrarily close to 1. (For a definition of the Dirichlet density over function fields the reader is again referred to appendix section B.5.)

Theorem 4.5.3. *Let q be a rational prime and let K be a function field over a finite field of constants of characteristic $p > 0$, $p \neq q$. Then for any $\varepsilon > 0$ there exists a set $\mathcal{V}$ of K-primes of density greater than $1 - \varepsilon$ such that the set*

$$A_{q,\mathcal{V}} = \{h \in K \mid \forall \mathfrak{p} \in \mathcal{V}, \operatorname{ord}_{\mathfrak{p}} h \geq 0 \vee \operatorname{ord}_{\mathfrak{p}} h \equiv 0 \bmod q\} \tag{4.5.3}$$

is Diophantine over K.

Proof. Let n be any positive integer. The field of constants of K, as any finite field, has an extension of degree q^n, and this extension is cyclic. (See Section 5, Chapter VII of [47].) Let $K_0 = K \subset K_1 \cdots \subset K_n$ be the corresponding tower of cyclic extensions with $[K_{i+1} : K_i] = q$. Let $\mathcal{W}_i$ be a set of primes of K_i not splitting in the extension K_{i+1}/K_i. Then the set

$$A_{\mathcal{W}_i,K_i} = \{h \in K_i \mid \forall \mathfrak{p} \in \mathcal{W}_{i,K_i}, \operatorname{ord}_{\mathfrak{p}} h \geq 0 \vee \operatorname{ord}_{\mathfrak{p}} h \equiv 0 \bmod q\} \tag{4.5.4}$$

is Diophantine over K_i. Let $\mathcal{W}_{i,K}$ be the set of K-primes lying below the primes in the set $\mathcal{W}_{i,K_i}$. Let

$$A_{\mathcal{W}_{i,K}} = \{h \in K \mid \forall \mathfrak{p} \in \mathcal{W}_{i,K}, \operatorname{ord}_{\mathfrak{p}} h \geq 0 \vee \operatorname{ord}_{\mathfrak{p}} h \equiv 0 \bmod q\}.$$

Then $A_{\mathcal{W}_{i,K}} = A_{\mathcal{W}_{i,K_i}} \cap K$ and, by the "going up and then down" method, $A_{\mathcal{W}_{i,K}}$ is Diophantine over K. Let $\mathcal{W}_K = \bigcup_{i=1}^n \mathcal{W}_{i,K}$. Then from the properties of Diophantine definitions we can conclude that

$$A = \bigcap_{i=1}^{n} A_{\mathcal{W}_i,K_i} = \{h \in K \mid \forall \mathfrak{p} \in \mathcal{W}_K, \operatorname{ord}_{\mathfrak{p}} h \geq 0 \vee \operatorname{ord}_{\mathfrak{p}} h \equiv 0 \bmod q\}$$

is also Diophantine over K. Further, by Lemma B.5.8, for n sufficiently large $\mathcal{W}_K$ has the desired Dirichlet density. □

5
Bound equations for number fields and their consequences

This chapter is devoted to the existential definability of bounds on the height of a number field element. We will consider two ways of bounding the height. The first method (due to Denef in [18]) relies on quadratic forms and is most effective over totally real fields. The second method relies on divisibility and thus depends on the choice of ring. The influence of divisibility depends on whether any primes are allowed to occur in the denominator of the divisors of elements of the ring. We will discuss this aspect of the matter in detail below. Denef was also the first to use the divisibility method for bounding height over the rings of integers in the context of existential definability (see [18] again). The present author extended Denef's divisibility method for use with bounds in rings of algebraic numbers. These kinds of bound were used in [99], [101], and [106] and in a slightly modified way in [73].

5.1 Real embeddings

In this section we show how to use quadratic forms to impose bounds on archimedean valuations when the corresponding completion of the number field is $\mathbb{R}$. The following lemma and its corollary constitute the technical foundation of the method. They are taken from [18].

Lemma 5.1.1. *Let K be a number field. Let $x \in K$. Let $x = x_1, \ldots, x_n$ be all the conjugates of x over $\mathbb{Q}$. Let $c \in K$, and let $c_1 = c, \ldots, c_n$ be all the conjugates of c over $\mathbb{Q}$. Suppose that $c_i < 0$ whenever $x_i < 0$. Then there exist $y_1, \ldots, y_4 \in K$ such that $x = y_1^2 + y_2^2 + cy_3^2 + y_4^2$.*

Proof. By the Hasse–Minkowskii theorem, a quadratic form represents an algebraic number over a number field if and only if it represents the number locally. (See [64] for more details on the Hasse–Minkowskii theorem.) Any quadratic form in four variables over a local field is universal, that is, it represents any element of the field; see [64]. So the real embeddings are the only ones we have to check. On the one hand, if for some i that $x_i \geq 0$, then $x_i = (\sqrt{x_i})^2 + 0^2 + c_i 0^2 + 0^2$. On the other hand, if $x_i < 0$ then $c_i < 0$. Thus x_i/c_i is a square in the completion of the corresponding embedding of K. □

Corollary 5.1.2. *Let K be a number field, and let $\mathcal{W}$ be any subset of its non-archimedean primes. Let $\sigma : K \to \mathbb{C}$ be any real embedding of K into $\mathbb{C}$. Then the set*

$$\{(x, y) \in K^2 \mid \sigma(x) \geq \sigma(y)\}$$

has a Diophantine definition over $O_{K,\mathcal{W}}$.

Proof. It is enough to consider the case $y = 0$. Let $c \in K$ be such that $\sigma(c) > 0$ and $\tau(c) < 0$ for any real embedding τ of K not equal to σ. (Such a c exists by the strong approximation theorem.) Then $\sigma(x) \geq 0$ if and only if $x = y_1^2 + y_2^2 + cy_3^2 + y_4^2$, by Lemma 5.1.1. Thus the corollary follows from Note 2.2.5.

5.2 Using divisibility in the rings of algebraic integers

In this section we start the discussion of divisibility as a method for bounding heights over number fields. As will become clear below, divisibility produces the best results when used over the rings of integers. Thus, it is over the rings of integers that we consider this method first. (We remind the reader that a discussion of the rings of integers and $\mathcal{S}$-integers of number fields can be found in Appendix B.)

It turns out that the foundation of the divisibility method can be found in solving some linear systems, as in the following lemma.

Lemma 5.2.1. *Let K be a number field of degree n over $\mathbb{Q}$. Let $x, y \in O_K \setminus \{0\}$ and assume that for $i = 0, \ldots, n$ we have that $(l_i - x) \mid y$ in the ring of integers of K, where $l_0 = 0, l_1, \ldots, l_n$ are distinct natural numbers. Then there exists a constant C, depending on $n, l_1, \ldots, l_n$ only, such that all the coefficients of the characteristic polynomial of x over $\mathbb{Q}$ are less than $C|\mathbf{N}_{K/\mathbb{Q}}(y)|$.*

Proof. If on the one hand $(l_i - x) \mid y$ then $\mathbf{N}_{K/\mathbb{Q}}(l_i - x) = c_i \mathbf{N}_{K/\mathbb{Q}}(y)$, where $c_i \in \mathbb{Q}$ and $|c_i| < 1$. On the other hand, if $(l_i - x) \mid y$ then

$$\mathbf{N}_{K/\mathbb{Q}}(l_i - x) = \prod_{j=1}^{n}(l_i - \sigma_j(x)) = a_0 + a_1 l_i + \cdots + l_i^n,$$

where $\sigma_1 = \mathrm{id}, \ldots, \sigma_n$ are all the embeddings of K into $\mathbb{C}$ and

$$F(T) = a_0 + a_1 T + \cdots + a_{n-1}T^{n-1} + T^n$$

is the characteristic polynomial of x over $\mathbb{Q}$. Thus, considered together, the $n+1$ equations $F(l_i) = c_i \mathbf{N}_{K/\mathbb{Q}}(y)$ can be viewed as a system of $n+1$ linear equations in the variables $a_0, \ldots, a_{n-1}, a_n = 1$. In matrix form this system looks as follows:

$$\begin{pmatrix} 1 & \cdots & 0 & 0 \\ 1 & \cdots & l_1^{n-1} & l_1^n \\ \vdots & \vdots & \vdots & \vdots \\ 1 & \cdots & l_{n-1}^{n-1} & l_{n-1}^n \\ 1 & \cdots & l_n^{n-1} & l_n^n \end{pmatrix} \begin{pmatrix} a_0 \\ a_1 \\ \vdots \\ a_{n-1} \\ 1 \end{pmatrix} = \begin{pmatrix} c_0 \mathbf{N}_{K/\mathbb{Q}}(y) \\ c_1 \mathbf{N}_{K/\mathbb{Q}}(y) \\ \vdots \\ c_{n-1}\mathbf{N}_{K/\mathbb{Q}}(y) \\ c_n \mathbf{N}_{K/\mathbb{Q}}(y) \end{pmatrix}. \tag{5.2.1}$$

Observe that the determinant of this system is a non-zero Vandermonde determinant whose value depends on the choice of constants $l_1, \ldots, l_n$ only. If we solve this system using Cramer's rule we will obtain the following expression for each a_i:

$$a_i = \frac{\begin{vmatrix} 1 & \cdots & c_0 \mathbf{N}_{K/\mathbb{Q}}(y) & \cdots & 0 \\ 1 & \cdots & c_1 \mathbf{N}_{K/\mathbb{Q}}(y) & \cdots & l_1^n \\ \vdots & \vdots & \vdots & \vdots & \vdots \\ 1 & \cdots & c_{n-1}\mathbf{N}_{K/\mathbb{Q}}(y) & \cdots & l_{n-1}^n \\ 1 & \cdots & c_n \mathbf{N}_{K/\mathbb{Q}}(y) & \cdots & l_n^n \end{vmatrix}}{\begin{vmatrix} 1 & \cdots & 0 & 0 \\ 1 & \cdots & l_1^{n-1} & l_1^n \\ \vdots & \vdots & \vdots & \vdots \\ 1 & \cdots & l_{n-1}^{n-1} & l_{n-1}^n \\ 1 & \cdots & l_n^{n-1} & l_n^n \end{vmatrix}}, \tag{5.2.2}$$

where the column $(c_j \mathbf{N}_{K/\mathbb{Q}}(y))$, $j = 0, \ldots, n$, has replaced the ith column of the original matrix, and therefore we have that

$$a_i = \mathbf{N}_{K/\mathbb{Q}}(y) \frac{\begin{vmatrix} 1 & \cdots & c_0 & \cdots & 0 \\ 1 & \cdots & c_1 & \cdots & l_1^n \\ \vdots & \vdots & \vdots & \vdots & \vdots \\ 1 & \cdots & c_{n-1} & \cdots & l_{n-1}^n \\ 1 & \cdots & c_n & \cdots & l_n^n \end{vmatrix}}{\begin{vmatrix} 1 & \cdots & 0 & 0 \\ 1 & \cdots & l_1^{n-1} & l_1^n \\ \vdots & \vdots & \vdots & \vdots \\ 1 & \cdots & l_{n-1}^{n-1} & l_{n-1}^n \\ 1 & \cdots & l_n^{n-1} & l_n^n \end{vmatrix}} \tag{5.2.3}$$

Since $|c_i| < 1$, we conclude that our lemma holds. □

A quick corollary of this lemma is the following fact.

Corollary 5.2.2. *Let K/E be an extension of number fields and let*

$$\Omega = \{\omega_1, \ldots, \omega_m\} \subset O_K$$

be a basis of K over E. Let $x, y \in O_K \setminus \{0\}$ and assume that for $i = 0, \ldots, n$, where $m \leq n = [K : \mathbb{Q}]$, we have that $(l_i - x) \mid y$ for distinct natural numbers $l_0 = 0, l_1, \ldots, l_n$. Then there exists a constant $\bar{C} \in \mathbb{Q}$, depending on $n, l_1, \ldots, l_n$, and Ω only, such that the following statements are true.

1. *If $\sigma : K \to \mathbb{C}$ is an embedding, then $|\sigma(x)| \leq \bar{C}|\mathbf{N}_{K/\mathbb{Q}}(y)|$.*
2. *We can write $x = \sum_{i=1}^m b_i \omega_i$, with $b_i \in E$, $|\sigma(b_i)| \leq \bar{C}|\mathbf{N}_{K/\mathbb{Q}}(y)|$ for any σ-embedding of K into $\mathbb{C}$.*

Proof. From the preceding lemma we can conclude that all the coefficients of the characteristic polynomial of x over $\mathbb{Q}$ are bounded by $C|\mathbf{N}_{K/\mathbb{Q}}(y)|$, where C is a positive constant depending on $n, l_1, \ldots, l_n$ only. Let

$$F(T) = a_0 + a_1 T + \cdots + a_{n-1}T^{n-1} + T^n$$

be this polynomial. Then, for all embeddings σ of K into $\mathbb{C}$, we have that

$$\sigma(x)^n = -a_0 - a_1\sigma(x) - \cdots - a_{n-1}\sigma(x)^{n-1}.$$

If $|\sigma(x)| > 1$ then

$$|\sigma(x)| \leq |a_0| + |a_1| + \cdots + |a_{n-1}| \leq nC|\mathbf{N}_{K/\mathbb{Q}}(y)|.$$

Thus, letting $\tilde{C} = \max(1, nC)$ we can conclude that $|\sigma(x)| \leq \tilde{C}|\mathbf{N}_{K/\mathbb{Q}}(y)|$.

Next let $x_1 = x, \ldots, x_n$ be all the conjugates of x over $\mathbb{Q}$. Without loss of generality assume that $x_1, \ldots, x_m$ are all the conjugates of x over E and let $\omega_{i,1} = \omega_i, \ldots, \omega_{i,m}$ be all the conjugates of ω_i over E. Then consider the following linear system of equations in variables $b_1, \ldots, b_m$:

$$x_j = \sum_{i=1}^{m} b_i \omega_{i,j}, \qquad j = 1, \ldots, m, \qquad b_i \in E.$$

If we solve this system using Cramer's rule as in Lemma 5.2.1 and keep in mind the bounds for $x_1, \ldots, x_m$, we will conclude that $|b_i| \leq \bar{C}|\mathbf{N}_{K/\mathbb{Q}}(y)|$ for some positive constant $\bar{C}$ depending on $\tilde{C}$ and Ω only, again exactly in the same fashion as in Lemma 5.2.1.

Next let σ be an embedding of K into $\mathbb{C}$. Applying σ to the system above we get

$$\sigma(x_j) = \sum_{i=1}^{m} \sigma(b_i)\sigma(\omega_{i,j}), \qquad j = 1, \ldots, m, \qquad b_i \in E.$$

Thus we can obtain a bound as above for $\sigma(b_i)$, for any σ-embedding of K into $\mathbb{C}$ and all $i = 1, \ldots, m$. □

Note 5.2.3. Before we finish this discussion of bounds, we should note that, assuming that y is not a unit, we can replace the bound $\bar{C}|\mathbf{N}_{K/\mathbb{Q}}(y)|$ in Lemma 5.2.1 and Corollary 5.2.2 by $|\mathbf{N}_{K/\mathbb{Q}}(y)|^c$, for sufficiently large positive $c \in \mathbb{N}$. It is often useful to have a bound not just for $|\sigma(b_i)|$ but also for $|\mathbf{N}_{K/\mathbb{Q}}(Db_i)|$, where D is a constant depending on the basis. It is clear that we can increase c to obtain the inequality $|\mathbf{N}_{K/\mathbb{Q}}(Db_i)| < |\mathbf{N}_{K/\mathbb{Q}}(y)^c|$.

5.3 Using divisibility in bigger rings

In this section we show how we can modify the divisibility method to make it work over some rings $O_{K,\mathcal{W}}$, where $\mathcal{W}$ is not empty.

5.3.1. When is division really division? Let $x, y \in O_K \setminus \{0\}$ and suppose that

$$x \mid y \text{ in } O_{K,\mathcal{W}}. \tag{5.3.1}$$

In general we cannot conclude that (5.3.1) implies that $x \mid y$ in O_K. However, if we assume that the divisor of x has no primes from $\mathcal{W}$ then the implication is valid.

From the observation above, it follows that we need a mechanism for producing elements of $O_{K,\mathcal{W}}$ without primes of $\mathcal{W}$ in the numerators of their divisors. Here is one way to accomplish this.

5.3.2. Avoiding primes allowed in the denominator of the divisors Let M/K be a finite extension of number fields. Let $\alpha \in O_K$ be a generator of this extension. Let $F(T)$ be the monic irreducible polynomial of α over K. Let $\mathcal{W}$ consist of all the primes which do not divide the discriminant of α and do not have relative-degree-1 factors in M. Then for all $\mathfrak{p} \in \mathcal{W}$ and for all $x \in K$ we have that $\text{ord}_{\mathfrak{p}} F(x) \leq 0$. (This follows from Lemma B.4.18.)

We are now ready to adjust our divisibility method for use in bigger rings.

Lemma 5.3.3. *Let K/E be a number field extension of degree s and let $[K : \mathbb{Q}] = n$. Let $\gamma \in O_K$ generate K over E. Let $G_u(T) \in O_K[T]$, $u = 1, \ldots, n+1$, be distinct monic polynomials that are irreducible over K. Assume also that $G_i(T)$ and $G_j(T)$ are not conjugates over $\mathbb{Q}$ for $i \neq j$. Let $\alpha/\beta \in K$, with $\alpha, \beta \in O_K$ and relatively prime to each other. Let $y \in O_K \setminus \{0\}$ be such that y is not an integral unit and*

$$\frac{y}{G_u(\alpha/\beta - l_i)} \in O_K, \qquad i = 0, \ldots, [n \deg(G_u) : \mathbb{Q}], \qquad u = 1, \ldots, n+1, \tag{5.3.2}$$

where $l_0 = 0, \ldots, l_z$, $z = \max_u([n \deg(G_u) : \mathbb{Q}])$ are distinct natural numbers. Let

$$\mathbf{N}_{K/\mathbb{Q}}(\beta)\alpha = e_0 + e_1\gamma + \cdots + e_{s-1}\gamma^{s-1}. \tag{5.3.3}$$

Then there exists a constant $c > 0$ depending on $l_0, \ldots, l_z, K, G_u(T)$ only such that

$$|\mathbf{N}_{K/\mathbb{Q}}(De_i)| < |\mathbf{N}_{K/\mathbb{Q}}(y)|^c, \qquad i = 0, \ldots, s-1, \tag{5.3.4}$$

where D is the discriminant of the power basis of γ.

Proof. Let C be defined as in Lemma B.10.8, $\bar{\tau}_u$ being the set of roots of G_u together with their $\mathbb{Q}$ conjugates. Since $G_1, \ldots, G_{n+1}$ are irreducible, distinct, and are not conjugates over $\mathbb{Q}$, the sets of roots of the polynomials $G_1, \ldots, G_{n+1}$ together with their $\mathbb{Q}$-conjugates are pairwise disjoint. Then $C > 0$ and by Lemma B.10.8 applied to conjugates of α/β over $\mathbb{Q}$ in place of $z_1, \ldots, z_n$, for some $u = 1, \ldots, n+1$ and for any τ, a root of $G_u(T)$, and any σ, an embedding of K into $\mathbb{C}$, we have that $|\sigma(\alpha/\beta) - \sigma(\tau) > C/2$. For this u, denote the field generated by a root τ of G_u by M, denote G_u by H, and let $q = \deg G_u$. Then in M we have that

$$\frac{y}{\prod_\tau(\alpha/\beta - l_i - \tau)} = \frac{y\beta^q}{\prod_\tau(\alpha - l_i\beta - \tau\beta)} \in O_M. \tag{5.3.5}$$

Since $(\alpha, \beta) = 1$ in O_K, it follows that $(\beta, \alpha - l_i\beta - \tau\beta) = 1$ in O_M. Thus for each i we have that

$$\frac{y}{(\alpha - l_i\beta - \tau\beta)} \in O_M.$$

Therefore

$$|\mathbf{N}_{M/\mathbb{Q}}(\alpha - l_i\beta - \tau\beta)| \leq |\mathbf{N}_{K/\mathbb{Q}}(y)^q|,$$
$$|\mathbf{N}_{M/\mathbb{Q}}(\beta)||\mathbf{N}_{M/\mathbb{Q}}(\alpha/\beta - l_i - \tau)| \leq |\mathbf{N}_{K/\mathbb{Q}}(y)^q|.$$

Using the fact that $|\mathbf{N}_{M/\mathbb{Q}}(\beta)| \geq 1$, we can conclude that

$$|\mathbf{N}_{M/\mathbb{Q}}(\alpha/\beta - l_i - \tau)| \leq |\mathbf{N}_{K/\mathbb{Q}}(y)^q|. \tag{5.3.6}$$

At the same time, using $i = 0$ and the inequality $|\sigma(\alpha/\beta) - \sigma(\tau)| > C/2$ applied to all embeddings σ of M into $\mathbb{C}$, we can conclude that

$$|\mathbf{N}_{K/\mathbb{Q}}(\beta)| = |\mathbf{N}_{M/\mathbb{Q}}(\beta)|^{1/q} \leq (C/2)^{-[K:\mathbb{Q}]}|\mathbf{N}_{K/\mathbb{Q}}(y)|.$$

From (5.3.6), by an argument similar to the one used to prove Lemma 5.2.1 and Corollary 5.2.2, we can conclude that there exists a positive constant $\bar{C}$, depending on $l_0, \ldots, l_z$, K, and $G_1(T), \ldots, G_{n+1}(T)$ only, such that

$$|\sigma(\alpha/\beta)| \leq \bar{C}|\mathbf{N}_{K/\mathbb{Q}}(y)^q| \tag{5.3.7}$$

for all embeddings σ of M into $\mathbb{C}$, and

$$\mathbf{N}_{K/\mathbb{Q}}(\beta)\alpha/\beta = e_0 + e_1\gamma + \cdots + e_{s-1}\gamma^{s-1} \tag{5.3.8}$$

where $e_0, e_1, \ldots, e_{s-1} \in E$. For all embeddings σ of K into $\mathbb{C}$, we can conclude that

$$|\sigma(e_j)| \leq \hat{C}|\mathbf{N}_{K/\mathbb{Q}}(y)^{q+1}|, \tag{5.3.9}$$

where $\hat{C}$ is a positive constant depending on $\gamma, l_0, \ldots, l_z, K$, and $G_1(T), \ldots, G_{n+1}(T)$ only. Since y is not an integral unit and $|\mathbf{N}_{K/\mathbb{Q}}(y)| \geq 2$, for some positive constant c depending on $\gamma, l_0, \ldots, l_z, K$, and $G_1(T), \ldots, G_{n+1}(T)$ only we have that

$$|\mathbf{N}_{K/\mathbb{Q}}(De_i)| < |\mathbf{N}_{K/\mathbb{Q}}(y)^c|, \tag{5.3.10}$$

where D, as above, is the discriminant of the power basis of γ over E. □

Corollary 5.3.4. *Let $K, E, s, n, \gamma, G_1(T), \ldots, G_{n+1}(T), z, \alpha, \beta, l_0, \ldots, l_z, e_0, \ldots, e_{s-1}$ be as above. Let $\mathcal{W}$ be a set of non-archimedean primes of K such that for all $u = 1, \ldots, n+1$, all $x \in K$, and all $\mathfrak{p} \in \mathcal{W}$ we have that $\mathrm{ord}_{\mathfrak{p}} G_u(x) \leq 0$. Let $v \in O_{K,\mathcal{W}} \setminus \{0\}$ be such that v is not a unit of $O_{K,\mathcal{W}}$ and satisfies*

$$\frac{v^{h_K}}{G_u(\alpha/\beta - l_i)} \in O_{K,\mathcal{W}}, \qquad i = 0, \ldots, n \deg G_u, \qquad u = 1, \ldots, n+1, \tag{5.3.11}$$

where h_K is the class number of K.

Then there exists a constant $c > 0$ depending on $\gamma, l_0, \ldots, l_z, K$, and $G_1(T), \ldots, G_{n+1}(T)$ only such that $v^{h_K} = yw$, $y \in O_K$, the divisor of y has no primes from $\mathcal{W}$, the divisor of w consists of primes in $\mathcal{W}$ only, and

$$|\mathbf{N}_{K/\mathbb{Q}}(De_i)| < |\mathbf{N}_{K/\mathbb{Q}}(y)|^c, \qquad i = 0, \ldots, s-1, \tag{5.3.12}$$

where D is the discriminant of the power basis of γ over E.

Proof. Given our assumptions on $\mathcal{W}$ and $G_1, \ldots, G_{n+1}$, for any prime $\mathfrak{p} \in \mathcal{W}$ we have that $\mathrm{ord}_{\mathfrak{p}} G_u(\alpha/\beta - l_i) \leq 0$. Thus, for all u, the primes in the numerator of the divisor of $G_u(\alpha/\beta - l_i)$ must be canceled by the numerator of the divisor of v^{h_K}. Note that $v^{h_K} = yw$, where the divisor of y is integral and is a product of powers of all the primes outside $\mathcal{W}$ occurring in the divisor of v^{h_K}, while w has a divisor composed of primes from $\mathcal{W}$ only. Then for all u we have that $\frac{y}{G_u(\alpha/\beta - l_i)} \in O_K$. Now the corollary follows from Lemma 5.3.3. □

We finish this section with a special case of the corollary above.

Corollary 5.3.5. *Let $K, E, s, n, \gamma, G_1(T), \ldots, G_{n+1}(T), z, \alpha, \beta, l_0, \ldots, l_z, e_0, \ldots, e_{s-1}$ be again as above. Assume also that K/E is Galois and that $G_1(T), \ldots, G_{n+1} \in O_E[T]$. Let $v \in E \cap O_{K,\mathcal{W}} \setminus \{0\}$ be such that v is not a unit of $\mathcal{W}$ and satisfies*

$$\frac{v^{h_E}}{G_u(\alpha/\beta - l_i)} \in O_{K,\mathcal{W}}, \qquad i = 0, \ldots, n \deg G_u, \qquad u = 1, \ldots, n+1, \tag{5.3.13}$$

where h_E is the class number of E.

Then there exists a constant $c > 0$, depending on $\gamma, l_0, \ldots, l_z, K$, and $G_1(T), \ldots, G_{n+1}$ only, such that $v^{h_E} = yw$, $y \in O_E$, the divisor of y has no primes from $\mathcal{W}$, the divisor of w consists of primes in $\mathcal{W}$ only, and

$$|\mathbf{N}_{K/\mathbb{Q}}(De_i)| < |\mathbf{N}_{K/\mathbb{Q}}(y)|^c, \qquad i = 0, \ldots, s-1, \tag{5.3.14}$$

Proof. Let $\bar{\mathcal{W}}$ be the closure of $\mathcal{W}$ over E. Given our assumptions on the extension K/E, $\mathcal{W}$, and $G_1, \ldots, G_{n+1}$, for any prime $\mathfrak{p} \in \bar{\mathcal{W}}$ we have that $\text{ord}_{\mathfrak{p}} G_u(\alpha/\beta - l_i) \leq 0$. Thus, as above, for all u the primes in the numerator of the divisor of $G_u(\alpha/\beta - l_i)$ must be canceled by the numerator of the divisor of v^{h_E}. However, since $v \in E$ we can write $v^{h_E} = yw$, with $y, w \in E$, where the divisor of y is integral and a product of powers of all the primes outside $\bar{\mathcal{W}}$ occurring in the divisor of v^{h_E} while w has a divisor composed of primes from $\bar{\mathcal{W}}$ only. Then, as before, for all u we have that

$$\frac{y}{G_u(\alpha/\beta - l_i)} \in O_K$$

and this corollary also follows from Lemma 5.3.3. □

6

Units of rings of $\mathcal{W}$-integers of norm 1

$\mathcal{W}$-units play an important role in the construction of Diophantine definitions over number fields. In this chapter we discuss some properties of these units.

6.1 What are the units of the rings of $\mathcal{W}$-integers?

Definition 6.1.1. Let K be a number field. Let $\mathcal{W}$ be a collection of its non-archimedean primes. Let $x \in K$ be such that its divisor is a product of powers of elements of $\mathcal{W}$. Then x is called a $\mathcal{W}$-unit. If $\mathcal{W}$ is empty then a $\mathcal{W}$-unit is just an integral unit of K, i.e. an algebraic integer whose multiplicative inverse is also an algebraic integer.

Next we list some useful properties of these units. The first one is a generalization of the well-known Dirichlet unit theorem.

Proposition 6.1.2. *Let K and $\mathcal{W}$ be as above. Then $\mathcal{W}$-units form a multiplicative group. If the number of primes in $\mathcal{W}$ is finite, then the rank of this group is equal to the rank of the integral unit group plus the number of elements in $\mathcal{W}$.* (See the generalized version of the Dirichlet unit theorem in [64].)

The next lemma is a direct consequence of the definition of the rings of $\mathcal{W}$-integers. (See Definition B.1.20.)

Lemma 6.1.3. *Let K and $\mathcal{W}$ be again as above. Then the only invertible elements of $O_{K,\mathcal{W}}$ (or the only units of $O_{K,\mathcal{W}}$) are the $\mathcal{W}$-units.*

The next lemma is important in making sure that the divisibility conditions discussed in Chapter 5 can be satisfied.

Lemma 6.1.4. *Let $\mathfrak{A}$ be any integral divisor of a number field K relatively prime to any prime in a set of non-archimedean K-primes $\mathcal{W}$. Then there exists a positive $k(\mathfrak{A}) = k \in \mathbb{N}$ such that for any natural number $l \equiv 0 \bmod k$ and for any $\mathcal{W}$-unit ε it follows that $\varepsilon^l - 1 \equiv 0 \bmod \mathfrak{A}$.*

Proof. The lemma follows from the fact that $O_{K,\mathcal{W}} \bmod \mathfrak{A}$ is a finite ring (by Proposition B.1.28) with a finite multiplicative group. We can set k equal to the order of this group. □

Using the proposition above we derive a corollary which will have a role in the "weak vertical method" to be described in Chapter 7.

Corollary 6.1.5. *Let E be a subfield of K. Let*

$$\Omega = \{1, \omega_2, \ldots, \omega_n\} \subset O_K$$

be a basis of K over E. Let $\mathcal{T}$ be a set of K-primes closed under conjugation over E. (By "closed under conjugation" we mean that, given a set of all K-factors of an E-prime, either the whole set is in $\mathcal{T}$ or no element of the set is in $\mathcal{T}$. In this way the expression "closed under conjugation" makes sense even if K/E is not Galois.) Let $\mathfrak{B}$ be any integral divisor of E relatively prime to any prime of $\mathcal{T}$. Then there exists a positive $l = l(\mathfrak{B}) \in \mathbb{N}$ such that for any positive $m \equiv 0 \bmod l$ and for any $\mathcal{T}$-unit ε we have that

$$\varepsilon^m = c_1 + c_2\omega_2 + \cdots + c_n\omega_n,$$

where for all $i = 1, \ldots, n$ it is the case that $c_i \in O_{K,\mathcal{T}} \cap E$, with $c_1 \equiv 1 \bmod \mathfrak{B}$ and $c_i \equiv 0 \bmod \mathfrak{B}$ for $i = 2, \ldots, n$.

Proof. Let $\mathfrak{D} = \mathfrak{D}_1\mathfrak{D}_2$ be the E-divisor of the discriminant of Ω, where $\mathfrak{D}_1$ is an integral divisor, prime to all the primes in $\mathcal{T}$, while all the primes comprising $\mathfrak{D}_2$ are in $\mathcal{T}$. (Such a factorization of $\mathfrak{D}$ exists given our assumptions on $\mathcal{T}$.) Let

$$l = k\big(\mathfrak{D}_1^{h_E}\mathfrak{B}^{h_E}\big)$$

where h_E is the class number of E. Then by the definition of $k(\mathfrak{D}_1^{h_E}\mathfrak{B}^{h_E})$ we have that for any $m \equiv 0 \bmod l$ it follows that

$$\varepsilon^m - 1 \equiv 0 \bmod \mathfrak{D}_1^{h_E}\mathfrak{B}^{h_E}.$$

Let $B \in E$ be an element whose divisor is $\mathfrak{B}^{h_E}$. Consider

$$(\varepsilon^m - 1)B^{-1} \equiv 0 \bmod \mathfrak{D}_1^{h_E}$$

in $O_{K,\mathcal{T}}$. By Lemma B.4.12,

$$(\varepsilon^m - 1)B^{-1} = a_1 + a_2\omega_2 + \cdots + a_n\omega_n,$$

where for any $\mathfrak{q} \notin \mathcal{T}$ we have that $\text{ord}_{\mathfrak{q}} a_i \geq 0$ for $i = 1, \ldots, n$. Further,

$$\varepsilon^m - 1 = Ba_1 + Ba_2\omega_2 + \cdots + Ba_n\omega_n$$

and thus

$$\varepsilon = c_1 + c_2\omega_2 + \cdots + c_n\omega_n,$$

where $c_i \in E$ and for any K-prime $\mathfrak{p} \notin \mathcal{T}$ we have that

$$\text{ord}_{\mathfrak{p}}(c_1 - 1) \geq \text{ord}_{\mathfrak{p}} B \geq \text{ord}_{\mathfrak{p}} \mathfrak{B},$$

while for $i = 2, \ldots, n$ we know that

$$\text{ord}_{\mathfrak{p}} c_i \geq \text{ord}_{\mathfrak{p}} B \geq \text{ord}_{\mathfrak{p}} \mathfrak{B}.$$

□

The next lemma discusses a property of integral units of some fields.

Lemma 6.1.6. *Let M/K be a number field extension of degree 2, where K is totally real and M is totally complex. Then there exists $m \in \mathbb{N}$ such that for every integral unit ε in M and every $k \equiv 0 \mod m$ we have that $\varepsilon^k \in O_K$.*

Proof. The Dirichlet unit theorem implies that the integral unit group of K is of the same rank as the integral unit group of M. Since the integral unit groups are finitely generated, this means that there exists $m \in \mathbb{N}$ such that, for every integral unit ε in M and every $k \equiv 0 \mod m$, we have that $\varepsilon^k \in O_K$. (See [37] for a discussion of the Dirichlet unit theorem.) □

6.2 Norm equations of units

In this section we again encounter norm equations, though we will now put them to a different use. (This section may be compared with Chapter 4.)

Proposition 6.2.1. *Let M/K be a number field extension. Let $\mathcal{W}_K$ be a subset of non-archimedean primes of K. Let $\mathcal{W}_M$ be the set of all the primes of M lying above the primes of $\mathcal{W}_K$. Consider the equation*

$$\mathbf{N}_{M/K}(x) = 1, \tag{6.2.1}$$

where $x \in M$. Then the following statements are true.

1. *The set*

$$\{x \in O_{M,\mathcal{W}_M} : \mathbf{N}_{M/K}(x) = 1\} \tag{6.2.2}$$

is a multiplicative group. In particular, if $\varepsilon \in O_{M,\mathcal{W}_M}$ *is a solution to (6.2.1) then* ε^n *is also a solution. Further,*

$$\frac{\varepsilon^n - 1}{\varepsilon - 1} \equiv n \bmod (\varepsilon - 1) \text{ in } \mathbb{Z}[\varepsilon]. \tag{6.2.3}$$

2. *If* $\mathcal{W}_K$ *is a finite set then the rank of the group defined in (6.2.2) is equal to the difference of the ranks of the* $\mathcal{W}_M$*-unit group in* M *and the* $\mathcal{W}_K$*-unit group in* K.
3. *If a* K*-prime* $\mathfrak{q}$ *does not split in the extension* M/K *and* x *is any solution of (6.2.1) then* $\text{ord}_{\mathfrak{Q}} x = 0$ *for the sole factor* $\mathfrak{Q}$ *of* $\mathfrak{q}$ *in* M.
4. *If* $\mathfrak{t}$ *is a prime of* K *and* $\mathfrak{T}_1, \mathfrak{T}_2$ *are two factors of* $\mathfrak{t}$ *in* M *then (6.2.1) has a solution* x *with a divisor of the form* $\mathfrak{T}_1^{a_1}/\mathfrak{T}_2^{a_2}$.
5. *Let* $\mathcal{V}_M$ *be the set of all the primes of* $\mathcal{W}_M$ *lying above primes of* $\mathcal{W}_K$ *splitting in the extension* M/K. *Then the only solutions to (6.2.1) in* $O_{M,\mathcal{W}_M}$ *are the* $\mathcal{V}_M$*-units. (In particular, if* $\mathcal{V}_M$ *is empty then all the solutions are integral.)*

Proof.

1. The first assertion of the lemma is true because non-zero $\mathcal{W}_M$-integers and solutions to the norm equation (6.2.1) form multiplicative groups.
2. The second assertion of the lemma is true because the norm map is a homomorphism from the group of $\mathcal{W}_M$-units to the group of $\mathcal{W}_K$-units. The image of the homomorphism is of the same rank as the group of $\mathcal{W}_K$-units. Indeed, if δ is a $\mathcal{W}_K$-unit then $\delta^{[M:K]}$ is in the image of the group of $\mathcal{W}_M$-units under the norm map. Thus the image of the norm map is of finite index in the group of $\mathcal{W}_K$-units.
3. The third assertion of the lemma can be derived from the following considerations. Let $\prod \mathfrak{P}_i^{a_i}$, $a_i \in \mathbb{Z}$, be the divisor of a solution in M to (6.2.1). Let $\mathfrak{p}_i$ be the prime below $\mathfrak{P}_i$ in K. Then $\prod \mathfrak{p}_i^{f_i a_i}$, where f_i is the relative degree of $\mathfrak{P}_i$ over $\mathfrak{p}_i$, is the trivial divisor. (See Section 8, Chapter I of [37].) Thus for each i there exists a j such that $\mathfrak{p}_i = \mathfrak{p}_j$. In other words, the prime $\mathfrak{p}_i = \mathfrak{p}_j$ splits in the extension M/K.
4. To see that the fourth statement of the lemma is true, note the following. If h_M is the class number of M then M has an element with a divisor of the form $\mathfrak{T}_1^{f_2 h_M}/\mathfrak{T}_2^{f_1 h_M}$, where f_1, f_2 are the relative degrees of $\mathfrak{T}_1$ and $\mathfrak{T}_2$ over K respectively. The norm of this divisor is trivial. So if $x \in M$ has the divisor of this form, the K-norm of x is an integral unit δ. Let $y = \delta^{-1}x^{[M:K]}$. Then $\mathbf{N}_{M/K}(y) = 1$.

5. The final statement of the lemma can be derived from Part 3 and the following fact. Any solution in $O_{M,\mathcal{W}_M}$ to (6.2.1) is a unit of the ring $O_{M,\mathcal{W}_M}$ and thus must be a $\mathcal{W}_M$-unit. □

Next we consider the norm equations of units in more specific situations.

Lemma 6.2.2. *Let K be a totally real field of degree n over $\mathbb{Q}$ and let M/K be a degree-2 extension. Then there exists $d \in O_K$ satisfying the following conditions:*

- *d is not a square of K;*
- *for all embeddings $\sigma : M \to \mathbb{C}$ such that $\sigma(M) \not\subset \mathbb{R}$ we have that $\sigma(d) > 0$;*
- *for all embeddings $\sigma : M \to \mathbb{C}$ such that $\sigma(M) \subset \mathbb{R}$ we have that $\sigma(d) < 0$.*

Further, the rank of the multiplicative group

$$\Xi_M = \left\{\varepsilon \in O_{M(\sqrt{d})} : \mathbf{N}_{M(\sqrt{d})/M}(\varepsilon) = 1\right\}$$

is equal to the rank of the multiplicative group

$$\Xi_K = \left\{\varepsilon \in O_{K(\sqrt{d})} : \mathbf{N}_{K(\sqrt{d})/K}(\varepsilon) = 1\right\},$$

and for some non-zero $k \in \mathbb{N}$, depending on M, K, and d only, for any $\varepsilon \in \Xi_M$ we have that $\varepsilon^k \in \Xi_K$.

Proof. First of all we observe the following. Since M/K is an extension of degree 2, for some $a \in K$ we have that $M = K(\sqrt{a})$. Therefore for any σ as above, $\sigma(M)$ is real if and only if $\sigma(a) > 0$. Hence, the last two conditions on d can be restated in the following form: for any embedding $\tau : K \to \mathbb{C}$ we have that $\tau(a)$ and $\tau(d)$ have different signs. Now the existence of d follows by the Strong Approximation Theorem (see Theorem B.2.1).

Next note that by the construction of d, any embedding of $M(\sqrt{d})$ into $\mathbb{C}$ will be non-real. Since $[M(\sqrt{d}) : \mathbb{C}] = 4n$, by the Dirichlet unit theorem, the rank of the integral unit group of $M(\sqrt{d})$ is $2n - 1$. Proceeding further, let r be the number of embeddings τ of K into $\mathbb{C}$ such that $\tau(a) > 0$ and let s be the number of embeddings τ such that $\tau(a) < 0$. Obviously $r + s = n$. Further, M will have $2r$ real and $2s$ non-real embeddings. Thus the rank of the integral unit group of M is $2r + s - 1$ and the rank of Ξ_M is $2n - 1 - 2r - s + 1 = s$.

At the same time, $K(\sqrt{d})$ will have $2s$ real and $2r$ non-real embeddings into $\mathbb{C}$, so that the rank of the integral unit group of $K(\sqrt{d})$ is $2s + r - 1$. Thus the rank of Ξ_K is $2s + r - 1 - r - s + 1 = s$ also, and the second assertion of the lemma holds.

To prove the last assertion by Lemma B.3.1 it is enough to show that the fields M and $K(\sqrt{d})$ are linearly disjoint over K. Indeed, linear disjointness of M and $K(\sqrt{d})$ over K would imply that any element $x \in K(\sqrt{d})$ satisfies the same irreducible polynomial over K as over M and therefore has the same conjugates over M as over K. The last statement implies that

$$\mathbf{N}_{M(\sqrt{d})/M}(x) = \mathbf{N}_{K(\sqrt{d})/K}(x),$$

and, therefore, $\Xi_K \subseteq \Xi_M$. Further, since the extensions M/K and $K(\sqrt{d})$ are both Galois, by Lemma B.3.3, in order to show that $K(\sqrt{d})$ and M are linearly disjoint over K it is enough to show that $M \cap K(\sqrt{d}) = K$ or, in other words that $K(\sqrt{d}) \neq M$. But by the construction of d one of these fields is real while the other is not. Therefore they cannot coincide. □

6.3 The Pell equation

Under certain circumstances we can say more about the integral solutions of unit norm equations. In this section we look at a very particular norm equation for an extension of degree 2: the Pell equation. More specifically, we are interested in equations of the following form:

$$X^2 - (a^2 - 1)Y^2 = 1. \tag{6.3.1}$$

If K is a number field and $X, Y, a \in K$ with $a^2 - 1$ not a square of K, then (6.3.1) asserts that an element $X + \sqrt{a^2-1}\,Y$ of $K(\sqrt{a^2-1})$ has K-norm equal to 1. Solutions to this equation have very nice properties, described below, and these properties serve as the foundation for the construction of a Diophantine definition of $\mathbb{Z}$ over the rings of integers of totally real number fields and fields with exactly one pair of non-real embeddings. The Pell equation also played a prominent role in a solution of the original Hilbert's Tenth Problem. For more details on the role of this remarkable equation see a helpful article by Martin Davis ([12]). Finally we note that the material presented in this section originally appeared in [15], [68], and [91].

Some properties of the Pell equation manifest themselves over any number field, while others depend on the field. We start with the first kind after introducing the initial notation set.

Notation 6.3.1.

- Let K denote a number field of degree n over $\mathbb{Q}$.
- Let $a \in O_K$ denote a K-integer such that $d = a^2 - 1$ is not a square in K.

- Let $a_1 = a, \ldots, a_n$ be all the conjugates of a over $\mathbb{Q}$.
- If $d = a^2 - 1$ is a positive real number then let $\sqrt{a^2 - 1}$ have the usual meaning, i.e. the positive real number whose square is $a^2 - 1$. Otherwise let $\sqrt{a^2 - 1}$ be one of the two complex numbers whose square is $a^2 - 1$.
- Let $\delta = \delta(a) = \sqrt{a^2 - 1}$.
- Let $\varepsilon = \varepsilon(a) = a + \delta(a)$.
- Let $x_m = x_m(a)$ and $y_m = y_m(a) \in O_K$, $m \in \mathbb{Z}$, be such that

$$x_m + \delta y_m = \varepsilon^m.$$

Now we list the more general properties of the Pell equation.

Lemma 6.3.2. *The following statements are true.*

1. $h \mid m \Rightarrow y_h \mid y_m$ *for all* $h, m \in \mathbb{Z} \setminus \{0\}$.
2. $x_{2km} \equiv \pm 1 \bmod x_m$ *for all* $k, m \in \mathbb{Z}$.
3. $x_m \mid y_{2km}$ *for all* $k, m \in \mathbb{Z}$.
4. $y_m(a) \equiv m \bmod (a - 1)$ *for all* $m \in \mathbb{Z}$.
5. *For* $k, l \in \mathbb{N}$ *we have that* $x_{k \pm l} = x_k x_l \pm (a^2 - 1) y_k y_l$.
6. *For* $k, l \in \mathbb{N}$ *we have that* $y_{k+l} = x_k y_l \pm x_l y_k$.
7. $x_{2m \pm j} \equiv \pm x_j \bmod x_m$ *for all* $j, m \in \mathbb{Z}$.
8. *If* $b, c \in O_K$, $\delta(b) \notin K$, *and* $a \equiv b \bmod c$ *then for all* $m \in \mathbb{Z}$ *we have*

$$x_m(a) \equiv x_m(b) \bmod c$$

and

$$y_m(a) \equiv y_m(b) \bmod c.$$

Proof. First of all, observe that for any $m \in \mathbb{Z}$ we have that $x_{-m} = x_m$ and $y_{-m} = -y_m$. So without loss of generality we can assume that all the indices are non-negative. Next we note that the properties listed above will follow from the formulas for $y_m(a)$ and $x_m(a)$ in terms of m and a or in terms of $x_h(a)$ and $y_h(a)$ when $h \mid m$, which we obtain simply by using the binomial theorem. Indeed let $h \mid m$, let $l = m/h$, and observe the following:

$$(a + \delta)^m = (x_h + \delta y_h)^l = \sum_{i=0}^{l} \binom{l}{i} x_h^{l-i} y_h^i \delta^i.$$

Thus, if l is even we have

$$(a + \delta)^m = \sum_{j=0}^{l/2} \binom{l}{2j} x_h^{l-2j} y_h^{2j} d^j + \delta \left(\sum_{j=0}^{(l-2)/2} \binom{l}{2j+1} x_h^{l-2j-1} y_h^{2j+1} d^j \right).$$

If l is odd then

$$(a+\delta)^m = \sum_{j=0}^{(l-1)/2} \binom{l}{2j} x_h^{l-2j} y_h^{2j} d^j + \delta \left(\sum_{j=0}^{(l-1)/2} \binom{l}{2j+1} x_h^{l-2j-1} y_h^{2j+1} d^j \right).$$

Hence, we have the formulas

$$x_m = \sum_{0 \le j \le l/2} \binom{l}{2j} x_h^{m-2j} y_h^{2j} d^j \tag{6.3.2}$$

and

$$y_m = \sum_{0 \le j \le (l-1)/2} \binom{l}{2j+1} x_h^{m-2j-1} y_h^{2j+1} d^j. \tag{6.3.3}$$

Thus Part 1 of the lemma follows directly from (6.3.3).

To see that Part 2 holds, observe that

$$x_{2mk} = x_m^{2k} + \binom{m}{2k-2} x_m^{2k-2} d y_m^2 + \cdots + \left(d y_m^2\right)^k \equiv (-1)^k \bmod x_m.$$

Similarly, we observe that

$$y_{2mk} = 2k x_m^{2k-1} y_m + \cdots + 2k x_m y_m^{2k-1} d^{k-1} \equiv 0 \bmod x_m,$$

and Part 3 holds.

Next, set $h = 1$, $x_h = a$, $y_h = 1$, and observe that

$$y_m = \sum_{0 \le j \le (m-1)/2}^{(m-2)/2} \binom{m}{2j+1} a^{m-2j-1} d^j. \tag{6.3.4}$$

Thus, modulo $a - 1$ we have that $y_m \equiv \binom{m}{1} = m$ and Part 4 is proved.

Now we note that Part 5 and Part 6 follow from the fact that for natural numbers l and k we have that $\varepsilon_{l \pm k} = \varepsilon_l \varepsilon_{\pm k}$. Thus, for Part 7,

$$\begin{aligned} x_{2m \pm j} &= x_{2m} x_j \pm (a^2 - 1) y_{2m} y_j \\ &= \left(x_m^2 + (a^2-1) y_m^2\right) x_j \pm (a^2 - 1)(2 x_m y_m) y_j \\ &= \left(2x_m^2 - 1\right) x_j \pm (a^2 - 1)(2 x_m y_m) y_j \equiv -x_j \bmod x_m. \end{aligned}$$

Part 8 follows by induction using addition formulas. □

Our next step is to specialize our discussion to a class of number fields. We will now need a second notation set.

Notation 6.3.3.

- From now on till the end of this section, let K denote a totally real (*tr*) number field or a number field with exactly one pair of non-real (*opnr*) embeddings.

- Let $\sigma_1 = \text{id}, \ldots, \sigma_n : K \to \mathbb{C}$ be all the embeddings of K into $\mathbb{C}$.
- If K has exactly one pair of non-real embeddings then assume that $K, \sigma_2(K) \not\subset \mathbb{R}$ and that for any $x \in K$ we have $\sigma_2(x)$ equal to $\bar{x}$, the complex conjugate of x.
- Let $I_0(K) = \{1\}$ if K is *tr*, and let $I_0(K) = \{1, 2\}$ if K is *opnr*.
- Let a be a K-element which satisfies the following inequalities:
  (a) $|a_i| > 2^{2n+2}$ for $i \in I_0$,
  (b) $0 < a_i < \frac{1}{2}$ for $i \notin I_0$,
  where $\sigma_i(a) = a_i$.
- Let $M = K(\delta)$. (Given our assumptions on a, it follows immediately that if K is *tr* then M has exactly two real embeddings, and if K is *opnr* then M is totally complex.)
- For $i = 1, \ldots, n$ and $j = 1, 2$, let $\sigma_{i,j}$ be one of the two extensions of σ_i to M. Also let $\sigma_{1,1} = \text{id}$.
- For $i = 1, \ldots, n$, let $\varepsilon_i = \sigma_{i,1}(\varepsilon)$ if $|\sigma_{i,1}(\varepsilon)| \geq 1$ and let $\varepsilon_i = \sigma_{i,2}(\varepsilon)$ if $|\sigma_{i,1}(\varepsilon)| < 1$.
- For $i = 1, \ldots, n$, let $\delta_i = \sigma_{i,1}(\delta)$. Then $\sigma_{i,2}(\delta) = -\delta_i$.
- Let C be a real constant, defined below in Lemma 6.3.6. Let $e \in \mathbb{N}$ be such that

$$|\varepsilon_1^e| > \frac{4|\delta_1|^{m_0}}{C^{n-m_0}},$$

  where $m_0 = |I_0|$.
- Let $G = \{u - \delta v : u, v \in O_K, u^2 - (a^2 - 1)v^2 = 1\}$.
- Let $H = \{\mu \in O_M : \mathbf{N}_{M/K}(\mu) = 1\}$.
- Let U_M, U_K be the groups of integral units of M and K respectively.

First we need to to establish some facts concerning the size of the absolute value of all the conjugates of $\delta(a)$ and $\varepsilon(a)$ in H.

Lemma 6.3.4. *The following statements hold.*

1. $\frac{1}{2}|a_i| < |\sigma_{i,j}(\delta)| < |a_i| + 1$, *where* $i \in I_0(K)$, $j = 1, 2$.
2. $\frac{1}{2} < |\sigma_{i,j}(\delta)| < 1$ *for* $i \notin I_0(K)$, $j = 1, 2$.
3. *Let* $\mu \in H$. *Then for* $i \notin I_0$ *we have that* $|\sigma_{i,j}(\mu)| = 1$. *Further, if* μ *is not a root of unity then either* $|\mu| > 1$ *or* $|\mu^{-1}| > 1$.
4. $|a| - \sqrt{|a^2 - 1|} < 1$.
5. *Assuming* $|\varepsilon| > 1$, *we have that* $|a| < |\varepsilon| < 2|a|$.
6. $\varepsilon(a)$ *is not a root of unity.*

Proof. The proof follows directly from our assumptions. Indeed, note that since $|a_i| > 2^{2n+2}$ for $i \in I_0(K)$ with $n \geq 2$, we have that $\frac{3}{4}|a_i|^2 > 1$ for such an i. Thus we conclude that

$$\tfrac{1}{4}|a_i|^2 < |a_i|^2 - 1 \leq |\sigma_{i,j}(\delta)|^2 = \left|a_i^2 - 1\right| \leq |a_i|^2 + 1 < (|a_i| + 1)^2,$$

and the first assertion of the lemma holds.

Next, let $i \notin I_0(K)$. Then

$$\tfrac{3}{4} < 1 - a_i^2 < 1,$$

and for $j = 1, 2$ we have $\frac{1}{2} < |\sigma_{i,j}(\delta)| < 1$. Now let $\mu \in H$. Then $\mu = x - y\delta$, with $x, y \in K$. If $i \notin I_0$ then $\sigma_{i,j}(\delta) = \pm\sqrt{\sigma_i(a)^2 - 1}$, where $\sigma_i(a) \in \mathbb{R}$ and $0 < \sigma_i(a) < \frac{1}{2}$. Thus for $i \notin I_0$ we have that $\sigma_i(a)^2 - 1 < 0$ and $\pm\delta \in i\mathbb{R}$. Since $\sigma_i(K) \subset \mathbb{R}$, we conclude that

$$\sigma_{i,1}(\mu) = \sigma_i(x) - \sigma_{i,1}(\delta)\sigma_i(y)$$

and

$$\sigma_{i,2}(\mu) = \sigma_i(x) - \sigma_{i,2}(\delta)\sigma_i(y) = \sigma_i(x) + \sigma_{i,1}(\delta)\sigma_i(y)$$

are complex conjugates whose product is 1. Therefore $|\sigma_{i,1}(\mu)| = |\sigma_{i,2}(\mu)| = 1$.

Suppose now that $\mu \in H$ and $|\mu| = 1$. Then given our assumptions on σ_1 and σ_2 for the case when K is *opnr*, for both kinds of fields we have $|\sigma_{i,j}(\mu)| = 1$ for *all* values of i and j. This would make μ a root of unity. Thus the third assertion of the lemma follows.

The next inequality,

$$|a| - \sqrt{|a^2 - 1|} < 1, \tag{6.3.5}$$

is equivalent to $|a|^2 - 2|a| + 1 < |a^2 - 1|$. But

$$|a|^2 + 1 < |a|^2 + 2|a| - 1 \leq |a^2 - 1| + 2|a|,$$

and therefore (6.3.5) holds. To see that if $|\varepsilon| > 1$ then $|\varepsilon| > |a|$, consider the following inequality:

$$|\varepsilon^2| = |a + \sqrt{a^2 - 1}|^2 \geq |2a^2 + 2a\sqrt{a^2 - 1}| - 1 = 2|\varepsilon||a| - 1$$

or, in other words,

$$|\varepsilon|^2 - 2|a||\varepsilon| + 1 = (|\varepsilon| - |a| - \sqrt{|a|^2 - 1})(|\varepsilon| - |a| + \sqrt{|a|^2 - 1}) \geq 0.$$

Thus, either

$$0 < |\varepsilon| \leq |a| - \sqrt{|a|^2 - 1} < 1, \tag{6.3.6}$$

by the argument above, or

$$|\varepsilon| \geq |a| + \sqrt{|a|^2 - 1} > |a|. \tag{6.3.7}$$

Now, to show that $|\varepsilon| = |a + \sqrt{a^2 - 1}| < 2|a|$ it is enough to observe that $|\sqrt{a^2 - 1}| < |a|$. Finally, from inequalities (6.3.6) and (6.3.7) it follows that $|\varepsilon| \neq 1$ and therefore it is not a root of unity. □

Our next job is to show that $\varepsilon(a)$ essentially generates the group G.

Lemma 6.3.5. *Let $\mu \in G$. Then for some $k \in \mathbb{Z}$ we have $\mu = \xi\varepsilon(a)^k$, where ξ is a root of unity.*

Proof. Consider $\mathbf{N}_{M/K} : U_M \to U_K$. Then H is the kernel of this map. Further, the image of the map is of finite index in U_K, since every square of an integral unit of K is in the image. Thus the rank of H is equal to rank U_M − rank U_K. Direct calculation using the Dirichlet unit theorem and our assumptions on a imply that for both types of K (*tr* and *opnr*) this difference is 1. Since $G \subseteq H$ the rank of G is at most 1, but since $a - \delta \in G$ we know that it is exactly 1. Further, the torsion-free part of G is isomorphic to a $\mathbb{Z}$-module and thus to a free module, and therefore we conclude that, modulo the roots of unity, every element of G is an integer power of some M-unit ν with a K-norm equal to 1. Without loss of generality, we can assume that $|\nu| > 1$ and $\varepsilon(a) = \xi\nu^l$, where $l \neq 0$ and ξ is a root of unity. We want to show that $|l| = 1$. To that effect let $\nu = x + \delta y$ with $x, y \in O_K$, and observe that $2\delta \mid (\nu - \nu^{-1})$. Consequently, on the one hand $|\mathbf{N}_{M/\mathbb{Q}}(2\delta)| \leq |\mathbf{N}_{M/\mathbb{Q}}(\nu - \nu^{-1})|$. On the other hand,

$$|\mathbf{N}_{M/\mathbb{Q}}(2\delta)| = 2^{2n} \prod_{i \in I_0, j=1,2} |\sigma_{i,j}(\delta)| \prod_{i \notin I_0, j=1,2} |\sigma_{i,j}(\delta)| > (4|\delta|^2)^{m_0} > (a^2)^{m_0},$$

by Lemma 6.3.4. At the same time, we have that

$$\begin{aligned} |\mathbf{N}_{M/\mathbb{Q}}(\nu - \nu^{-1})| &= \prod_{i \in I_0, j=1,2} |\sigma_{i,j}(\nu) - \sigma_{i,j}(\nu^{-1})| \prod_{i \notin I_0, j=1,2} |\sigma_{i,j}(\nu) - \sigma_{i,j}(\nu^{-1})| \\ &< 2^{2n-2m_0}|\nu - \nu^{-1}|^{2m_0} \leq 2^{2n-2m_0}(|\nu| + 1)^{2m_0} < 2^{2n}|\nu|^{2m_0}. \end{aligned}$$

Thus we obtain the inequality $|a|^{2m_0} < 2^{2n}|\nu|^{2m_0}$. Suppose now that $\varepsilon_1 = \xi\nu^l$, where $l \geq 2$ and ξ is a root of unity. Then $|a|^2 < 2^{2n}|\varepsilon_1| < 2^{2n+1}|a|$ and we have a contradiction of one of our assumptions on a. □

Having established the nature of elements of G, we continue to explore the properties of powers of $\varepsilon = \varepsilon(a)$.

Lemma 6.3.6. *The following statements hold.*

1. $k < |\sigma_i(x_k(a))|$ *for* $i \in I_0$ *and any* $k \in \mathbb{N}$.
2. *There exists a constant* C, *depending on* a *and* K *only, such that for any* $k \in \mathbb{N} \setminus \{0\}$ *there exist* $m, h \in k\mathbb{N}$ *with the properties*

$$|\sigma_i(x_m)| > \tfrac{1}{2}, \qquad i \notin I_0, \tag{6.3.8}$$

and

$$|\sigma_i(y_h)| > C, \qquad i \notin I_0. \tag{6.3.9}$$

3. *If* $\sigma_i(y_{eh})$ *satisfies (6.3.9) for* $i \notin I_0$ *then*
   (a) $y_{eh} | y_{em} \Rightarrow h \,|\, m$,
   (b) $y_{eh}^2 | y_{em} \Rightarrow h y_{eh} \,|\, m$,
   where the integer constant e *was defined in Notation 6.3.3.*
4. *If* $\sigma_i(x_m)$ *satisfies (6.3.8) for* $i \notin I_0$ *then*

$$x_k \equiv \pm x_j \bmod x_m \quad \Rightarrow \quad k \equiv \pm j \bmod m.$$

5. *For any non-zero* $m \in \mathbb{Z}$ *and* $C_1, C_2 \in \mathbb{R}^+$, *there exists a positive* $s \in \mathbb{N}$ *such that*

$$b = \left(x_m^2 + y_m^2(a^2-1)\right)^{2s}\left(x_m^4 + a\left(1-x_m^2\right)^2\right)$$

has the following properties:
   (a) $b \equiv 1 \bmod y_m(a)$;
   (b) $b \equiv a \bmod x_m(a)$;
   (c) $|\sigma_i(b)| > C_1$ *for* $i \in I_0$ *and* $0 < \sigma_i(b) < C_2$ *for* $i \notin I_0$. *In particular, it can be arranged that* b *satisfies the requirements listed for* a *in Notations 6.3.1 and 6.3.3.*

Proof.
1. On the one hand, since ε is not a root of unity by Lemma 6.3.4, either $|\varepsilon| > 1$ or $|\varepsilon|^{-1} > 1$. Thus, by the same lemma, either ε or ε^{-1} is of absolute value greater than a. On the other hand,

$$|x_m| = \frac{|\varepsilon^m + \varepsilon^{-m}|}{2} \geq \frac{|a|^m - 1}{2} > \frac{2^{2nm} - 1}{2} > m$$

since $n \geq 2$.

2. Let $k \in \mathbb{N}$ be given. Consider the set $\{\varepsilon_j, j \notin I_0\}$. By Lemma 6.3.4, $|\varepsilon_j| = 1$. Thus for each $j \notin I_0$ we have that $\varepsilon_j = e^{i\beta_j}$, where $\beta_j/\pi \notin \mathbb{Q}$. (Otherwise, ε_j would be a root of unity.) Let $\alpha_j = \beta_j/\pi$ and let $A = \{\alpha_{j_1}, \ldots, \alpha_{j_s}\}$ be a maximal subset of $\{\alpha_j, j \in I_0\}$ with respect to the linear independence of the set $\{1, \alpha_{j_1}, \ldots \notin, \alpha_{j_s}\}$ over $\mathbb{Q}$. Since, by a preceding observation, for all $j \notin I_0$

we have $\alpha_j \notin \mathbb{Q}$, we know that A is not empty. Let $J_0 = \{j_1, \ldots, j_s\}$. Then for any $r \notin I_0$ and for some $b_r, b_{r,j} \in \mathbb{Z}$, we have that $\varepsilon_r^{b_r} = \prod_{j \in J_0} \varepsilon_j^{b_{r,j}}$. Thus, if we let $b = \prod_{r \notin I_0} b_r$ then for all $r \notin I_0$ and for some $a_{r,j} \in \mathbb{Z}$, we have that $\varepsilon_r^b = \prod_{j \in J_0} \varepsilon_j^{a_{r,j}}$. Consequently, we conclude that for all $r \notin I_0$ and for any $l \in k\mathbb{Z}$ we also have that

$$\varepsilon_r^{lb} = \prod_{j \in J_0} \varepsilon_j^{la_{r,j}} = \prod_{j \in J_0} \mathrm{e}^{i\pi l a_{r,j}\alpha_j} = \mathrm{e}^{i\pi \sum_{j \in J_0} l a_{r,j}\alpha_j}$$

$$= \cos\left(\pi \sum_{j \in J_0} l a_{r,j}\alpha_j\right) + i \sin\left(\pi \sum_{j \in J_0} l a_{r,j}\alpha_j\right) = \sigma_r(x_{lb}) \pm \sigma_{r,1}(\delta)\sigma_r(y_{lb}).$$

Let $m = \max_{r,j}\{|a_{r,j}|\}$. Since $|\cos \pi\theta|$ is a continuous function, for any $\lambda_1 > 0$ there exists $\lambda_2 > 0$ such that $|\theta| < \lambda_2 \Rightarrow 1 - |\cos \pi\theta| < \lambda_1$. By Kronecker's theorem (see Theorem 442, Chapter XXIII of [36]), there exists $l \in k\mathbb{N}$ such that for all $j \in J_0$ there exists $l_j \in \mathbb{Z}$ with the property that $|l\alpha_j - l_j| < \lambda_2/mn$. Then

$$\left| \sum_{j \in J_0} l a_{r,j}\alpha_j - \sum_{j \in J_0} a_{r,j} l_j \right| < \lambda_2,$$

and therefore

$$1 - |\sigma_r(x_{lb})| = 1 - \left| \pm \cos\left(\pi \sum_{j \in J_0} l a_{r,j}\alpha_j - \pi \sum_{j \in J_0} a_{r,j} l_j\right) \right| < \lambda_1.$$

If we select $\lambda_1 < \frac{1}{2}$ then for all $r \notin I_0$ we have that $|\sigma_r(x_{lb})| > \frac{1}{2}$.

Next, for $r \notin I_0$ let $A_r = \sum_{j \in J_0} a_{r,j}$. Let $A \in \mathbb{N}$ be such that $A > \max_{r \notin I_0}\{|A_r|\}$ and $\mathrm{ord}_2 A_r/A \leq 0$ for all $r \notin I_0$. Since $|\sin \pi\theta|$ is a continuous function, it is equicontinuous on any closed and bounded interval. Thus for any $\lambda_1 > 0$ there exists $0 < \lambda_2 < \frac{1}{4}$ such that for any $\theta_1, \theta_2 \in [-1, 1]$ with $|\theta_1 - \theta_2| < \lambda_2$ we have that $||\sin(\pi\theta_1)| - |\sin(\pi\theta_2)|| < \lambda_1$. Next using Kronecker's theorem again choose $l \in k\mathbb{Z}$ such that for all $j \in J_0$ and for some $l_j \in \mathbb{Z}$ we have

$$\left| l\alpha_j - l_j - \frac{1}{2A} \right| < \frac{\lambda_2}{An}.$$

Then for all $r \notin I_0$ we have that

$$\left| \sum_{j \in J_0} \left(l a_{r,j}\alpha_j - l_j a_{r,j} - \frac{a_{r,j}}{2A} \right) \right| < \lambda_2.$$

Thus

$$\left| \sum_{j \in J_0} (l a_{r,j}\alpha_j - l_j a_{r,j}) - \frac{A_r}{2A} \right| < \lambda_2$$

and, therefore, for $\lambda_1 < \left|\frac{1}{2}\sin(\pi A_r/2A)\right|$ we can conclude that

$$
\begin{aligned}
|\sigma_{r,1}(\delta)\sigma_r(y_{lb})| &= \left|\sin\left(\pi\sum_{j\in J_0} la_{r,j}\alpha_j\right)\right| = \left|\sin\left(\pi\sum_{j\in J_0} la_{r,j}\alpha_j - l_j a_{r,j}\right)\right| \\
&> \frac{1}{2}\left|\sin\left(\frac{\pi A_r}{2A}\right)\right|,
\end{aligned}
$$

where $\sin(\pi A_r/2A) \neq 0$, since $A_r/2A \notin \mathbb{Z}$ by the construction of A. Thus we can set

$$
C = \min_r \frac{|\sin(\pi A_r/2A)|}{2|\sigma_{r,1}(\delta)|} > 0.
$$

3. Let $r, q \in \mathbb{N}$ and $0 < r < q$. Assume also that $|\sigma_i(y_{eq})| > C$ for all $i \notin I_0$. Then

$$
\begin{aligned}
\left|\mathbf{N}_{M/\mathbb{Q}}(y_{er})\right| &= \prod_{i=1}^{n}\prod_{j=1}^{2}\left|\sigma_{i,j}\left(\frac{\varepsilon^{er} - \varepsilon^{-er}}{2\delta}\right)\right| \\
&\leq \prod_{i\in I_0}\frac{\left|\left(\varepsilon_i^{er} - \varepsilon_i^{-er}\right)\left(\varepsilon_i^{-er} - \varepsilon_i^{er}\right)\right|}{4\left|\delta_i^2\right|}\prod_{i\notin I_0}\frac{1}{|\delta_i|^2} \\
&\leq \frac{4|\varepsilon_1|^{2em_0r}}{4^{m_0}|\mathbf{N}_{K/\mathbb{Q}}(a^2-1)|} < \frac{|\varepsilon_1|^{2em_0r}}{4^{m_0-1}}.
\end{aligned}
$$

So, finally,

$$
\left|\mathbf{N}_{K/\mathbb{Q}}(y_{er})\right| \leq \frac{|\varepsilon_1|^{em_0r}}{2^{m_0-1}}.
$$

At the same time,

$$
\begin{aligned}
\left|\mathbf{N}_{K/\mathbb{Q}}(y_{eq})\right| &= \prod_{i=1}^{n}|\sigma_i(y_{eq})| \geq C^{n-m_0}\prod_{i\in I_0}|\sigma_i(y_{eq})| \\
&\geq C^{n-m_0}\prod_{i\in I_0}\frac{\left|\varepsilon_i^{eq} - \varepsilon_i^{-eq}\right|}{2|\delta_i|} \geq C^{n-m_0}\frac{(|\varepsilon_1|^{eq}-1)^{m_0}}{2^{m_0}|\delta_1|^{m_0}} \\
&> C^{n-m_0}\frac{|\varepsilon_1|^{m_0eq}}{2^{m_0+1}|\delta_1|^{m_0}}.
\end{aligned}
$$

Thus, since by our assumption on e we have that

$$
|\varepsilon_1|^e > \frac{4|\delta_1|^{m_0}}{C^{n-m_0}},
$$

we can conclude that $|\mathbf{N}_{K/\mathbb{Q}}(y_{eq})| > |\mathbf{N}_{K/\mathbb{Q}}(y_{er})|$.

Suppose now that $y_{eq} \mid y_{el}$. Write $l = sq + r$, where $0 \leq r < q$. Assume that $r > 0$. Then $y_{el} = y_{seq+er} = y_{seq}x_{er} + x_{seq}y_{er}$. By Lemma 6.3.2 we know that $y_{eq} \mid y_{seq}$ and thus $y_{eq} \mid y_{seq}x_{er}$. But since $(y_{seq}, x_{seq}) = 1$ we conclude that

$(y_{eq}, x_{seq}) = 1$, so $y_{eq} \mid y_{er}$. But this is impossible by the argument above, and consequently $r = 0$.

Let us examine now the second divisibility condition, $y_{eq}^2 \mid y_{el}$. Now we know that $q \mid l$ and thus $l = qs$. Therefore, using the binomial theorem,

$$y_{eqs} = \sum_{0 \le j \le (l-1)/2} \binom{s}{2j+1} x_{eq}^{s-2j-1} y_{eq}^{2j+1} d^j$$

is established. Thus

$$\frac{y_{eqs}}{y_{eq}} = s x_{eq}^{s-1} + y_{eq}(\ldots),$$

and, since $(y_{eq}, x_{eq}) = 1$, we conclude that $y_{eq} \mid s$.

4. As above we have to start with some inequalities concerning norms. Let m be a positive integer and assume that x_m satisfies inequality (6.3.8). Let r_1, r_2 be positive integers such that $r_1 < m$ and $r_2 < m$. We claim that under these assumptions

$$|\mathbf{N}_{K/\mathbb{Q}}(x_{r_1} \pm x_{r_2})| < |\mathbf{N}_{K/\mathbb{Q}}(x_m)| \tag{6.3.10}$$

and

$$|\mathbf{N}_{K/\mathbb{Q}}(x_{r_1})| < |\mathbf{N}_{K/\mathbb{Q}}(x_m).| \tag{6.3.11}$$

In estimating the norms we will proceed pretty much in the same manner as above, by expressing x_m, x_{r_1}, and x_{r_2} as sums of powers of ε. Thus

$$\begin{aligned} |\mathbf{N}_{K/\mathbb{Q}}(x_m)| &= \prod_{i \in I_0} |\sigma_i(x_m)| \prod_{i \notin I_0} |\sigma_i(x_m)| \\ &\ge \frac{1}{2^{n-m_0}} \prod_{i \in I_0} \left| \frac{\varepsilon_i^m + \varepsilon_i^{-m}}{2} \right| \ge \frac{(|\varepsilon_1|^m - 1)^{m_0}}{2^n}. \end{aligned}$$

Without loss of generality we can assume that we have $r_1 \le r_2 < m$ and the following upper bound for the norm of $x_{r_1} \pm x_{r_2}$,

$$\begin{aligned} |\mathbf{N}_{K/\mathbb{Q}}(x_{r_1} \pm x_{r_2})| &= \prod_{i \in I_0} |\sigma_i(x_{r_1} \pm x_{r_2})| \prod_{i \notin I_0} |\sigma_i(x_{r_1} \pm x_{r_2})| \\ &\le 2^{n-m_0} \prod_{i \in I_0} \frac{|\varepsilon_i|^{r_1} + |\varepsilon_i|^{r_2} + |\varepsilon_i|^{-r_1} + |\varepsilon_i|^{-r_2}}{2} \\ &\le 2^{n-m_0} (|\varepsilon_1|^{r_2} + 1)^{m_0}. \end{aligned}$$

Thus, in this case it is enough to show that

$$2^{n-m_0} (|\varepsilon_1|^{r_2} + 1)^{m_0} < \frac{(|\varepsilon_1|^m - 1)^{m_0}}{2^n}. \tag{6.3.12}$$

Next we note that $|\varepsilon_1|^{r_2} + 1 < 2|\varepsilon_1|^{r_2}$ and $|\varepsilon_1|^m - 1 > \frac{1}{2}|\varepsilon_1|^m$, while $m_0 \geq 1$. Therefore to show that (6.3.12) holds, it is enough to show that

$$2^{2n+1}|\varepsilon_1|^{r_2} < |\varepsilon_1|^m \quad \Leftrightarrow \quad 2^{2n+1} < |\varepsilon_1|^{m-r_2}. \tag{6.3.13}$$

But, by Lemma 6.3.4 and the assumptions on a, we have the following inequalities:

$$2^{2n+1} < 2^{2n+2} < |a| < |\varepsilon_1| \leq \left|\varepsilon_1^{m-r_2}\right|.$$

Thus (6.3.12) holds.

Suppose now that $x_j \equiv \pm x_l \bmod x_m$. We can write $j = 2mj_1 \pm r_1$, $l = 2ml_1 \pm r_2$, where $|r_1| \leq m$, $|r_2| \leq m$. Then $x_j = x_{r_1}x_{2mj_1} \pm dy_{r_1}y_{2mj_1} \equiv \pm x_{r_1}$ by Lemma 6.3.2. Similarly, $x_l \equiv \pm x_{r_2} \bmod x_m$. Consequently we have $x_{r_1} \equiv \pm x_{r_2} \bmod x_m$. Thus if $r_1 \neq r_2$ then we have a contradiction with either inequality (6.3.10) or inequality (6.3.11).

5. First of all we observe that, for all $m \in \mathbb{Z}$, we have that $|\sigma_i(x_m)| < 1$ for $i \notin I_0$ and $|\sigma(x_m)| > 1$ for $i \in I_0$. Indeed,

$$|\sigma_i(x_m)| = \left|\frac{\sigma_{i,1}(\varepsilon^m + \varepsilon^{-m})}{2}\right|,$$

where for $i \notin I_0$, we know that

$$|\sigma_{i,1}(\varepsilon^m)| = |\sigma_{i,1}(\varepsilon^{-m})| = 1.$$

The only way in which the absolute value of the sum of two complex numbers can be equal to the sum of their absolute values is for both numbers to have the same argument. In our case this would require $\varepsilon^m = \pm 1$, contradicting our assumptions on a and m. Thus $|\sigma_i(x_m)| < 1$ for $i \notin I_0$ and all $m \neq 0$. Similarly, for $i \in I_0$ we have by Lemma 6.3.4 that

$$|\sigma_i(x_m)| = \left|\frac{\sigma_{i,1}(\varepsilon^m + \varepsilon^{-m})}{2}\right| > \frac{1}{2}\left|\varepsilon_1^m\right| - 1 > \frac{1}{2}|a_i| - 1 > 2^{2n+1} - 1 > 1.$$

Next we have that

$$\sigma_i(b) = \left(\sigma_i(x_m)^2 + \sigma_i(y_m)^2\left(a_i^2 - 1\right)\right)^{2s}\left(\sigma_i(x_m)^4 + a_i\left(1 - \sigma_i\left(x_m^2\right)\right)^2\right),$$

where for $i \in I_0$ we have that $a_i > 1$ and for $i \notin I_0$ we have that $0 < a_i < 1$. Consequently, for $i \in I_0$ we have that

$$\left|\sigma_i(x_m)^2 + \sigma_i(y_m)^2\left(a_i^2 - 1\right)\right| = \left|2\sigma_i(x_m)^2 - 1\right| \geq 2\,|\sigma_i(x_m)|^2 - 1 > 1,$$

and for $i \notin I_0$ we have that $\sigma_i(x_m) \in \mathbb{R}$ and

$$\sigma_i(x_m)^2 - \sigma_i(y_m)^2\left(1 - a_i^2\right) < \sigma_i(x_m)^2 < 1,$$

while at the same time

$$\sigma_i(x_m)^2 - \sigma_i(y_m)^2(1 - a_i^2) = 2\sigma(x_m)^2 - 1 > -1.$$

Hence for $i \notin I_0$ it is the case that

$$\left|\sigma_i(x_m)^2 + \sigma_i(y_m)^2(a^2 - 1)\right| < 1.$$

Thus, since m is fixed we can choose s to make $|\sigma_i(b)|$ arbitrarily large for $i \in I_0$ and arbitrarily small for $i \notin I_0$. In particular we can arrange that $|\sigma_i(b)| > 2^{2n+2}$ for $i \in I_0$ and $|\sigma_i(b)| < \frac{1}{2}$ for $i \notin I_0$.

Note also that for all i if $\sigma_i(b) \in \mathbb{R}$ we have that $\sigma_i(b) > 0$. Indeed, the first term in the product is a square and the second term is the sum of a square and a positive number times a square.

Further, for $i \notin I_0$ we have that $\sigma_i(b^2 - 1) < 0$ cannot be a square in $\sigma_i(K)$ and therefore $b^2 - 1$ is not a square in K. Finally, the equivalences in the statement of the lemma will hold simply because

$$\begin{aligned} x_m^2 &\equiv 1 \bmod y_m, \\ (a^2 - 1)y_m^2 &\equiv -1 \bmod x_m, \end{aligned}$$

and

$$1 - x_m^2 \equiv 1 \bmod x_m^2.$$

□

6.4 Non-integral solutions of some unit norm equations

In this section we will consider solutions to norm equations which are not integral. We will continue to call these solutions units because they will be units of some rings of $\mathcal{W}$-integers. (We remind the reader that we use the term "$\mathcal{W}$-integer" for the cases where $\mathcal{W}$ is infinite as well as for the cases where $\mathcal{W}$ is finite.) If we select the set $\mathcal{W}$ properly with respect to the extensions under consideration, the set of solutions to our norm equations will once again serve as a basis for the Diophantine definitions we seek.

Proposition 6.4.1. *Let K be a totally real number field. Let F be a subextension of K such that the extension K/F is cyclic. Let L be a totally complex extension of degree 2 of $\mathbb{Q}$ and let E be a totally real cyclic extension of $\mathbb{Q}$ of prime degree $p > 2$. Assume that p is relatively prime to $[K : \mathbb{Q}]$. These extensions are described in the following diagram:*

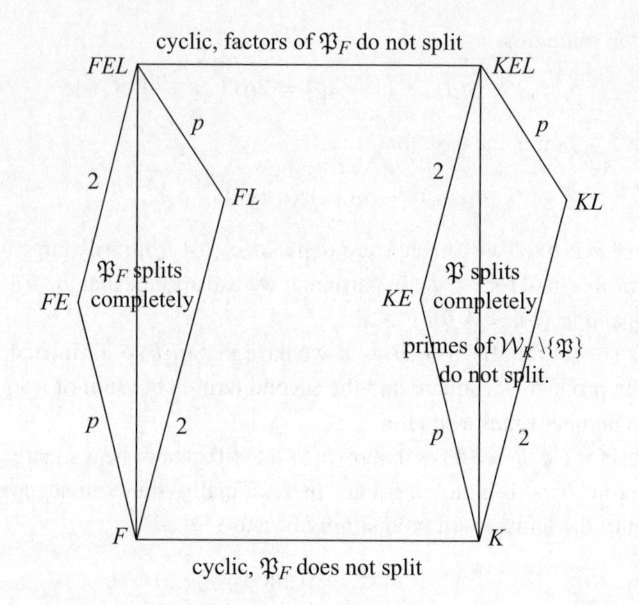

Next consider the system of norm equations

$$\begin{cases}\mathbf{N}_{LEK/EK}(x) = 1,\\ \mathbf{N}_{LEK/LK}(x) = 1.\end{cases} \tag{6.4.1}$$

Let $\mathcal{V}_K$ consist of primes of K not splitting in the extension KE/K. Let $\mathfrak{P}$ be a prime of K satisfying the following conditions:

1. $\mathfrak{P}$ *must lie above an F-prime* $\mathfrak{P}_F$ not splitting *in the extension K/F.* (This is where we need K/F to be cyclic. Otherwise we might have no primes which do not split.)
2. $\mathfrak{P}_F$ *should split completely in the extension FEL/F.*
3. $\mathfrak{P}$ *should split completely in the extension EKL/K.*

Let $\mathcal{W}_K = \mathcal{V}_K \cup \{\mathfrak{P}\}$. Let $\mathcal{W}_{KLE}$ be the set of primes of KLE lying above the primes of $\mathcal{W}_K$. Let $r \in \mathbb{N}$ satisfy the following requirements:

1. *Let $m \in \mathbb{N}$ be such that, for any root of unity $\xi \in LEK$, we have $\xi^m = 1$. Then $r \equiv 0 \bmod 2m$.*
2. *Also, $r/m \equiv 0 \bmod h_{LEF}$, where h_{LEF} is the class number of FEL.*

Let $x \in O_{KLE,W_{KLE}}$ be a solution to system (6.4.1). *Then $x^r \in LEF$. Further, the set of non-root-of-unity solutions to the system in $O_{KLE,W_{KLE}}$ is not empty.*

Proof. First of all, by Lemma B.3.3 we have that $[LKE : EK] = 2$, $[LEK : LK] = p$, $[LEF : EF] = 2$, $[LEF : LF] = p$. Further, the fields KE, FE

are totally real and the fields LK, LKE, FL, FLE are totally complex. From Proposition 6.2.1 we can conclude that if $x \in LEK$ is a solution to this system then the divisor of x must be composed of the primes lying above primes of EK and LK splitting in the extensions LEK/EK and LEK/LK respectively. Given the fact that both extensions are cyclic of distinct prime degrees, we can conclude that LEK-primes occurring in the divisor of x lie above K-primes splitting completely in the extension LEK/K. Further, since LEK/EK is a totally complex extension of degree 2 of a totally real field, all the integral solutions to this system of norm equations have to be roots of unity. Absence of non-root-of-unity integral solutions leads to the following consequence. Let x_1, x_2 be two solutions to the first equation of the norm system above such that x_1 and x_2 have the same divisor. Then $x_1^r = x_2^r$.

Given our assumptions on $\mathfrak{P}_F$ and $\mathfrak{P}$ and on the degrees of the extensions, by Lemma B.4.8 no factor of $\mathfrak{P}_F$ in FEL will split in the extension KEL/FEL. By our choice for the membership in $\mathcal{W}_K$, we have that x is a solution to our system while being an element of the integral closure of $O_{K,\mathcal{W}}$ in LEK only if the divisor of x consists of the factors of $\mathfrak{P}$ in LKE.

Let $\mathfrak{p}$ be a factor of $\mathfrak{P}$ in LEK. Note that by Lemmas B.3.5 and B.3.6, we have that

$$\begin{aligned}
\mathrm{Gal}(LEK/LK) &\cong \mathrm{Gal}(EK/K) \cong \mathrm{Gal}(E/\mathbb{Q}),\\
\mathrm{Gal}(LEK/EK) &\cong \mathrm{Gal}(LK/K) \cong \mathrm{Gal}(L/\mathbb{Q}),\\
\mathrm{Gal}(LEK/K) &\cong \mathrm{Gal}(LEK/EK) \times \mathrm{Gal}(LEK/LK).
\end{aligned}$$

Now let σ_L be a generator of $G(LEK/EK)$ and let σ_E be a generator of $G(LEK/LK)$. Then $\sigma_L\sigma_E = \sigma_E\sigma_L$ will generate $G(LEK/K)$. Since $\mathfrak{P}$ splits completely in the extension LEK/K, if $\tau_1, \tau_2 \in G(LEK/K)$ are such that $\tau_1(\mathfrak{p}) = \tau_2(\mathfrak{p})$ for some KLE-factor $\mathfrak{p}$ of $\mathfrak{P}$ then $\tau_1 = \tau_2$. Denote $\sigma_L^i\sigma_E^j(\mathfrak{p})$ by $\mathfrak{p}_{i,j}$, where $i = 1, 2$ and $j = 1, \ldots, p$. Suppose now that $z \in O_{KLE,\mathcal{W}_{KLE}}$ is a solution to the norm system. Then the divisor of z is of the form $\mathfrak{Z} = \prod \mathfrak{p}_{i,j}^{a_{i,j}}$, with

$$\sum_{i,j} a_{i,j} = 0. \tag{6.4.2}$$

Since each $\mathfrak{p}_{i,j}$ is the only unramified factor of the prime below it in FLE, we can consider $\mathfrak{Z}$ as a divisor of FLE. Let w be an element of FLE whose divisor is $\mathfrak{Z}^{h_{FLE}}$. Then, as in the argument used in the proof of Proposition 6.2.1, we have that (6.4.2) implies that $\mathbf{N}_{FLE/FE}(w) = \nu$, where $\nu \in FE$ is an integral unit and $\mathbf{N}_{FLE/FE}(\nu^{-1}w^2) = 1$.

Next we observe that $FLE \cap KE = FE$, since FE is the largest totally real subfield contained in FLE. Thus FLE and KE are linearly disjoint over

FE by Lemma B.3.3. Therefore $\mathbf{N}_{KLE/KE}(v^{-1}w^2) = \mathbf{N}_{FLE/FE}(v^{-1}w^2) = 1$. The divisor of $v^{-1}w^2$ is the same as the divisor of $z^{2h_{FLE}}$, however. Therefore $z^{2h_{FLE}m} = (v^{-1}w^2)^m \in FLE$.

Are we guaranteed to have such solutions? Indeed we are. Let $y \in LEK$ be such that its divisor is $\mathfrak{p}^a$, where $a \in \mathbb{N}$. (Such a y certainly exists if $a \equiv 0 \bmod h_{LEK}$, where h_{LEK} is the class number of LEK.) Next, consider

$$x = \frac{y\sigma_E\sigma_L(y)}{\sigma_L(y)\sigma_E(y)} = \frac{y/\sigma_L(y)}{\sigma_E(y)/\sigma_E\sigma_L(y)} = \frac{y/\sigma_E(y)}{\sigma_L(y)/\sigma_L\sigma_E(y)}.$$

We claim that x is not a root of unity and satisfies the norm system above. First of all note that since $y = u/\sigma_L(u) = v/\sigma_E(v)$, the EK-norm and LK-norm of y are equal to 1. Second, note that the divisor of x is of the form

$$\left(\frac{\mathfrak{p}\sigma_E\sigma_L(\mathfrak{p})}{\sigma_L(\mathfrak{p})\sigma_E(\mathfrak{p})}\right)^a.$$

But, by the argument above, the primes $\mathfrak{p}, \sigma_L(\mathfrak{p}), \sigma_E(\mathfrak{p}), \sigma_E\sigma_L(\mathfrak{p})$ are all distinct. Thus the divisor of x is not trivial and x is not a root of unity. □

The next proposition addresses the question of the existence of a prime $\mathfrak{P}$ satisfying all the requirements listed above.

Proposition 6.4.2. *Let K, F, L, E, p be as above. Then there are infinitely many primes $\mathfrak{P}_F$ satisfying the following conditions:*

1. *$\mathfrak{P}$ lies above an F-prime $\mathfrak{P}_F$ not splitting in the extension K/F.*
2. *$\mathfrak{P}_F$ splits completely in the extension FEL/F.*
3. *$\mathfrak{P}$ splits completely in the extension EKL/K.*

Proof. Consider the extension KEL/F. By Lemma B.3.3 the fields K, FE, LF are pairwise Galois extensions of F that are linearly disjoint over F, and therefore by Lemma B.3.6 we have that

$$\mathrm{Gal}(KLE/F) \cong \mathrm{Gal}(K/F) \times \mathrm{Gal}(LF/F) \times \mathrm{Gal}(EF/F).$$

Let σ_K be the generator of $\mathrm{Gal}(KLE/FLE) \cong \mathrm{Gal}(K/F)$, where the congruence holds by Lemma B.3.5. Further, by the same lemma, the restriction of σ_K to K generates $\mathrm{Gal}(K/F)$. Now let $\mathfrak{p}_{KLE}$ be a prime whose Frobenius automorphism over F is σ_K. Then $\mathfrak{p}_K = \mathfrak{p}_{KLE} \cap K$ is not moved by the restriction of σ_K to K. Hence $\mathfrak{p}_K$ is the only factor above $\mathfrak{p}_F = \mathfrak{p}_K \cap F$. The decomposition group of $\mathfrak{p}_{KLE}$ over K is equal to $\mathrm{Gal}(KLE/K) \cap G(\mathfrak{p}_{KLE})$, however, where $G(\mathfrak{p}_{KLE})$ is the decomposition group of $\mathfrak{p}_{KLE}$ over F. But this intersection contains the identity only. Therefore $\mathfrak{p}_K$ splits completely in the extension KLE/K. Let $\mathfrak{p}_{FLE} = \mathfrak{p}_{KLE} \cap FLE$. Then

$$G_{\text{over}F}(\mathfrak{p}_{FLE}) \cong G_{\text{over}F}(\mathfrak{p}_{KLE})/G_{\text{over } FLE}(\mathfrak{p}_{KLE}) = \{\text{id}\},$$

where $G_{\text{over}F}(\mathfrak{p}_{FLE})$ is the decomposition group of $\mathfrak{p}_{FLE}$ over F, $G_{\text{over}F}(\mathfrak{p}_{KLE})$ is the decomposition group of $\mathfrak{p}_{KLE}$ over F, and $G_{\text{over } FLE}(\mathfrak{p}_{KLE})$ is the decomposition group of $\mathfrak{p}_{KLE}$ over *FLE*. Thus $\mathfrak{p}_F$ splits completely in the extension FLE/F. Now our assertion follows by the Chebotarev density theorem. □

The next proposition is a generalization of Lemma 6.1.6 for the case of $\mathcal{W}$-units.

Proposition 6.4.3. *Let L be a totally real field. Let $d \in L$ be such that $L(\sqrt{d})$ is a totally complex extension of $\mathbb{Q}$. Let K be a totally real cyclic extension of L of odd prime degree p. Let $\mathcal{W}_{L(\sqrt{d})}$ be a set of primes of $L(\sqrt{d})$ not splitting in the extension $K(\sqrt{d})/L(\sqrt{d})$. Let ε be an element of the integral closure of $O_{L(\sqrt{d}),\mathcal{W}_{L(\sqrt{d})}}$ in $K(\sqrt{d})$. Then the following statements are true.*

- *If ε satisfies*

$$\mathbf{N}_{K(\sqrt{d})/L(\sqrt{d})}(\varepsilon) = 1 \tag{6.4.3}$$

  then $\varepsilon \in O_{K(\sqrt{d})}$.
- *There exists a natural number m, depending on K, L, d only, such that $\varepsilon^m \in K$.*

Proof. Since ε is an element of the integral closure of $O_{L(\sqrt{d}),\mathcal{W}_{L(\sqrt{d})}}$ in $K(\sqrt{d})$, by Proposition 6.2.1, the only primes which can appear in the denominator of its divisor are the factors of the primes in $\mathcal{W}_{L(\sqrt{d})}$. Since $\mathbf{N}_{K(\sqrt{d})/L(\sqrt{d})}(\varepsilon) = 1$, however, we must conclude that every prime appearing in the divisor of ε must have a distinct conjugate over $L(\sqrt{d})$. This is not true of the primes in $\mathcal{W}_{L(\sqrt{d})}$, by construction. Therefore, the only elements of the integral closure of $O_{L(\sqrt{d}),\mathcal{W}_{L(\sqrt{d})}}$ in $K(\sqrt{d})$ satisfying (6.4.3) are integral units. At the same time, since the integral unit group of $K(\sqrt{d})$ has the same rank as the integral unit group of K, there exists a natural number $m > 0$ such that the mth power of any integral unit in $K(\sqrt{d})$ is in K. Hence the lemma is true. □

7

Diophantine classes over number fields

In this chapter we prove the main known results concerning the Diophantine classes of the rings of integers and $\mathcal{W}$-integers of number fields. We start by constructing Diophantine definitions of $\mathbb{Z}$ over some of these rings. Next we use these definitions to put together parts of the big picture of the Diophantine classes of the rings of $\mathcal{W}$-integers of number fields, discussed in Chapter 1. Most of the chapter is taken up with proving vertical results, i.e. resolving problems of the following nature. *Let $R_1 \subset R_2$ be integral domains with quotient fields F_1, F_2 respectively, such that R_2 is the integral closure of R_1 in F_2 and F_2/F_1 is a non-trivial finite field extension. Then give a Diophantine definition of R_1 over R_2 or alternatively show that $R_1 \leq_{Dioph} R_2$.*

The proofs of all the vertical results presented in this book can be classified as being done by one of two vertical methods, which we name "weak" and "strong." These methods were developed by Denef and Lipshitz in [15], [19], and [18] and consequently used by Pheidas in [68] and by the present author in [91], [99], [101], [106], [93], and [103].

Before presenting the details of the constructions for particular rings, we describe the main features of the weak and strong vertical methods.

7.1 Vertical methods of Denef and Lipshitz

7.1.1. The weak and strong methods As we have noted above, the method used to solve vertical problems over number fields has a "weak" version and a "strong" version. In the weak version, the norm equations in combination with some bound equations assert that if solutions can be found in the ring above, for a given value of a parameter, then this value is in the ring below. And, conversely, if the parameter is equal to a rational integer then the solutions will be found in the ring above. In the strong version, the norm equations in combination with the bound equations assert that the solutions can be found in the ring above if and only if the parameter is equal to a specific kind of element below, for example, a rational integer.

Which method is used depends on the kind of control that we exercise over the solutions of the norm equations, as will be demonstrated by the examples below. We will start with a formal description of the weak version of the vertical method.

Lemma 7.1.2. The weak vertical method *Let K/F be a number field extension with a basis $\Lambda = \{1, \alpha, \ldots, \alpha^{m-1}\} \subset O_K$. Let $x \in O_K$ and $w, y \in O_F$. Assume that y is not zero and is not an integral unit. Let $c \in \mathbb{N}$ be fixed and let $n = [K : \mathbb{Q}]$. Suppose that the following equalities and inequalities hold:*

$$x = \sum_{i=0}^{m-1} a_i \alpha^i, \qquad a_i \in F, \tag{7.1.1}$$

$$|\mathbf{N}_{K/\mathbb{Q}}(Da_i)| \leq |\mathbf{N}_{K/\mathbb{Q}}(y)^c|, \tag{7.1.2}$$

where D is the discriminant of Λ, and

$$x \equiv w \bmod y^{2c}. \tag{7.1.3}$$

Then $x \in O_F$.

Proof. From (7.1.1) and (7.1.3) we conclude that

$$x - w = (a_0 - w) + a_1\alpha + \cdots + a_{n-1}\alpha^{m-1} \equiv 0 \bmod y^{2c}.$$

Thus

$$\frac{x-w}{y^{2c}} = \frac{a_0 - w}{y^{2c}} + \frac{a_1}{y^{2c}}\alpha + \cdots + \frac{a_{m-1}}{y^{2c}}\alpha^{m-1} \in O_K.$$

By Lemma B.4.12 for $i = 1, \ldots, m-1$ we have that $Da_i/y^{2c} \in O_F$ and therefore that on the one hand $|\mathbf{N}_{K/\mathbb{Q}}(Da_i)| \geq \mathbf{N}_{K/\mathbb{Q}}(y^{2c})$ or $|\mathbf{N}_{K/\mathbb{Q}}(a_i)| = 0$. On the other hand, from (7.1.2) we conclude that

$$|\mathbf{N}_{K/\mathbb{Q}}(Da_i)| \leq |\mathbf{N}_{K/\mathbb{Q}}(y)|^c < \mathbf{N}_{K/\mathbb{Q}}(y)^{2c},$$

since y is not an integral unit. Hence for $i = 1, \ldots, m$ we have that $|\mathbf{N}_{K/\mathbb{Q}}(a_i)| = 0$ and therefore that $a_i = 0$ for $i = 1, \ldots, m-1$. Consequently $x \in O_F$.

Lemma 7.1.3. The strong vertical method *Let K/F be a number field extension. Let $x, y \in O_K$, $w \in O_F$. Assume that y is not zero and is not an integral unit. Let $c \in \mathbb{N}$ be fixed. Assume that the following equations and* *inequalities are satisfied. For all embeddings σ of K into $\mathbb{C}$,*

$$|\sigma(x)| \leq |\mathbf{N}_{K/\mathbb{Q}}(y^c)|, \tag{7.1.4}$$
$$|\sigma(w)| \leq |\mathbf{N}_{K/\mathbb{Q}}(y^c)|, \tag{7.1.5}$$
$$x \equiv w \bmod 2y^{2cn} \text{ in } O_K, \tag{7.1.6}$$

where $n = [K : \mathbb{Q}]$. Then $x = w \in O_F$.

Proof. From (7.1.6) we can conclude that either $x = w$ or

$$|\mathbf{N}_{K/\mathbb{Q}}(x - w)| \geq 2^n \mathbf{N}_{K/\mathbb{Q}}(y^{2cn}).$$

At the same time, from inequalities (7.1.4) and (7.1.5) we conclude that

$$|\mathbf{N}_{K/\mathbb{Q}}(x - w)| \leq 2^n \mathbf{N}_{K/\mathbb{Q}}(y^{nc}).$$

Since y is not an integral unit, $|\mathbf{N}_{K/\mathbb{Q}}(y^{nc})| < \mathbf{N}_{K/\mathbb{Q}}(y^{2cn})$. Thus we must conclude that $x = w$. □

Remark 7.1.4. The difference between the weak and the strong vertical methods ultimately boils down to the fact that in using the weak method one does not have to have a bound on the element of the smaller field in the congruence, while such a bound is necessary for the stronger method.

7.2 Integers of totally real number fields and fields with exactly one pair of non-real embeddings

In this section we discuss the construction of a Diophantine definition of $\mathbb{Z}$ over the rings of algebraic integers of totally real number fields and of fields with exactly one pair of non-real embeddings. The result concerning totally real number fields was originally proved by Denef in [18]. The construction for fields with exactly one pair of non-real embeddings was first carried out independently by Pheidas in [68] and the present author in [91]. We will closely follow these constructions, which are examples of the strong vertical method.

In what follows $K, a, e, n, C, \sigma_i, \sigma_{i,j}, d, \delta_i, \varepsilon_i, I_0, m_0$ will be defined as in notation lists 6.3.1 and 6.3.3. Further, let c be the constant from Note 5.2. Now consider the following statements, in which all the variables range over O_K:

$$x^2 - (a^2 - 1)y^2 = 1, \tag{7.2.1}$$
$$\bar{w}^2 - (a^2 - 1)\bar{z}^2 = 1, \tag{7.2.2}$$
$$w - \delta z = (\bar{w} - \delta\bar{z})^e, \tag{7.2.3}$$

$$u^2 - (a^2 - 1)v^2 = 1, \tag{7.2.4}$$

$$s^2 - (b^2 - 1)y^2 = 1, \tag{7.2.5}$$

$$0 < \sigma_i(b) \le 2^{-16}, \quad |\sigma_i(z)| \ge C, \quad |\sigma_i(u)| \ge \tfrac{1}{2}, \qquad i \notin I_0, \tag{7.2.6}$$

$$v \neq 0, \tag{7.2.7}$$

$$z^2 \mid v, \tag{7.2.8}$$

$$b \equiv 1 \bmod z, \qquad b \equiv a \bmod u, \tag{7.2.9}$$

$$s \equiv x \bmod u, \tag{7.2.10}$$

$$t \equiv \xi \bmod z, \tag{7.2.11}$$

$$\prod_{i=0}^{n}(i - \xi)(i - x) \mid f, \qquad \text{where } f \neq 0 \text{ and } (2^n f^{cn}) \mid z, \tag{7.2.12}$$

Before we explore the meaning of these equations for the variables involved, we should note that (7.2.6) can be rewritten in a Diophantine form by Corollary 5.1.2. Further, condition (7.2.7) is Diophantine by Proposition 2.2.4.

Now assume that the equations in (7.2.1)–(7.2.12) have solutions in O_K. We show that this assumption implies that $\xi \in \mathbb{Z}$. Indeed, using Lemma 6.3.5, our assumptions on a, and the part of (7.2.6) which has to do with $\sigma_i(b)$, we can conclude the following from (7.2.1)–(7.2.5):

$$\begin{aligned} x &= \pm x_k(a), & y &= \pm y_k(a), \\ w &= \pm x_{eh}(a), & z &= \pm y_{eh}(a), \\ u &= \pm x_m(a), & v &= \pm y_m(a), \\ s &= \pm x_j(b), & t &= \pm y_j(b) \end{aligned}$$

for some $k, h, m, j \in \mathbb{N}$, where we again use Notation 6.3.1. (To see that b satisfies the appropriate assumptions, note the following. Since $b \in O_K$ we have that $\prod_{i\in I_0} |\sigma_i(b)| \prod_{i\notin I_0} |\sigma_i(b)| \ge 1$. Thus (7.2.6) implies that $\prod_{i\in I_0} |\sigma_i(b)| \ge 2^{16(n-m_0)}$. If $m_0 = 1$ we have that $|b| > 2^{16(n-1)} > 2^{2n+2}$ for $n \ge 2$. If $m_0 = 2$ then $n \ge 3$ and $|b_1| > 2^{8(n-2)}$; in this case we have $2^{8(n-2)} \ge 2^{2n+2}$ and therefore the size requirement for b is satisfied also.) Thus (7.2.6)–(7.2.11) can be rewritten as the following equations:

$$|\sigma_i(y_{eh}(a))| \ge C, \quad |\sigma_i(x_m(a))| \ge \tfrac{1}{2}, \qquad i \notin I_0, \tag{7.2.13}$$

$$y_m(a) \neq 0, \tag{7.2.14}$$

$$y_{eh}^2(a) \mid y_m(a), \tag{7.2.15}$$

$$b \equiv 1 \bmod y_{eh}(a), \qquad b \equiv a \bmod x_m(a), \tag{7.2.16}$$

$$x_j(b) \equiv \pm x_k(a) \bmod x_m(a), \tag{7.2.17}$$

$$\pm y_j(b) \equiv \xi \bmod y_{eh}(a). \tag{7.2.18}$$

From Lemma 6.3.6 and (7.2.13) we also have the following congruences:

$$y_j(b) \equiv j \bmod (b-1) \quad \Rightarrow \quad y_j(b) \equiv j \bmod y_{eh}(a) \qquad \text{by (7.2.16)}, \tag{7.2.19}$$

$$j \equiv \pm\xi \bmod y_{eh}(a) \qquad \text{by (7.2.18)}, \tag{7.2.20}$$

$$x_j(b) \equiv \pm x_j(a) \bmod x_m(a) \qquad \text{by (7.2.16)}, \tag{7.2.21}$$

$$x_j(a) \equiv \pm x_k(a) \bmod x_m(a) \qquad \text{by (7.2.17)}, \tag{7.2.22}$$

$$k \equiv \pm j \bmod m \qquad \text{by (7.2.22)}, \tag{7.2.23}$$

$$y_{eh}(a) \mid m \qquad \text{by (7.2.15)}, \tag{7.2.24}$$

$$k \equiv \pm j \bmod y_{eh}(a) \qquad \text{by (7.2.23) and (7.2.24)}, \tag{7.2.25}$$

$$k \equiv \pm\xi \bmod 2^n f^{cn} \qquad \text{by (7.2.12) and (7.2.20)}. \tag{7.2.26}$$

From Lemma 6.3.6 we also have that $k < |x_k(a)|$. Thus, by Lemma 5.2.1 and (7.2.12) we conclude that $k < |x_k(a)| \leq |\mathbf{N}_{K/\mathbb{Q}}(f)^c|$. Similarly, for $i = 1, \ldots, n$ we have that $|\xi_i| \leq |\mathbf{N}_{K/\mathbb{Q}}(f)|^c$. Therefore, by the strong version of the vertical method, $\xi = \pm k \in \mathbb{Z}$.

Next we show that if $\xi \in \mathbb{N}$, $\xi > n$ (so that $\prod_{i=0}^{n}(i - \xi) \neq 0$), then (7.2.1)–(7.2.12) can be satisfied in all other variables in O_K. Set $k = \xi$, $x = x_k(a)$, $y = y_k(a)$. Then (7.2.1) is satisfied. By the properties of $\mathcal{W}$-units (see Definition 6.1.1) and by the properties of the solutions to Pell equations (see Section 6.3), we can find $h \in \mathbb{N}$ such that (7.2.12) is satisfied by $z = y_{eh}(a)$ with $|\sigma_i(y_{eh}(a))| \geq C$ for all $i \notin I_0$. Set $w = x_{eh}(a)$. Then (7.2.3) and the z-part of (7.2.6) are satisfied. By the properties of $\mathcal{W}$-units and again by the properties of solutions to Pell equations, we can find $m \in \mathbb{N}$ such that $y_{eh}^2(a) \mid y_m(a)$ and $|\sigma_i(x_m)| > \frac{1}{2}$ for $i \notin I_0$. Set $u = x_m(a)$, $v = y_m(a)$. Then (7.2.4), the u-part of (7.2.6), and (7.2.8) are satisfied. By Lemma 6.3.2, there exists $b \in O_K$ satisfying the b-part of (7.2.6) and (7.2.9). Set $s = x_k(b)$, $t = y_k(b)$. Then (7.2.5) is satisfied. Finally set $\bar{w} = x_h(a)$, $\bar{z} = y_h(a)$, and the properties of solutions to Pell equations will ensure that the remaining conditions are also satisfied.

We summarize the discussion in this section in the following theorem.

Theorem 7.2.1. *Let K be a number field which is totally real or has exactly two conjugate non-real embeddings. Then $\mathbb{Z} \equiv_{Dioph} O_K$.*

7.3 Integers of extensions of degree 2 of totally real number fields

In this section we present a construction of a Diophantine definition of O_K over O_M, where K is a totally real number field and M is an extension of degree 2 of K. This result was originally proved by Jan Denef and Leonard Lipshitz in [19]. We will use the weak vertical method 7.1.2 to reach our goal.

Let $a \in O_K$ be such that $M = K(\sqrt{a})$. Let d be defined as in Lemma 6.2.2. Let $n = [K : \mathbb{Q}]$. Let c be the constant from Note 5.2.3 adjusted for $D = 4a$. Now consider the following equations, in which all the variables range over the specified domains and k is defined as in Lemma 6.2.2:

$$\mathbf{N}_{M(\sqrt{d})/M}(\varepsilon) = 1, \qquad \varepsilon \in O_M[\sqrt{d}], \tag{7.3.1}$$

$$\mathbf{N}_{M(\sqrt{d})/M}(\delta) = 1, \qquad \delta \in O_M[\sqrt{d}], \tag{7.3.2}$$

$$\mathbf{N}_{M(\sqrt{d})/M}(\gamma) = 1, \qquad \gamma \in O_M[\sqrt{d}], \tag{7.3.3}$$

$$\varepsilon^k - 1 = w(\delta^k - 1), \qquad w \in O_M[\sqrt{d}], \qquad \delta^k - 1 \neq 0, \tag{7.3.4}$$

$$x - w = (\delta^k - 1)u, \qquad u \in O_M[\sqrt{d}], \qquad x \in O_M, \tag{7.3.5}$$

$$(\gamma^k - 1)^{2c}v = \delta^k - 1, \qquad v \in O_M[\sqrt{d}], \tag{7.3.6}$$

$$2\prod_{i=0}^{2n}(i - x)y = \gamma^k - 1, \qquad y \in O_M[\sqrt{d}], \qquad \gamma^k - 1 \neq 0. \tag{7.3.7}$$

We claim the following.

1. *If these equations are satisfied by variables ranging over the specified domains then $x \in O_K$.*

Proof. From (7.3.1)–(7.3.3), by Lemma 6.2.2 we conclude that $\varepsilon^k, \delta^k, \gamma^k \in O_{K(\sqrt{d})}$. Therefore (7.3.4) implies that $w \in O_{K(\sqrt{d})}$ also. Further, from (7.3.7) we also have that by Lemma 5.3.3

$$x = a_0 + a_1\sqrt{a}, \qquad \text{where } a_0, a_1 \in K(\sqrt{d}) \tag{7.3.8}$$

and

$$\mathbf{N}_{M(\sqrt{d})/\mathbb{Q}}(4aa_1) \leq \mathbf{N}_{M(\sqrt{d})/\mathbb{Q}}(\gamma^k - 1)^c. \tag{7.3.9}$$

(Note that we have included a factor 2 in the product dividing $\gamma^k - 1$ to make it obvious that $\gamma^k - 1$ is not an integral unit.) Now, by the weak vertical method applied to (7.3.5) we conclude that $x \in O_{K(\sqrt{d})}$. Since by assumption x is also in M, and $K(\sqrt{d}) \cap M = K$ as in Lemma 6.2.2, we conclude that $x \in O_K$. □

2. *If not all conjugates of a over $\mathbb{Q}$ are real then equations (7.3.1)–(7.3.7) can be satisfied for any $x \in \mathbb{N}$, $x > 2n$.* (If all the conjugates of a are positive then $M = K(\sqrt{a})$ is a totally real field. We took care of totally real fields in Section 7.2.)

Proof. Indeed, if not all conjugates of a over $\mathbb{Q}$ are positive then $K(\sqrt{d})$ is not a degree-2 non-real extension of $\mathbb{Q}$ and therefore $K(\sqrt{d})$ has integral units which are not roots of unity. Further, $K(\sqrt{d})$ has real embeddings and therefore it does not have roots beyond 1 and -1. (We remind the reader that by the construction of d we have that, for every embedding σ of $M(\sqrt{d})$ into its algebraic closure, $\sigma(K(\sqrt{d}))$ is a real field if and only if $\sigma(M)$ is not a real field.) Let $\lambda \neq \pm 1$ be an integral unit of $K(\sqrt{d})$. From the discussion in Section 6.1 we can assume $\lambda \in O_K[\sqrt{d}]$. Let $x \in \mathbb{N}$ be given. From the discussion in Section 6.1 on $\mathcal{W}$-units, it is also clear that for some positive $l \in \mathbb{N}$ we have that $2\prod_{i=0}^{2n}(i-x) \mid (\lambda^l - 1)$. (The assumption that $x > 2n$ ensures that the product is not zero.) Let $\gamma = \lambda^l$. Then (7.3.7) is satisfied. By the same argument as above, there exists a non-zero $r \in \mathbb{N}$ such that $(\gamma - 1)^{2c} \mid (\lambda^r - 1)$. Let $\delta = \lambda^r$. Finally, let $\varepsilon = \delta^x$. Thus all the remaining equations are satisfied. Since O_K has an integral basis over $\mathbb{Z}$, the last observation concludes our proof. □

The only remaining task is to rewrite the norm equations in polynomial form and adjust all the other equations to make sure that the variables range over O_M as opposed to $O_M[\sqrt{d}]$. The first task has been discussed already in subsection 4.1.1 and the second can be performed using coordinate polynomials (see appendix section B.7).

Taking into account the transitivity of Diophantine generation and Theorem 7.2.1 we see that we have proved the following.

Theorem 7.3.1. *Let K be any extension of degree 2 of a totally real field. Then $O_K \equiv_{Dioph} \mathbb{Z}$.*

This theorem has an easy corollary, which is of some interest.

Corollary 7.3.2. *Let K be an abelian number field. Then $O_K \equiv_{Dioph} \mathbb{Z}$.*

Proof. First of all we observe that any abelian number field is either totally real or an extension of degree 2 of a totally real number field. Indeed, if a Galois extension is not totally real then it is not a subfield of $\mathbb{R}$ and therefore complex conjugation must be an element of its Galois group over $\mathbb{Q}$. In this case the fixed field of complex conjugation must be a real abelian extension of $\mathbb{Q}$ and therefore totally real. This totally real subfield will of course be a subfield of degree 2. (An alternative approach to proving this fact would involve the Kronecker–Weber theorem (Theorem 5.9, Chapter V of [37]), stating that all abelian extensions are subfields of cyclotomics, which are extensions of degree 2 of totally real fields. To finish the job we would then need to apply the transitivity of Dioph-generation.) □

7.4 The main results for the rings of $\mathcal{W}$-integers and an overview of the proof

Below we state the strongest results concerning the definability of $\mathbb{Z}$ over large rings of $\mathcal{W}$-integers. They provide a measure of how close we have come to proving the Diophantine undecidability of a number field and how far we have yet to travel. Poonen's results will bring us considerably closer.

Before we proceed we would like to remind the reader that an introductory discussion of rings of $\mathcal{W}$-integers can be found in Section B.1 of the number theory appendix.

Theorem 7.4.1. (See Theorem 7.9.4) *Let M be a totally real field or a totally complex extension of degree 2 of a totally real field. Then for any $\varepsilon > 0$ there exists a set $\mathcal{W}_M$ of primes of M whose Dirichlet density is greater than $1 - [M : \mathbb{Q}]^{-1} - \varepsilon$ and such that $\mathbb{Z}$ has a Diophantine definition over $O_{M,\mathcal{W}_M}$.*

Theorem 7.4.2. (See Theorem 7.9.4) *Let M be as above and let $\varepsilon > 0$ be given. Let $\mathcal{S}_{\mathbb{Q}}$ be the set of all the rational primes splitting in M. (If the extension is Galois but not cyclic, $\mathcal{S}_{\mathbb{Q}}$ contains all the rational primes.) Then there exists a set of M-primes $\mathcal{W}_M$ such that the set of rational primes $\mathcal{W}_{\mathbb{Q}}$ below $\mathcal{W}_M$ differs from $\mathcal{S}_{\mathbb{Q}}$ by a set contained in a set of Dirichlet density less than ε and such that $\mathbb{Z}$ is existentially definable over $O_{M,\mathcal{W}_M}$.*

In this section we give an overview of the proofs of the above theorems. We start with a diagram and a series of observations:

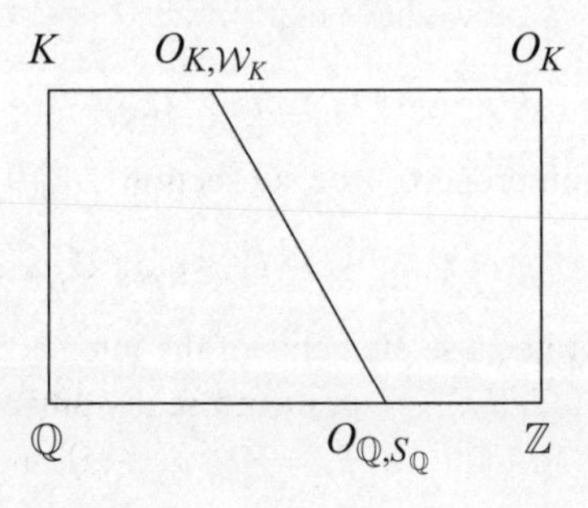

Let $\mathcal{W}_K$ be a set of primes of a number field K. Let $\mathfrak{P}_K$ be a prime of K such that some conjugate of $\mathfrak{P}_K$ over $\mathbb{Q}$ is not in $\mathcal{W}_K$. Let $\mathfrak{P}_{\mathbb{Q}}$ be the rational prime below $\mathfrak{P}_K$. Let $x \in \mathbb{Q}$ be such that $\text{ord}_{\mathfrak{P}_{\mathbb{Q}}} x < 0$. Then $x \notin O_{K,\mathcal{W}_K}$. Indeed, let $\mathfrak{Q}_K$ be a conjugate of $\mathfrak{P}_K$ over $\mathbb{Q}$ that is not in $\mathcal{W}_K$. Then $\mathfrak{Q}_K$ is a factor of $\mathfrak{P}_{\mathbb{Q}}$ in K and therefore $\text{ord}_{\mathfrak{Q}_K} x < 0$. Since x has a pole at a prime outside $\mathcal{W}_K$, x is not an element of $O_{K,\mathcal{W}_K}$.

Suppose now that $\mathcal{W}_K = \mathcal{V}_K \cup \mathcal{S}_K$, where $\mathcal{V}_K$ consists exclusively of primes $\mathfrak{P}_K$ such that $\mathfrak{P}_K$ has a $\mathbb{Q}$-conjugate not in $\mathcal{W}_K$, $\mathcal{S}_K$ consists of primes all of whose $\mathbb{Q}$-conjugates are in $\mathcal{W}_K$, and $\mathcal{S}_K$ is finite. Then $O_{K,\mathcal{W}_K} \cap \mathbb{Q} = O_{\mathbb{Q},\mathcal{S}_{\mathbb{Q}}}$, where $\mathcal{S}_{\mathbb{Q}}$, the set of rational primes below the primes of $\mathcal{S}_K$, is a finite set. If we were able to show that

$$O_{K,\mathcal{W}_K} \cap \mathbb{Q} \leq_{Dioph} O_{K,\mathcal{W}_K} \tag{7.4.1}$$

then, using the fact that $\mathbb{Z} \leq_{Dioph} O_{\mathbb{Q},\mathcal{S}_{\mathbb{Q}}}$ and the transitivity of Dioph-generation, we could conclude that $\mathbb{Z} \leq_{Dioph} O_{K,\mathcal{W}_K}$. Thus the problem would be reduced to a vertical question and we would have to determine for which $\mathcal{W}_K$ we can show that (7.4.1) holds. However, there is a price to pay for this reduction: the density of $\mathcal{W}_K$, as we will see below, cannot in general exceed $1 - [K : \mathbb{Q}]^{-1}$.

Let us begin with the class of fields about which we know the most: totally real number fields. First we consider the following. Let K be a totally real number field and let M be its Galois closure over $\mathbb{Q}$. Then M is also a totally real number field. Further, let $\mathcal{W}_K$ be a set of primes of K and $\mathcal{W}_M$ the set of primes of M above the primes of $\mathcal{W}_K$. Then $O_{M,\mathcal{W}_M}$ is the integral closure of $O_{K,\mathcal{W}_K}$ in M and $O_{M,\mathcal{W}_M} \leq_{Dioph} O_{K,\mathcal{W}_K}$ by Proposition 2.2.1. Suppose that we could show that (7.4.1) holds for M and $\mathcal{W}_M$ instead of for K and $\mathcal{W}_K$. Then, by the transitivity of Dioph-generation, $O_{M,\mathcal{W}_M} \cap \mathbb{Q} \leq_{Dioph} O_{K,\mathcal{W}_K}$. But

$$O_{M,\mathcal{W}_M} \cap \mathbb{Q} = (O_{M,\mathcal{W}_M} \cap K) \cap \mathbb{Q} = O_{K,\mathcal{W}_K} \cap \mathbb{Q}.$$

Thus (7.4.1) holds in this case. Consequently, without loss of generality we can assume that $K/\mathbb{Q}$ is a Galois extension.

Now let $F_1, \ldots, F_r$ be all the cyclic subextensions of K over $\mathbb{Q}$. Suppose further that for $i = 1, \ldots, r$ we have that

$$O_{K,\mathcal{W}_K} \cap F_i \leq_{Dioph} O_{K,\mathcal{W}_K}.$$

By the finite intersection property (see subsection 2.1.19), we then have

$$O_{K,\mathcal{W}_K} \cap F_1 \cap \cdots \cap F_r \leq_{Dioph} O_{K,\mathcal{W}_K}.$$

But $F_1 \cap \cdots \cap F_r = \mathbb{Q}$ because elements of the intersection are not moved by any cyclic subgroup of $\text{Gal}(K/\mathbb{Q})$ or, therefore, by any element of $\text{Gal}(K/\mathbb{Q})$. Consequently $O_{K,\mathcal{W}_K} \cap F_1 \cap \cdots \cap F_r = O_{K,\mathcal{W}_K} \cap \mathbb{Q}$ and we have the desired result. Thus it is enough to solve the vertical problem for the case of cyclic extensions of totally real number fields.

So, to solve the problem in the cyclic case we will apply the weak vertical method, which requires bounds on the elements of the ring. To impose bounds we will use a technique from Chapter 5: we will require that $\mathcal{W}_K$ consists of primes not splitting in certain extensions of K. This requirement will impose an independent constraint on the densities of the prime sets that we will consider for the vertical problem: these densities will have to be strictly less than 1, though they can come arbitrarily close to 1. If we combine the constraint on density coming from the bound equations and the constraint on density coming from the method for reducing the problem of defining $\mathbb{Z}$ to a vertical problem then we can conclude that we will be able to consider a prime set $\mathcal{W}_K$ of density strictly less than $1 - [K : \mathbb{Q}]^{-1}$ but arbitrarily close to $1 - [K : \mathbb{Q}]^{-1}$.

7.5 The main vertical definability results for rings of $\mathcal{W}$-integers in totally real number fields

Most of this section is devoted to proving the vertical result (Theorem 7.5.6) described above, which will provide a route towards a Diophantine definition of $\mathbb{Z}$ over "large" rings of $\mathcal{W}$-integers. As we observed earlier, it is sufficient to prove a special case of the problem, the case of a cyclic extension, and this can be done by the weak vertical method. We start with a notation list for this section.

Notation 7.5.1.

- Let K be a totally real number field with $[K : \mathbb{Q}] = n$.
- Let F be a subextension of K such that the extension K/F is cyclic and $[K : F] = m$.
- Let L be a totally complex extension of degree 2 of $\mathbb{Q}$.
- Let E be a totally real cyclic extension of $\mathbb{Q}$ of degree $p > 2$ with $(p, n) = 1$.
- Let $\delta \in O_{EL}$ be a generator of EL over $\mathbb{Q}$.
- Let $G_0(T)$ be the monic irreducible polynomial of δ over $\mathbb{Q}$.
- For $i = 1, \ldots, n$, let $G_i(T) = G_0(T + i)$.
- Let $\mathcal{V}_K$ be a set of primes of K satisfying the following requirements.
  (a) No prime of $\mathcal{V}_K$ splits in the extension KE/K. Since KLE/K is a Galois extension, this assumption will also imply, by Proposition B.1.11, that no prime of $\mathcal{V}_K$ has a relative-degree-1 factor in the extension KLE/K.
  (b) No prime of $\mathcal{V}_K$ divides the discriminant of G_i for any i.

- Let $l_0 = 0, \ldots, l_s$, where $s = pn$, be distinct natural numbers.
- Let $\mathfrak{P}_K$ be a prime of K satisfying the following requirements.
  (a) $\mathfrak{P}_K$ must lie above an F-prime $\mathfrak{P}_F$ not splitting in the extension K/F.
  (b) $\mathfrak{P}_F$ must split completely in the extension FEL/F.
  (c) $\mathfrak{P}_K$ must split completely in the extension EKL/K.
  (d) $\mathfrak{P}_K$ does not divide the free term of any G_i.
- Let $\mathcal{W}_K = \mathcal{V}_K \cup \{\mathfrak{P}_K\}$.
- Let $\bar{\mathcal{W}}_K$ be the closure of $\mathcal{W}_K$ with respect to conjugation over F.
- Let $\mathcal{W}_{KLE}$ be the set of primes of KLE lying above the primes of $\mathcal{W}_K$.
- Let r be defined as in Proposition 6.4.1.
- Let $\gamma \in O_K$ generate K over F.
- Let P be a rational prime with all its K-factors outside $\mathcal{W}_K$. (Such a P can always be found, by Lemma B.4.7.)
- Let Q be a rational prime below $\mathfrak{P}_K$.
- Let h_K, h_F be the class numbers of K and F respectively.

Before we proceed with the main theorems of this section, we need to note several technical points.

Lemma 7.5.2. *The following statements are true.*

1. *For all $i = 0, \ldots, n$ we have that G_i is an irreducible polynomial over K.*
2. *For any pair $i \neq j$ we have that G_i and G_j do not have any common roots.*
3. *Let l, h be positive integers. Assume also that $l > e(\mathfrak{P}_K/\mathbb{Q})$, the ramification degree of $\mathfrak{P}_K$ over $\mathbb{Q}$. Then for any $x \in K$, and any $i = 0, \ldots, n$, any $\mathfrak{q} \in \mathcal{W}_K$, and any $j = 0, \ldots, s$ we have that $\text{ord}_{\mathfrak{q}} G_i((Qx^l)^h + Ql_j) \leq 0$.*
4. *$\mathcal{V}_K$ can contain all but finitely many primes of K not splitting in the extension EK/K.*
5. *There are infinitely many $\mathfrak{P}_K$ satisfying the requirements listed in Notation 7.5.1.*

Proof. 1. First of all, we observe that by Lemma B.3.3, our assumptions on L, and our assumption on p and n we have that K and E are linearly disjoint over $\mathbb{Q}$. Therefore $[EK : K] = [E : \mathbb{Q}] = p$ and G_0 is irreducible over K. Consequently, $G_i(T) = G_0(T + i)$ is also irreducible over K for all $i = 1, \ldots, n$.

2. Since for all i, j we have that G_i and G_j are irreducible over $\mathbb{Q}$ and of the same degree, the only way in which they can have a common root is if $G_i(T) = cG_j(T)$, where $c \in \mathbb{Q}$. But both polynomials are monic and thus would have to be equal. If $G_i = G_j$, however, then we have that all the roots of G differ from each other by an integer and hence that $G'(\delta) \in \mathbb{Z}$, where $0 <$

$\deg(G') = \deg(G) - 1$. This would of course contradict the fact that G is the monic irreducible polynomial of δ over $\mathbb{Q}$.

3. First note that for all $i = 0, \ldots, n$ we have that $\delta - i \in O_{KLE}$ generates KLE over K. Thus, since no prime of $\mathcal{V}_K$ will have a degree-1 factor in the extension KLE/K by Lemma B.4.18, for all $\mathfrak{Q} \in \mathcal{V}_K$, for all G_i, $i = 0, \ldots, n$, and for all $y \in K$ we know that $\text{ord}_{\mathfrak{Q}} G_i(y) \leq 0$. Next let $y = (Qx^l)^h + Ql_j$ for some $j = 0, \ldots, s$. We have to consider two cases. If $\text{ord}_{\mathfrak{P}_K} x < 0$ then, by our assumption on l, we conclude that $\text{ord}_{\mathfrak{P}_K}((Qx^l)^h + l_j) < 0$ and, since $G_i(T)$ is monic with integral coefficients, $\text{ord}_{\mathfrak{P}_K} G_i((Qx^l)^h + Ql_j) < 0$. If $\text{ord}_{\mathfrak{P}_K} x \geq 0$, however, then $\text{ord}_{\mathfrak{P}_K}((Qx^l)^h + Ql_j) > 0$ and therefore $\text{ord}_{\mathfrak{P}_K} G_i((Qx^l)^h + Ql_j) = 0$, since $\mathfrak{P}_K$ does not divide the free term of any G_i.

4. This part follows from the fact that only finitely many primes can divide the discriminants of any G_i, $i = 0, \ldots, n$.

5. This assertion follows from Proposition 6.4.2. □

We are now ready to apply the weak vertical method.

Theorem 7.5.3. Applying the weak vertical method to the case of cyclic extensions of totally real number fields *Let $x_1, x_2, x_3 \in O_{K,\mathcal{W}_K}[\delta] \subseteq O_{KLE,\mathcal{W}_{KLE}}$ be solutions to (6.4.1) such that x_1, x_2, x_3 are not roots of unity; let $z, c_{1,0}, \ldots, c_{1,2p-1} \in O_{K,\mathcal{W}_K}$, let $w_j \in O_{K,\mathcal{W}_K}[\delta] \subseteq O_{KLE,\mathcal{W}_{KLE}}$, $j = 1, \ldots, 2p-1$, and assume that*

$$\prod_{j=1}^{2p-1}\left(z - \frac{x_3^r - 1}{x_2^r - 1} - c_{1,j}^{ch_F} w_j\right) = 0, \qquad j = 1, \ldots, 2p-1, \tag{7.5.1}$$

where c is the constant defined in Corollary 5.3.5, adjusted for the case where D is equal to the discriminant of the power basis of γ,

$$\frac{x_3^r - 1}{x_2^r - 1} \in O_{K,\mathcal{W}_K}[\delta], \tag{7.5.2}$$

$$x_1^r = c_{1,0} + c_{1,1}\delta + \cdots + c_{1,2p-1}\delta^{2p-1}, \tag{7.5.3}$$

$$\frac{c_{1,j}^{h_F}}{PG_u(z - Ql_i)} \in O_{K,\mathcal{W}_K},$$

$$i = 0, \ldots, s, \qquad u = 0, \ldots, n, \qquad j = 1, \ldots, 2p-1. \tag{7.5.4}$$

Assume further that $z = (Qx^l)^{h_K} = \alpha/\beta$, $\alpha, \beta \in O_K$, and α and β are relatively prime, while $l > e(\mathfrak{P}_K/\mathbb{Q})$, as in Lemma 7.5.2. Then $z \in F \cap O_{K,\mathcal{W}_K}$.

Conversely, if $x \in \mathbb{N}$ *then there exist* $x_1, x_2, x_3 \in O_{K,\mathcal{W}_K}[\delta] \subseteq O_{KLE,\mathcal{W}_{KLE}}$, $c_{1,0}, \ldots, c_{1,2p-1} \in O_{K,\mathcal{W}_K}$, *and* $w_j \in O_{K,\mathcal{W}_K}[\delta] \subseteq O_{KLE,\mathcal{W}_{KLE}}$, $j = 1, \ldots, 2p-1$, *such that* x_1, x_2, x_3 *are not roots of unity and (6.4.1) and (7.5.1)–(7.5.4) are satisfied.*

Proof. By Proposition 6.4.1, we have that $x_1^r, x_2^r, x_3^r \in FLE$. Therefore $c_{1,j} \in K \cap FLE = F$ for all $j = 0, \ldots, 2p-1$. Next suppose that all the product terms in (7.5.1) which are zero have $c_{1,j} = 0$. Then $z = (x_3^r - 1)/(x_2^r - 1) \in K \cap FLE = F$ and we are done. Therefore, without loss of generality we can assume that for some $j^* \in \{1, \ldots, 2p-1\}$ we have that $c_{1,j^*} \neq 0$ and

$$z - \frac{x_3^r - 1}{x_2^r - 1} - c_{1,j^*}^{h_{FC}} w_{j^*} = 0. \tag{7.5.5}$$

Since P has all its factors outside $\mathcal{W}_K$, they must be outside $\bar{\mathcal{W}}_K$ also. Indeed, if $\mathfrak{t}$ is a factor of P in $\bar{\mathcal{W}}_K \setminus \mathcal{W}_K$ then $\mathfrak{t}$ must have an F-conjugate $\bar{\mathfrak{t}} \in \mathcal{W}_K$. But if $\mathfrak{t}$ and $\bar{\mathfrak{t}}$ are conjugates over F, they are also conjugates over $\mathbb{Q}$ and $\bar{\mathfrak{t}}$ must be a factor of P, contradicting the choice of P. Thus c_{1,j^*} is divisible by at least one prime outside $\bar{\mathcal{W}}_K$ and therefore is not a unit of $O_{K,\bar{\mathcal{W}}_K}$. Hence, by Corollary 5.3.5, $c_{1,j^*}^{h_F} = yv$ where $y \in O_F$, y is not an integral unit, y is not divisible by any prime of $\bar{\mathcal{W}}_K$, $v \in F$ is divisible by primes of $\bar{\mathcal{W}}_K$ only, and

$$\left|\mathbf{N}_{K/\mathbb{Q}}(De_i)\right| < \left|\mathbf{N}_{K/\mathbb{Q}}(y)\right|^c, \qquad i = 0, \ldots, m-1 \tag{7.5.6}$$

with

$$\mathbf{N}_{K/\mathbb{Q}}(\beta)z = e_0 + e_1\gamma + \cdots + e_{m-1}\gamma^{m-1}, \qquad e_i \in F. \tag{7.5.7}$$

Next we turn our attention to the equations in (7.5.1) and (7.5.2). First of all let

$$\frac{x_3^r - 1}{x_2^r - 1} = f_0 + f_1\delta + \cdots + f_{2p-1}\delta^{2p-1}, \qquad f_i \in O_{K,\mathcal{W}_K} \cap F,$$

$$w_{j^*} = g_{0,j^*} + g_{1,j^*}\delta + \cdots + g_{2p-1,j^*}\delta^{2p-1},$$

where $g_{0,j^*}, \ldots, g_{2p-1,j^*} \in O_{K,\mathcal{W}_K}$. Then (7.5.1) can be rewritten as

$$z - (f_0 + f_1\delta + \cdots + f_{2p-1}\delta^{2p-1}) = c_{1,j^*}^{ch_F}(g_{0,j^*} + g_{1,j^*}\delta + \cdots + g_{2p-1,j^*}\delta^{2p-1}).$$

Since δ is of degree $2p$ over K, we can conclude that

$$z - f_0 = c_{1,j^*}^{ch_F} g_{0,j^*}.$$

Further, $\mathbf{N}_{K/\mathbb{Q}}(\beta)z - \mathbf{N}_{K/\mathbb{Q}}(\beta)f_0 = y^c\bar{g}$, with $\bar{g} \in O_{K,\mathcal{W}_K}$. Since $y \in O_F$ is not divisible by any prime in $\bar{\mathcal{W}}_K$ and $f_0 \in F \cap O_{K,\mathcal{W}_K}$, by the strong approximation theorem, for some $\bar{f} \in O_F$ we have that $(\mathbf{N}_{K/\mathbb{Q}}(\beta)f_0 - \bar{f})/y^c \in O_{K,\mathcal{W}_K}$. Thus

$$\mathbf{N}_{K/\mathbb{Q}}(\beta)z - \bar{f} = y^c\hat{g}, \qquad \hat{g} \in O_{K,\mathcal{W}_K}.$$

But $(\mathbf{N}_{K/\mathbb{Q}}(\beta)z - \bar{f})/y^c$ has a non-negative order at every element of $\mathcal{W}_K$. Thus $\hat{g} \in O_K$. Now by the weak vertical method (Theorem 7.1.2), we can conclude that $\mathbf{N}_{K/\mathbb{Q}}(\beta)z \in F$. Hence $z \in F$.

Conversely, let $x \in \mathbb{N}$. By Proposition 6.4.1 applied to the extension KLE/K, we can find a solution $y \in O_{KLE,\mathcal{W}_{KLE}}$ to (6.4.1) such that y is not a root of unity and the divisor of y consists of factors of $\mathfrak{P}_K$ only. By Corollary 6.1.5, also applied to the extension KLE/K, there exists $l_1 \in \mathbb{N} \setminus \{0\}$ such that $x_1 = y^{l_1} \in O_{K,\mathcal{W}_K}[\delta]$ and (7.5.3), (7.5.4) can be satisfied by $c_{1,0}, \ldots, c_{1,2p-1} \in O_{K,\mathcal{W}_K}$. Next, we apply Corollary 6.1.5 again to find $l_2 \in \mathbb{N} \setminus \{0\}$ such that $x_2 = y^{l_2} \in O_{K,\mathcal{W}_K}[\delta]$ and $x_2 - 1 \equiv 0 \bmod \mathfrak{D}$, where $\mathfrak{D}$ is the product of the non-$\mathcal{W}_K$ parts of the divisors of $c_{1,j}^{ch_F}$ for all $j = 1, \ldots, 2p-1$ such that $c_{1,j} \neq 0$. (For some $j = 1, \ldots, 2p-1$, we do have that $c_{1,j} \neq 0$. Indeed, suppose that $c_{1,1} = \cdots = c_{1,2p-1} = 0$. Then $x_1^r = c_{1,0} \in K$ and $1 = \mathbf{N}_{KLE/KE}(x_1^r) = x_1^{2r}$, making x_1 a root of unity contrary to our assumptions.) Finally, let $x_3 = x_2^z$. Then x_3 is also a solution to (6.4.1). At the same time,

$$\frac{x_3^r - 1}{x_2^r - 2} - z \equiv 0 \bmod x_2^r - 1 \equiv 0 \bmod x_2 - 1 \qquad \text{in } \mathbb{Z}[x_2] \subset O_{K,\mathcal{W}_K}[\delta].$$

Since $x_2 - 1 \equiv 0 \bmod c_{1,j}^{ch_F}$ in $O_{K,\mathcal{W}_K}[\delta]$ for all j such that $c_{1,j} \neq 0$, we can find a $j^* \in \{1, \ldots, 2p-1\}$ such that $w_j \in O_{K,\mathcal{W}_K}[\delta]$ satisfies (7.5.5). □

Corollary 7.5.4. $O_{K,\mathcal{W}_K} \cap F \leq_{Dioph} O_{K,\mathcal{W}_K}$.

Proof. First of all we need to rewrite system (6.4.1) as a system over K with solutions in K. Let $\sigma_1 = \text{id}, \ldots, \sigma_p$ be all the elements of $\text{Gal}(LEK/LK)$ and let $\tau_1 = \text{id}$ and τ_2 be the elements of $\text{Gal}(LEK/EK)$. Then let $x = \sum_{i=0}^{2p-1} c_i\delta^i$, $c_i \in K$, and consider the system

$$\begin{cases} \prod_{j=1}^{p}\left(\sum_{i=0}^{2p-1} c_i\sigma_j(\delta^i)\right) = 1, \\ \prod_{j=1}^{2}\left(\sum_{i=0}^{2p-1} c_i\tau_j(\delta^i)\right) = 1. \end{cases} \tag{7.5.8}$$

It is clear that $x \in O_{K,\mathcal{W}_K}[\delta]$ is a solution to (6.4.1) if and only if $c_0, \ldots, c_{2p-1} \in O_{K,\mathcal{W}_K}$ constitute a solution to (7.5.8).

Similarly, we can set $x_j = \sum_{i=0}^{2p-1} c_{j,i}\delta^i$, $w_l = \sum_{i=0}^{2p-1} f_{i,l}\delta^i$ and rewrite (7.5.1) as

$$\prod_{l=1}^{2p-1}\left(z - \frac{\left(\sum_{i=0}^{2p-1} c_{3,i}\delta^i\right)^r - 1}{\left(\sum_{i=0}^{2p-1} c_{2,i}\delta^i\right)^r - 1} - c_{1,l}^{ch_F}\sum_{i=0}^{2p-1} f_{i,l}\delta^i\right) = 0. \qquad (7.5.9)$$

Then, if we let $\Delta = \{1, \delta, \dots, \delta^{2p-1}\}$, we can use coordinate polynomials with respect to this basis to rewrite (7.5.8) and (7.5.9) as polynomial equations with coefficients in O_K and all the variables ranging over $O_{K,\mathcal{W}_K}$. (See appendix section B.7 for a discussion of coordinate polynomials.)

Next we address the issue of making sure that x_1, x_2, x_3 are not roots of unity. By our choice of r, it is enough to make sure that $x_j^r - 1 \neq 0$. To accomplish this we can again use coordinate polynomials and the fact that $O_{K,\mathcal{W}_K}$ is Diophregular. We leave the details to the reader.

Finally, given any $x \in O_{K,\mathcal{W}_K}$ and any positive integer l, if we force $x^{lh_K}, (x+1)^{lh_K}, \dots, (x+lh_K)^{lh_K}$ into F then by Corollary B.10.10 we can conclude that $x \in F$. This observation completes the proof of the corollary. $\square$

We are almost ready for our main vertical result, but we need to add items to our notation list before we proceed.

Notation 7.5.5.

- Let $F_1, \dots, F_r$ be all the cyclic subextensions of K.
- For each $j = 1, \dots, r$, let the pair of primes $\mathfrak{P}_{F_j}, \mathfrak{P}_j$ in F_j and K respectively satisfy the requirements listed in Notation 7.5.1, but with respect to the extension K/F_j rather than the extension K/F.
- Let $\mathcal{W}_{K,j} = \mathcal{V}_K \cup \{\mathfrak{P}_j\}$.
- Let $\mathcal{U}_K = \mathcal{V}_K \cup \{\mathfrak{P}_1\} \cup \dots \cup \{\mathfrak{P}_r\}$.

We are now ready to prove the main theorem for this section.

Theorem 7.5.6. $O_{K,\mathcal{U}_K} \cap \mathbb{Q} \leq_{Dioph} O_{K,\mathcal{U}_K}$.

Proof. By Corollary 7.5.4, we have that $O_{K,\mathcal{W}_{K,i}} \cap F_i \leq_{Dioph} O_{K,\mathcal{W}_{K,i}}$. Further, from Chapter 4 we also know that $O_{K,\mathcal{W}_{K,i}} \leq_{Dioph} O_{K,\mathcal{U}_K}$. Thus, by the transitivity of Dioph-generation, $O_{K,\mathcal{W}_{K,i}} \cap F_i \leq_{Dioph} O_{K,\mathcal{U}_K}$. Next, by the finite intersection property of Dioph-generation, we also have that

$$R = \left(\bigcap_{i=1}^{r} O_{K,\mathcal{W}_{K,i}}\right) \cap \mathbb{Q} = \left(\bigcap_{i=1}^{r} O_{K,\mathcal{W}_{K,i}}\right) \cap \left(\bigcap_{i=1}^{r} F_i\right) \leq_{Dioph} O_{K,\mathcal{U}_K}.$$

Finally, using the Dioph-regularity of R, which is a ring of $\mathcal{S}$-integers of $\mathbb{Q}$ with possibly infinite $\mathcal{S}$, together with the transitivity of Dioph-generation, we get $\mathbb{Q} \leq_{Dioph} O_{K,\mathcal{U}_K}$ and at last, by the finite intersection property again, $\mathbb{Q} \cap O_{K,\mathcal{U}_K} \leq_{Dioph} O_{K,\mathcal{U}_K}$. □

7.6 Consequences for vertical definability over totally real fields

In this section we will discuss several useful consequences of Theorem 7.5.6. We will start with a corollary which follows immediately from this theorem.

Corollary 7.6.1. *Let $K/\mathbb{Q}$ be a Galois extension of degree n with K a totally real number field. Let E be a cyclic totally real extension of $\mathbb{Q}$ of degree $p > n$. Let $\mathcal{V}_K$ be a set of primes of K not splitting in the extension EK/K. Then there exists a set of K-primes $\bar{\mathcal{V}}_K$ such that $(\bar{\mathcal{V}}_K \setminus \mathcal{V}_K) \cup (\mathcal{V}_K \setminus \bar{\mathcal{V}}_K)$ is a finite set and $O_{K,\bar{\mathcal{V}}_K} \cap \mathbb{Q} \leq_{Dioph} O_{K,\bar{\mathcal{V}}_K}$.*

Our next step is to observe that we actually have a slightly stronger corollary than this.

Corollary 7.6.2. *Let $K/\mathbb{Q}$ be a Galois extension of degree n with K a totally real number field. Let E be a cyclic totally real extension of $\mathbb{Q}$ of degree $p > n$. Let $\mathcal{V}_K$ be a set of primes of K such that all but finitely many of $\mathcal{V}_K$ do not split in the extension EK/K. Then there exists a set of K-primes $\bar{\mathcal{V}}_K$ containing $\mathcal{V}_K$ such that $\bar{\mathcal{V}}_K \setminus \mathcal{V}_K$ is a finite set and $O_{K,\bar{\mathcal{V}}_K} \cap \mathbb{Q} \leq_{Dioph} O_{K,\bar{\mathcal{V}}_K}$.*

The proof of this statement is very similar to the proof of Theorem 7.5.6 and we leave the details to the reader. Our next step is to drop the assumption that the field in question is Galois.

Corollary 7.6.3. *Let $K/\mathbb{Q}$ be a finite extension with K a totally real number field. Let K_G be the Galois closure of K over $\mathbb{Q}$. Let $n = [K_G : \mathbb{Q}]$. Let E be a totally real cyclic extension of $\mathbb{Q}$ of degree $p > n$. Let $\mathcal{U}_K$ be a set of primes of K such that all but finitely many of these primes do not split in the extension EK/K. Then there exists a set of K-primes $\bar{\mathcal{U}}_K$ containing $\mathcal{U}_K$ such that $\bar{\mathcal{U}}_K \setminus \mathcal{U}_K$ is a finite set and $O_{K,\bar{\mathcal{U}}_K} \cap \mathbb{Q} \leq_{Dioph} O_{K,\bar{\mathcal{U}}_K}$.*

Proof. First of all, we observe that K_G is also a totally real number field. Second, we note that, by Lemma B.3.3 and our assumption on the degrees of

E and K_G, we have that K_G and E are linearly disjoint over $\mathbb{Q}$ and therefore so are K and E. Let $\mathfrak{P}_K$ be a prime of K not splitting in the extension EK/K. Let $\mathfrak{P}_{\mathbb{Q}}$ be the prime below $\mathfrak{P}_K$ in $\mathbb{Q}$, and let $\mathfrak{P}_{K_G}$ be a prime above $\mathfrak{P}_{\mathbb{Q}}$ in K_G. Then, by Lemma B.4.7, we have that $\mathfrak{P}_{\mathbb{Q}}$ does not split in the extension $E/\mathbb{Q}$. Consequently, by Lemma B.4.8, $\mathfrak{P}_{K_G}$ does not split in the extension EK_G/K_G. Let $\mathcal{W}_{K_G}$ be the set of K_G-primes above the primes of $\mathcal{W}_K$. Then primes of $\mathcal{W}_{K_G}$ satisfy the conditions of Corollary 7.6.2. Hence there exists a set of K_G-primes $\bar{\mathcal{W}}_{K_G}$ containing $\mathcal{W}_{K_G}$ such that $\bar{\mathcal{W}}_{K_G} \setminus \mathcal{W}_{K_G}$ is a finite set and

$$O_{K_G,\bar{\mathcal{W}}_{K_G}} \cap \mathbb{Q} \leq_{Dioph} O_{K_G,\bar{\mathcal{W}}_{K_G}}. \tag{7.6.1}$$

Let $\tilde{\mathcal{W}}_{K_G}$ be the closure of $\bar{\mathcal{W}}_{K_G}$ with respect to conjugation over K. Since $\mathcal{W}_{K_G}$ is closed with respect to conjugation over K and $\bar{\mathcal{W}}_{K_G}$ is larger than $\mathcal{W}_{K_G}$ by finitely many primes only, $\tilde{\mathcal{W}}_{K_G} \setminus \mathcal{W}_{K_G}$ is still finite. Let $\bar{\mathcal{W}}_K$ be the set of all K-primes below the primes of $\tilde{\mathcal{W}}_{K_G}$ and observe that $O_{K_G,\tilde{\mathcal{W}}_{K_G}}$ is the integral closure of $O_{K,\bar{\mathcal{W}}_K}$ in K_G. Note also that the set $\bar{\mathcal{W}}_K \setminus \mathcal{W}_K$ is finite. From Proposition 2.2.1 and Theorem 4.2.4 we have the following:

$$O_{K,\bar{\mathcal{W}}_{K_G}} \leq_{Dioph} O_{K_G,\tilde{\mathcal{W}}_{K_G}}, \tag{7.6.2}$$

$$O_{K_G,\tilde{\mathcal{W}}_{K_G}} \leq_{Dioph} O_{K,\bar{\mathcal{W}}_K}. \tag{7.6.3}$$

If we now combine the equations (7.6.1)–(7.6.3) and use the transitivity of Diophantine generation, we may conclude that

$$\mathbb{Q} \cap O_{K_G,\bar{\mathcal{W}}_{K_G}} \leq_{Dioph} O_{K,\bar{\mathcal{W}}_K}. \tag{7.6.4}$$

But $\mathbb{Q} \cap O_{K_G,\bar{\mathcal{W}}_{K_G}}$ is Dioph-regular by Proposition 2.2.4 and therefore, again by transitivity, $\mathbb{Q} \leq_{Dioph} O_{K,\bar{\mathcal{W}}_K}$. Since $O_{K,\bar{\mathcal{W}}_K} \leq_{Dioph} O_{K,\bar{\mathcal{W}}_K}$, by the finite intersection property 2.1.19,

$$\mathbb{Q} \cap O_{K,\bar{\mathcal{W}}_K} \leq_{Dioph} O_{K,\bar{\mathcal{W}}_K}. \tag{7.6.5}$$

This concludes the proof of the corollary. □

We can call the preceding results a "view from above," i.e. from the point of view of the primes of K. Now we consider the same situation but from the point of view of the primes of $\mathbb{Q}$: a "view from below."

Corollary 7.6.4. *Let $K/\mathbb{Q}$ be a finite extension with K a totally real number field. Let K_G be the Galois closure of K over $\mathbb{Q}$. Let $n = [K_G : \mathbb{Q}]$. Let E be a totally real cyclic extension of $\mathbb{Q}$ of degree $p > n$. Let $\mathcal{W}_{\mathbb{Q}}$ be a set of primes of $\mathbb{Q}$ such that all but finitely many primes of $\mathcal{W}_{\mathbb{Q}}$ do not split in the extension $E/\mathbb{Q}$.*

Then there exists a set of $\mathbb{Q}$-primes $\bar{\mathcal{W}}_{\mathbb{Q}}$ containing $\mathcal{W}_{\mathbb{Q}}$ such that $\bar{\mathcal{W}}_{\mathbb{Q}} \setminus \mathcal{W}_{\mathbb{Q}}$ is a finite set and $O_{\mathbb{Q},\bar{\mathcal{W}}_{\mathbb{Q}}}$ has a Diophantine definition over its integral closure in K.

Proof. Let $\mathfrak{P}_{\mathbb{Q}} \in \mathcal{W}_{\mathbb{Q}}$ and let $\mathfrak{P}_K$ be a prime above it in K. Then, by Lemma B.4.8, $\mathfrak{P}_K$ does not split in the extension EK/K. Let $\mathcal{W}_K$ be the set of all K-factors of primes in $\mathcal{W}_{\mathbb{Q}}$. Then by Corollary 7.6.3 there exists a set of K-primes $\bar{\mathcal{W}}_K$ containing $\mathcal{W}_K$ such that $\bar{\mathcal{W}}_K \setminus \mathcal{W}_K$ is a finite set and $O_{K,\bar{\mathcal{W}}_K} \cap \mathbb{Q} \leq_{Dioph} O_{K,\bar{\mathcal{W}}_K}$. Let $\tilde{\mathcal{W}}_K$ be the closure of $\bar{\mathcal{W}}_K$ under conjugation over $\mathbb{Q}$. Then, since $\mathcal{W}_K$ is closed under conjugation over $\mathbb{Q}$, we have that $\tilde{\mathcal{W}}_K \setminus \bar{\mathcal{W}}_K$ is finite. Let $\bar{\mathcal{W}}_{\mathbb{Q}}$ be the set of rational primes below $\tilde{\mathcal{W}}_K$. Then $\bar{\mathcal{W}}_{\mathbb{Q}} \setminus \mathcal{W}_{\mathbb{Q}}$ is finite and $O_{K,\tilde{\mathcal{W}}_K}$ is the integral closure of $O_{\mathbb{Q},\bar{\mathcal{W}}_{\mathbb{Q}}}$ in K. Thus, as above, we have the following relations:

$$O_{K,\bar{\mathcal{W}}_K} \leq_{Dioph} O_{K,\tilde{\mathcal{W}}_K},$$
$$O_{K,\bar{\mathcal{W}}_K} \cap \mathbb{Q} \leq_{Dioph} O_{K,\bar{\mathcal{W}}_K},$$

and therefore

$$O_{K,\bar{\mathcal{W}}_K} \cap \mathbb{Q} \leq_{Dioph} O_{K,\tilde{\mathcal{W}}_K}.$$

At the same time, $\mathbb{Q} \leq_{Dioph} O_{K,\bar{\mathcal{W}}_K} \cap \mathbb{Q}$ by Proposition 2.2.4 as before. Therefore $\mathbb{Q} \leq_{Dioph} O_{K,\tilde{\mathcal{W}}_K}$. Finally, we certainly have $O_{K,\tilde{\mathcal{W}}_K} \leq_{Dioph} O_{K,\tilde{\mathcal{W}}_K}$. Thus, by the finite intersection property 2.1.19, $O_{\mathbb{Q},\bar{\mathcal{W}}_{\mathbb{Q}}} = \mathbb{Q} \cap O_{K,\tilde{\mathcal{W}}_K} \leq_{Dioph} O_{K,\tilde{\mathcal{W}}_K}$. This concludes the proof of the corollary. □

We finish this section with two more vertical definability results whose proofs are analogous to the proofs above and are left to the reader.

Corollary 7.6.5. *Let K/U be a finite extension of totally real number fields. Let K_G be the Galois closure of K over $\mathbb{Q}$. Let $n = [K_G : \mathbb{Q}]$. Let E be a totally real cyclic extension of $\mathbb{Q}$ of degree $p > n$. Let $\mathcal{W}_K$ be a set of primes of K such that all but finitely many primes of $\mathcal{W}_K$ do not split in the extension EK/K. Then there exists a set of K-primes $\bar{\mathcal{W}}_K$ containing $\mathcal{W}_K$ such that $\bar{\mathcal{W}}_K \setminus \mathcal{W}_K$ is a finite set and $O_{K,\bar{\mathcal{W}}_K} \cap U \leq_{Dioph} O_{K,\bar{\mathcal{W}}_K}$.*

Corollary 7.6.6. *Let K/U be a finite extension of totally real number fields. Let K_G be the Galois closure of K over $\mathbb{Q}$. Let $n = [K_G : \mathbb{Q}]$. Let E be a totally real cyclic extension of $\mathbb{Q}$ of degree $p > n$. Let $\mathcal{W}_U$ be a set of primes of U such that all but finitely many primes of $\mathcal{W}_U$ do not split in the extension EU/U. Then there exists a set of U-primes $\bar{\mathcal{W}}_U$ containing $\mathcal{W}_U$ such that $\bar{\mathcal{W}}_U \setminus \mathcal{W}_U$*

is a finite set and $O_{U,\bar{\mathcal{W}}_U}$ *has a Diophantine definition over its integral closure in* K.

7.7 Horizontal definability for rings of $\mathcal{W}$-integers of totally real number fields and Diophantine undecidability for these rings

In this section we convert the vertical definability results into horizontal definability results and into a result asserting the Diophantine undecidability of some rings of $\mathcal{W}$-integers. We start with an undecidability result.

Theorem 7.7.1. *Let* $K/\mathbb{Q}$ *be a finite extension with* K *a totally real number field. Let* K_G *be the Galois closure of* K *over* $\mathbb{Q}$. *Let* $n = [K_G : \mathbb{Q}]$. *Let* E *be a totally real cyclic extension of* $\mathbb{Q}$ *of degree* $p > n$. *Let* $\mathcal{W}_K$ *be a set of primes of* K *such that all but finitely many primes of* $\mathcal{W}_K$ *do not split in the extension* EK/K. *Assume further that all but possibly finitely many primes of* $\mathcal{W}_K$ *have a conjugate over* $\mathbb{Q}$ *which is not in* $\mathcal{W}_K$. *Then there exists a set of* K*-primes* $\bar{\mathcal{W}}_K$ *containing* $\mathcal{W}$ *such that* $\bar{\mathcal{W}}_K \setminus \mathcal{W}_K$ *is a finite set and* $\mathbb{Z} \leq_{Dioph} O_{K,\bar{\mathcal{W}}_K}$. *Thus HTP is undecidable over* $O_{K,\bar{\mathcal{W}}_K}$.

Proof. By Corollary 7.6.3 there exists a set of K-primes $\bar{\mathcal{W}}_K$ containing $\mathcal{W}_K$ and such that $\bar{\mathcal{W}}_K \setminus \mathcal{W}_K$ is a finite set and $\mathbb{Q} \cap O_{K,\bar{\mathcal{W}}_K} \leq_{Dioph} O_{K,\bar{\mathcal{W}}_K}$. Since $\bar{\mathcal{W}}_K$ differs from $\mathcal{W}_K$ by at most finitely many primes, all but possibly finitely many primes of $\bar{\mathcal{W}}_K$ have a $\mathbb{Q}$-conjugate in K which is not in K. By Lemma B.4.20 we have that $\mathbb{Q} \cap O_{K,\bar{\mathcal{W}}_K} = O_{\mathbb{Q},\mathcal{V}_\mathbb{Q}}$, where $\mathcal{V}_\mathbb{Q}$ is a finite, possibly empty, set. From Theorem 4.2.4 we know that $\mathbb{Z} \leq_{Dioph} O_{\mathbb{Q},\mathcal{V}_\mathbb{Q}}$, and the theorem now follows by the transitivity of Diophantine generation. □

Now we state the horizontal definability result.

Corollary 7.7.2. *Let* $K/\mathbb{Q}$ *be a finite extension with* K *a totally real number field. Let* K_G *be the Galois closure of* K *over* $\mathbb{Q}$. *Let* $n = [K_G : \mathbb{Q}]$. *Let* E *be a totally real cyclic extension of* $\mathbb{Q}$ *of degree* $p > n$. *Let* $\mathcal{W}_K$ *be a set of primes of* K *such that all but finitely many primes of* $\mathcal{W}_K$ *do not split in the extension* EK/K. *Assume further that all but possibly finitely many primes of* $\mathcal{W}_K$ *have a conjugate over* $\mathbb{Q}$ *which is not in* $\mathcal{W}_K$. *Then there exists a set of* K*-primes* $\bar{\mathcal{W}}_K$ *containing* $\mathcal{W}_K$ *such that* $\bar{\mathcal{W}}_K \setminus \mathcal{W}_K$ *is a finite set and* $O_K \leq_{Dioph} O_{K,\bar{\mathcal{W}}_K}$.

Proof. From Theorem 7.7.1 we have that $\mathbb{Z} \leq_{Dioph} O_{K,\bar{\mathcal{W}}_K}$. By Proposition 2.2.1 $O_K \leq_{Dioph} \mathbb{Z}$. Therefore by transitivity $O_K \leq_{Dioph} O_{K,\bar{\mathcal{W}}_K}$. □

7.8 Vertical definability results for rings of $\mathcal{W}$-integers of the totally complex extensions of degree 2 of totally real number fields

In this section we will obtain results, analogous to those obtained in the preceding section for totally real fields, for totally complex extensions of degree 2 of totally real fields. We obtain these further results, as in the case of the ring of integers, using the fact that a totally real field and its totally complex extensions of degree 2 have integral unit groups of the same rank. We will use this property of these fields in combination with the weak vertical method to give a Diophantine definition of a ring of $\mathcal{W}$-integers of a totally real field over its integral closure in a totally complex extension of degree 2. As usual, we start with the notation list to be used in this section.

Notation 7.8.1.

- Let K be a totally real field of degree n over $\mathbb{Q}$.
- Let $d \in K$ be such that $M = K(\sqrt{d})$ is a totally complex extension of $\mathbb{Q}$.
- Let E be a totally real cyclic extension of $\mathbb{Q}$ of odd prime degree $p > 2n!$.
- Let $m \in \mathbb{N}$ be as in Lemma 6.4.3.
- Let δ_E be an integral generator of E over $\mathbb{Q}$ (and therefore a generator of ME over M).
- Let D be the discriminant of δ_E.
- Let $G(T) = G_0$ be the monic irreducible polynomial of δ_E over $\mathbb{Q}$.
- Let $G_i(T) = G(T + i)$, $i = 1, \ldots, 2n$.
- Let $\mathcal{W}_M$ be a set of primes of M not splitting in the extension ME/M and not dividing the discriminant of $G_i(T)$ for any $i = 0, \ldots, 2n$.
- Let $\mathcal{V}_M$ be a set of primes of M containing $\mathcal{W}_M$ and such that the difference between the sets is finite.
- Let $\bar{\mathcal{W}}_M$ be the closure of $\mathcal{W}_M$ under conjugation over K.
- Let $c = c(d)$ be the constant from Corollary 5.3.5 applied to the extension M/K and the basis $\{1, \sqrt{d}\}$.
- Let $l_0 = 0, \ldots, l_z$ be distinct integers, with $z = 2pn$.
- Let $\mathcal{W}_K$ be the set of K-primes below $\mathcal{W}_M$.
- Let h_K, h_M denote the class numbers of K and M respectively.
- Let P be a fixed rational prime such that all its M-factors are outside $\bar{\mathcal{W}}_M$. (We can again use Lemma B.4.7 to find such a P.)
- Let $\{\omega_1, \ldots, \omega_n\}$ be a basis of K over $\mathbb{Q}$.
- Let $\sigma_1 = \text{id}, \ldots, \sigma_n : K \to \mathbb{C}$ be all the embeddings of K into $\mathbb{C}$.

Our next step is a lemma which will use norm equations to force some elements of M into K.

Lemma 7.8.2. *Let ε be an element of the integral closure of $O_{M,\mathcal{W}_M}$ in EM such that*

$$\mathbf{N}_{ME/M}(\varepsilon) = 1, \tag{7.8.1}$$

and assume that $\varepsilon^m = \sum_{i=0}^{p-1} a_i \delta_E^i$, $a_i \in O_{M,\mathcal{W}_M}$. Then ε is an integral unit of O_{EM} and $a_0, \ldots, a_{p-1} \in O_{M,\mathcal{W}_M} \cap K$.

Proof. By Lemma 6.4.3 we have that $\varepsilon \in O_{ME}$ and $\varepsilon^m \in KE$. Since δ_E also generates KE/K and $[KE : K] = [ME : M]$, for some $b_i \in K$ we have that $\varepsilon^m = \sum_{i=0}^{p-1} b_i \delta_E^i$, where $b_i \in K \subset M$. Since b_i must be equal to a_i, the assertion of the lemma follows. □

The next lemma makes use of the norm equation to construct a Diophantine definition. As the reader will see, no doubt, this lemma is very similar in flavor to Theorem 7.5.3, but some details are a bit different.

Lemma 7.8.3. *Suppose that the following equations are satisfied by the variables $a_{0,j}, \ldots, a_{p-1,j}$, $b_{0,j}, \ldots, b_{p-1,j}$, x, x_r, $f_{s,r,0}, \ldots, f_{s,r,z}$, $U_{0,r}, \ldots, U_{p-1,r}$, $v_{0,r}, \ldots, v_{p-1,r}$ ranging over $O_{M,\mathcal{W}_M}$ for some $s = 1, \ldots, p-1$, for all $r = 0, \ldots, h_M$, and for all $j = 1, \ldots, h_M + 2$:*

$$\nu_j = \sum_{i=0}^{p-1} a_{i,j}\delta_E^i, \tag{7.8.2}$$

$$\rho_j = \sum_{i=0}^{p-1} b_{i,j}\delta_E^i, \tag{7.8.3}$$

$$\mathbf{N}_{ME/M}(\rho_j) = 1, \tag{7.8.4}$$

$$\nu_j = \rho_j^m, \tag{7.8.5}$$

$$x_r = (x+r)^{h_M}, \tag{7.8.6}$$

$$a_{s,1} \equiv 0 \bmod P, \tag{7.8.7}$$

$$f_{s,r,i} = \frac{a_{s,1}^{h_K}}{\prod_{u=0}^{2n} G_u(x_r - l_i)}, \qquad i = 0, \ldots, z, \tag{7.8.8}$$

$$U_{0,r} + U_{1,r}\delta_E + \cdots + U_{p-1,r}\delta_E^{p-1} = \frac{\nu_{r+2} - 1}{\nu_2 - 1}, \tag{7.8.9}$$

$$\begin{aligned} x_r &- U_{0,r} - U_{1,r}\delta_E - \cdots - U_{p-1,r}\delta_E^{p-1} \\ &= a_{s,1}^{ch_K}\left(v_{s,0,r} + v_{s,1,r}\delta_E + \cdots + v_{s,p-1,r}\delta_E^{p-1}\right). \end{aligned} \tag{7.8.10}$$

Then $x \in K$.

Proof. By Lemma 7.8.2, for all $j = 1, \ldots, h_M + 2$ we have that v_j is an integral unit of KE and $a_{i,j} \in O_{K,\mathcal{W}_K}$. Next, note the following. Let s^* be the value of s for which the equations hold, and assume that $a_{s^*,1} = 0$. Then we are done, because (7.8.10) in this case will force x_r into $M \cap K(\delta_E) = K$ and Corollary B.10.10 will force x into K. Thus without loss of generality we can assume that the equations hold for some value of s with $a_{s,1} \neq 0$. We fix this value of s for the remainder of the discussion.

We now observe that by Lemma B.4.18, for all $u = 0, \ldots, n$, for all $x \in M$, and for all $\mathfrak{P} \in \mathcal{W}_M$, we have that $\mathrm{ord}_{\mathfrak{P}} G_u(x) \leq 0$. Further, by Corollary 5.3.5 we have that $a_{s,1}^{h_K} = z_s y_s$, where $z_s \in K$, $y_s \in O_K$, $y_s \neq 0$, $y_s \equiv 0 \bmod P$, is a prime not lying below any prime of M in $\mathcal{W}_M$, the M-divisor of y_s consists of primes outside $\bar{\mathcal{W}}_M$, the M-divisor of z_s consists of primes from $\bar{\mathcal{W}}_M$, and

$$x_r = \alpha_r/\beta_r, \qquad \alpha_r, \beta_r \in O_M, \tag{7.8.11}$$

$$\mathbf{N}_{M/\mathbb{Q}}(\beta_r)x_r = e_{0,r} + e_{1,r}\sqrt{d}, \tag{7.8.12}$$

$$\left|\mathbf{N}_{K/\mathbb{Q}}(4de_{i,r})\right| < \left|\mathbf{N}_{K/\mathbb{Q}}(y_s^c)\right|, \qquad i = 0, 1. \tag{7.8.13}$$

From (7.8.9), (7.8.10) we get $x_r - U_{0,r} = a_{s,1}^{ch_K} v_{s,0,r}$, where $U_{0,r} \in O_{K,\mathcal{W}_K}$. Further,

$$\mathbf{N}_{M/\mathbb{Q}}(\beta_r)x_r - \mathbf{N}_{M/\mathbb{Q}}(\beta_r)U_{0,r} = \mathbf{N}_{M/\mathbb{Q}}(\beta_r)a_{s,1}^{ch_K} v_{0,r} = y^c\bar{v},$$

where $\bar{v} \in O_{M,\mathcal{W}_M}$. Since $y \in O_K$ is not divisible by any prime of $\mathcal{W}_M$ and $\mathbf{N}_{M/\mathbb{Q}}(\beta_r)U_{0,r} \in K \cap O_{M,\mathcal{W}_M}$, by the Strong Approximation Theorem there exists $\bar{U}_r \in O_K$ such that $(\mathbf{N}_{M/\mathbb{Q}}(\beta_r)U_{0,r} - \bar{U}_r)/y^c \in O_{M,\mathcal{W}_M}$. Hence

$$\mathbf{N}_{M/\mathbb{Q}}(\beta_r)x_r - \bar{U}_r = y^c C_r,$$

where $C_r \in O_{M,\mathcal{W}_M}$. At the same time, $(\mathbf{N}_{M/\mathbb{Q}}(\beta_r)x_r - \bar{U}_r)/y^c$ has a non-negative order at all the primes at which elements of the ring $O_{M,\mathcal{W}_M}$ are allowed to have a negative order. Therefore $C_r \in O_M$ and, by the weak vertical method 7.1.2, $x_r \in K$. By Corollary B.10.10, having $x_r \in K$ for $r = 1, \ldots, h_M$ implies that $x \in K$ as above. □

Lemma 7.8.4. *Let $x \in \mathbb{N}$, $x \neq 0$. Then all the equations (7.8.2)–(7.8.10) can be satisfied in all the other variables over $O_{M,\mathcal{W}_M}$.*

Proof. Let x be a non-zero natural number. Then for all r let $x_r = (x + r)^{h_M}$ to satisfy (7.8.6) and note that x_r will also be a non-zero natural number. Next, let μ be an integral unit of K such that $\mathbf{N}_{KE/K}(\mu) = 1$ and such that μ is not a root of unity. Then $\mathbf{N}_{ME/M}(\mu) = 1$ also. Let

$$B = P\prod_{r=0}^{h_M}\prod_{i=0}^{z}\prod_{u=1}^{2n} G_u(x_r - l_i).$$

By Corollary 6.1.5 applied to the extension KE/K, there exists a positive natural number $l(B)$ such that for any positive integer k we have that

$$\mu^{kl(B)} = \sum_{i=0}^{p-1} c_i \delta_E^i, \qquad c_i \in O_K,$$

and for $i \geq 1$ it is the case that

$$c_i \equiv 0 \bmod P \prod_{r=0}^{h_M} \prod_{i=0}^{z} \prod_{u=1}^{2n} G_u(x_r - l_i).$$

Let $\rho_1 = \mu^{l(B)}$ and $\nu_1 = \rho_1^m = \mu^{ml(B)}$. Then (7.8.2)–(7.8.5) can be satisfied for ν_1 and ρ_1 by $\nu_1 = \sum_{i=0}^{p-1} a_{i,1}\delta_E^i$, $a_{i,1} \in O_K$, where, by the construction of $l(B)$, for $i \geq 1$ we have that

$$a_{i,1} \equiv 0 \bmod P \prod_{r=0}^{h_M} \prod_{i=0}^{z} \prod_{u=1}^{2n} G_u(x_r - l_i),$$

and for some $i \geq 1$ it is the case that $a_{i,1} \neq 0$. (Otherwise, $\nu_1 \in K$ would have K-norm 1 and thus would be a root of unity.) Choose $i \geq 1$ such that $a_{i,1} \neq 0$ and set $s = i$. Then (7.8.7) and (7.8.8) can be satisfied. Next let

$$\rho_2 = \mu^l, \qquad \nu_2 = \rho_2^m, \qquad \rho_{r+2} = \mu^{lx_r}, \qquad \nu_{r+2} = \rho_{r+2}^m,$$

where $l \equiv l(a_{s,1}^{ch_K})$ is defined as above using Corollary 6.1.5. At this point we can also conclude that (7.8.2)–(7.8.5) will be satisfied for all j. Further, we see that $\nu_{r+2} = \nu_2^{x_r}$, so that

$$\frac{\nu_{r+2} - 1}{\nu_2 - 1} \in \mathbb{Z}[\nu_2] \subset O_K[\delta_E],$$

and hence (7.8.9) can be satisfied by $U_{s,r} \in O_K$. Now we also note that $\nu_2 - 1 \equiv 0 \bmod a_{s,1}^{ch_K}$ in $O_K[\delta_E]$. Since

$$\frac{\nu_{r+2} - 1}{\nu_2 - 1} \equiv x_r \bmod \nu_2 - 1$$

in $O_K[\delta_E]$, it is the case that

$$\frac{\nu_{r+2} - 1}{\nu_2 - 1} \equiv x_r \bmod a_{s,1}^{ch_K} \tag{7.8.14}$$

in $O_K[\delta_E]$. Consequently, (7.8.10) is also satisfied. This concludes the proof of the lemma. □

We now have all the necessary ingredients to prove the following.

Theorem 7.8.5. $O_{M,\mathcal{W}_M} \cap K \leq_{Dioph} O_{M,\mathcal{W}_M}$.

Proof. Consider the equation

$$x = \sum \pm \frac{y_i}{y_{n+1}} \omega_i, \qquad y_{n+1} \neq 0, \tag{7.8.15}$$

where for each $i = 1, \ldots, n$, we have that (7.8.2)–(7.8.10) can be satisfied for $x = y_i$ with all the other variables in $O_{M,\mathcal{W}_M}$. By Lemma 7.8.3 we know that if the equations are satisfied then $x \in K \cap O_{M,\mathcal{W}_M}$. Further, from Lemma 7.8.4 we can conclude that every element x of $K \cap O_{M,\mathcal{W}_M}$ has a representation (7.8.15), since every element of K has rational coordinates with respect to the basis $\{\omega_1, \ldots, \omega_n\}$. □

The only remaining task is to rewrite all the equations in polynomial form and so that elements of ME do not occur in the coefficients. This can be done, as before, using coordinate polynomials (see Section B.7).

We will now strengthen the above result in the same fashion as we did for totally real number fields.

Corollary 7.8.6. $O_{M,\mathcal{V}_M} \cap K \leq_{Dioph} O_{M,\mathcal{V}_M}$.

Proof. The proof follows a familiar outline. As before, we have the following sequence of relations:

$$O_{M,\mathcal{W}_M} \leq_{Dioph} O_{M,\mathcal{V}_M},$$
$$O_{M,\mathcal{W}_M} \cap K \leq_{Dioph} O_{M,\mathcal{W}_M},$$
$$K \leq_{Dioph} O_{M,\mathcal{W}_M},$$

and therefore

$$O_{M,\mathcal{V}_M} \cap K \leq_{Dioph} O_{M,\mathcal{V}_M}.$$

□

Now we make use of the Diophantine definitions we have constructed over totally real fields.

Theorem 7.8.7. *There exists a set of M-primes $\bar{\mathcal{W}}_M$ containing $\mathcal{W}_M$ such that the set $\bar{\mathcal{W}}_M \setminus \mathcal{W}_M$ is finite and $O_{M,\bar{\mathcal{W}}_M} \cap \mathbb{Q} \leq_{Dioph} O_{M,\bar{\mathcal{W}}_M}$.*

Proof. Let $\tilde{\mathcal{W}}_M \subseteq \mathcal{W}_M$ be the largest subset of $\mathcal{W}_M$ closed under conjugation over K. Let $\tilde{\mathcal{W}}_K$ be the set of primes of K below $\tilde{\mathcal{W}}_M$, so that

$$O_{K,\mathcal{W}_M} \cap K = O_{K,\tilde{\mathcal{W}}_M} \cap K = O_{K,\tilde{\mathcal{W}}_K}.$$

Then the primes in $\tilde{\mathcal{W}}_K$ do not split in the extension KE/K. (This follows from Lemma B.4.7.) Therefore, by Corollary 7.6.3 there exists a set of K-primes $\bar{\mathcal{W}}_K$, containing $\tilde{\mathcal{W}}_K$, such that $\bar{\mathcal{W}}_K \setminus \tilde{\mathcal{W}}_K$ is a finite set and $O_{K,\bar{\mathcal{W}}_K} \cap \mathbb{Q} \leq_{Dioph} O_{K,\bar{\mathcal{W}}_K}$. Let $\{\mathfrak{P}_1, \ldots, \mathfrak{P}_r\} = \bar{\mathcal{W}}_K \setminus \tilde{\mathcal{W}}_K$. Let $\bar{\mathcal{W}}_M$ be the result of adding to $\mathcal{W}_M$ all the factors of $\{\mathfrak{P}_1, \ldots, \mathfrak{P}_r\}$ in M. Note that $\bar{\mathcal{W}}_M \setminus \mathcal{W}_M$ is finite, since we have added factors of finitely many primes in K. Note further that $O_{M,\bar{\mathcal{W}}_M} \cap K = O_{K,\bar{\mathcal{W}}_K}$ and, by Corollary 7.8.6, $O_{M,\bar{\mathcal{W}}_M} \cap K \leq_{Dioph} O_{M,\bar{\mathcal{W}}_M}$. Thus the assertion of the theorem follows by the transitivity of Diophantine generation. □

The proofs of the following results are almost identical to the proofs of the analogous results from Sections 7.6 and 7.7.

Corollary 7.8.8. *Let $\mathcal{W}_{\mathbb{Q}}$ be a set of primes of $\mathbb{Q}$ not splitting in the extension $E/\mathbb{Q}$. Then there exists a set of $\mathbb{Q}$-primes $\bar{\mathcal{W}}_{\mathbb{Q}}$ containing $\mathcal{W}_{\mathbb{Q}}$ such that $\bar{\mathcal{W}}_{\mathbb{Q}} \setminus \mathcal{W}_{\mathbb{Q}}$ is a finite set and $O_{\mathbb{Q},\bar{\mathcal{W}}_{\mathbb{Q}}}$ has a Diophantine definition over its integral closure in M.*

Corollary 7.8.9. *Let M/F be a finite extension of number fields with M a totally complex extension of degree 2 of a totally real number field. Then there exists a set of M-primes $\bar{\mathcal{W}}_M$ containing $\mathcal{W}_M$ such that $\bar{\mathcal{W}}_M \setminus \mathcal{W}_M$ is a finite set and $O_{M,\bar{\mathcal{W}}_M} \cap F \leq_{Dioph} O_{M,\bar{\mathcal{W}}_M}$.*

Theorem 7.8.10. *Assume that all but possibly finitely many primes of $\mathcal{W}_M$ have a conjugate over $\mathbb{Q}$ that is not in $\mathcal{W}_M$. Then there exists a set of M-primes $\bar{\mathcal{W}}_M$ containing $\mathcal{W}_M$ such that $\bar{\mathcal{W}}_M \setminus \mathcal{W}_M$ is a finite set and $\mathbb{Z} \leq_{Dioph} O_{M,\bar{\mathcal{W}}_M}$. Thus HTP is undecidable over $O_{M,\bar{\mathcal{W}}_M}$.*

Corollary 7.8.11. *Assume that all but possibly finitely many primes of $\mathcal{W}_M$ have a conjugate over $\mathbb{Q}$ that is not in $\mathcal{W}_M$. Then there exists a set of M-primes $\bar{\mathcal{W}}_M$ containing $\mathcal{W}_M$ such that $\bar{\mathcal{W}}_M \setminus \mathcal{W}_M$ is a finite set and $O_M \leq_{Dioph} O_{M,\bar{\mathcal{W}}_M}$.*

7.9 Some consequences

In this section we examine in detail various consequences of the above results. Among other things, we will analyze the maximum possible density of prime sets which can be allowed in the denominator of the elements of our rings such that the necessary conditions for solutions of the vertical and horizontal problems are satisfied. We will consider the vertical definability first.

Theorem 7.9.1. *Let K be any totally real field or a totally complex extension of degree 2 of a totally real field. Let $\mathcal{W}_K$ be any set of primes of K. Then for any $\varepsilon > 0$ there exists a set of K-primes $\bar{\mathcal{W}}_K$ such that $\mathcal{W}_K \setminus \bar{\mathcal{W}}_K$ is contained in a set of Dirichlet density less than ε, $\bar{\mathcal{W}}_K \setminus \mathcal{W}_K$ is finite, and $O_{K,\bar{\mathcal{W}}_K} \cap \mathbb{Q}$ has a Diophantine definition over $O_{K,\bar{\mathcal{W}}_K}$.*

Proof. Let K_G be the Galois closure of K over $\mathbb{Q}$. Let $n = [K_G : \mathbb{Q}]$. Let $p > n$ be a prime number and let $E/\mathbb{Q}$ be a totally real cyclic extension of $\mathbb{Q}$ of degree p. Let

$$\tilde{\mathcal{W}}_K = \{\mathfrak{p} \in \mathcal{W}_K \mid \mathfrak{p} \text{ does not split in the extension } EK/K\}.$$

Then by Corollary 7.6.3 and Theorem 7.8.5 there exists a set $\bar{\mathcal{W}}_K$ containing $\tilde{\mathcal{W}}_K$ such that $\bar{\mathcal{W}}_K \setminus \tilde{\mathcal{W}}_K$ is a finite set and $O_{K,\bar{\mathcal{W}}_K} \cap \mathbb{Q} \leq_{Dioph} O_{K,\bar{\mathcal{W}}_K}$. Next consider the set $\mathcal{W}_K \setminus \tilde{\mathcal{W}}_K$. This is a subset of the set of all primes of K splitting in the extension KE/K. As before, by Lemma B.3.3, E and K are linearly disjoint over $\mathbb{Q}$ and therefore EK/K is a cyclic extension of degree p. The only non-ramified primes splitting in this extension are the primes with EK-factors whose Frobenius automorphism over K is identity. (See Lemma B.4.1 for a discussion of Frobenius automorphism.) By Lemma B.5.2, the density of this set is

$$\frac{1}{[KE : K]} = \frac{1}{p}.$$

Since $\bar{\mathcal{W}}_K$ contains $\tilde{\mathcal{W}}_K$, we have that $\mathcal{W}_K \setminus \bar{\mathcal{W}}_K$ is also contained in a set whose Dirichlet density is $1/p$. Thus, by selecting $p > 1/\varepsilon$ we will satisfy the requirement of the theorem. (Note that the required extension exists by Lemma B.3.9.) □

Using the same methodology as in the proofs of Theorem 7.9.1 and Corollaries 7.6.4, 7.6.5, 7.8.8, and 7.8.9, we can also prove the following two theorems.

Theorem 7.9.2. *Let $\mathcal{W}_{\mathbb{Q}}$ be any set of rational primes. Then, for any $\varepsilon > 0$ and any number field K that is totally real or a totally complex extension of degree 2 of a totally real field, there exists a set of rational primes $\bar{\mathcal{W}}_{\mathbb{Q}}$ such that $\bar{\mathcal{W}}_{\mathbb{Q}} \setminus \mathcal{W}_{\mathbb{Q}}$ is finite, $\mathcal{W}_{\mathbb{Q}} \setminus \bar{\mathcal{W}}_{\mathbb{Q}}$ is contained in a set of primes of Dirichlet density less than ε, and $O_{\mathbb{Q},\bar{\mathcal{W}}_{\mathbb{Q}}}$ has a Diophantine definition in its integral closure in K.*

Theorem 7.9.3. *Let K/F be an extension of number fields, where K is totally real or a totally complex extension of degree 2 of a totally real field. Let $\mathcal{W}_F$ be any set of primes of F. Then for any $\varepsilon > 0$ there exists a set of F-primes*

$\bar{\mathcal{W}}_F$ such that $\bar{\mathcal{W}}_F \setminus \mathcal{W}_F$ is finite, $\mathcal{W}_F \setminus \bar{\mathcal{W}}_F$ is contained in a set of primes of Dirichlet density less than ε, and $O_{F,\bar{\mathcal{W}}_F}$ has a Diophantine definition in its integral closure in K.

Next we have some undecidability and horizontal definability results.

Theorem 7.9.4. *Let K be a totally real field or a totally complex extension of degree 2 of a totally real field that is a non-trivial extension of $\mathbb{Q}$. Then for any $\varepsilon > 0$ there exists a set $\bar{\mathcal{W}}_K$ of primes of K whose Dirichlet density is greater than $1 - [K : \mathbb{Q}]^{-1} - \varepsilon$ and such that $\mathbb{Z}$ has a Diophantine definition over $O_{K,\bar{\mathcal{W}}_K}$. (Thus HTP is undecidable in $O_{K,\bar{\mathcal{W}}_K}$.)*

Proof. Let K_G be the Galois closure of K over $\mathbb{Q}$. Let $n = [K_G : \mathbb{Q}]$. Let $p > n$ be a prime number and let $E/\mathbb{Q}$ be a totally real cyclic extension of $\mathbb{Q}$ of degree p. (Again we remind the reader that such an extension exists by Lemma B.3.9.) Let $\mathcal{W}_K$ be a set of primes of K constructed in the following fashion. For every rational prime $\mathfrak{p}_{\mathbb{Q}}$ consider all the primes of K above it and remove a prime with the highest possible degree. Out of the remaining set of primes remove all the primes splitting in the extension EK/K. Then by Lemma B.5.5 the Dirichlet density of $\mathcal{W}_K$, denoted as before by $\delta(\mathcal{W}_K)$, exists and $\delta(\mathcal{W}_K) > 1 - [K : \mathbb{Q}]^{-1} - 1/p$. Further, by Theorems 7.7.1 and 7.8.10 there exists a set of K-primes $\bar{\mathcal{W}}_K$ containing $\mathcal{W}_K$ such that $\bar{\mathcal{W}}_K \setminus \mathcal{W}_K$ is a finite set and $\mathbb{Z} \leq_{Dioph} O_{K,\bar{\mathcal{W}}_K}$. By Proposition 4.6, Chapter IV of [37], we have $\delta(\mathcal{W}_K) = \delta(\bar{\mathcal{W}}_K)$ and the theorem holds for sufficiently large p. □

Theorem 7.9.5. *Let K be any totally real number field or a totally complex extension of degree 2 of a totally real number field, and let $\varepsilon > 0$ be given. Let $\mathcal{S}_{\mathbb{Q}}$ be the set of all rational primes splitting in K. Then there exists a set of K-primes $\bar{\mathcal{W}}_K$ with the property that the set of rational primes $\bar{\mathcal{W}}_{\mathbb{Q}}$ below $\bar{\mathcal{W}}_K$ is such that $\bar{\mathcal{W}}_{\mathbb{Q}} \setminus \mathcal{S}_{\mathbb{Q}}$ is contained in a set of Dirichlet density less than ε, $\mathcal{S}_{\mathbb{Q}} \setminus \bar{\mathcal{W}}_{\mathbb{Q}}$ is finite, and $\mathbb{Z}$ is definable over $O_{K,\bar{\mathcal{W}}_K}$.*

Proof. Let K_G be the Galois closure of K over $\mathbb{Q}$. Let $n = [K_G : \mathbb{Q}]$. Let $p > n$ be a prime number and let $E/\mathbb{Q}$ be a totally real cyclic extension of $\mathbb{Q}$. Let $\mathcal{W}_K$ be a set of primes of K constructed in the following fashion. For every rational prime $\mathfrak{p}_{\mathbb{Q}}$ consider all the primes of K above it and remove one prime from the set. Out of the remaining set of primes remove all the primes splitting in the extension EK/K.

Note that by Lemma B.3.5 the extension EK/K is a cyclic extension of degree p. Next note that, by Lemmas B.4.7 and B.4.8, a prime $\mathfrak{p}_{\mathbb{Q}}$ splits in the extension $E/\mathbb{Q}$ if and only if all the primes above it in K_G split in the extension EK_G/K_G. However, by the same lemmas, a prime of K_G splits in the extension EK_G/K_G if and only if the prime below it in K splits in the extension EK/K. Thus a rational prime splits in the extension $E/\mathbb{Q}$ if and only if any prime above it splits in the extension EK/K.

Let $\mathfrak{p}_{\mathbb{Q}}$ be a prime splitting in the extension $K/\mathbb{Q}$ but having no factors in $\mathcal{W}_K$; then at least one factor of $\mathfrak{p}_{\mathbb{Q}}$ in K splits in the extension EK/K. But, by the argument above, $\mathfrak{p}_{\mathbb{Q}}$ must split in the extension $E/\mathbb{Q}$. Let $\mathcal{W}_{\mathbb{Q}}$ be the set of rational primes below $\mathcal{W}_K$. Then $\mathcal{S}_{\mathbb{Q}} \setminus \mathcal{W}_{\mathbb{Q}}$ consists of rational primes splitting in the extension $E/\mathbb{Q}$. However, by Lemma B.5.2, the density of the set of such rational primes is less than $1/p$ and therefore less than ε for sufficiently large p. Finally, by Theorems 7.7.1 and 7.8.10, there exists a set of K primes $\bar{\mathcal{W}}_K$ containing $\mathcal{W}_K$ such that $\bar{\mathcal{W}}_K \setminus \mathcal{W}_K$ is a finite set of K-primes and $\mathbb{Z} \leq_{Dioph} O_{K,\bar{\mathcal{W}}_K}$. Let $\bar{\mathcal{W}}_{\mathbb{Q}}$ be the set of rational primes below $\bar{\mathcal{W}}_K$. Then $\bar{\mathcal{W}}_{\mathbb{Q}} \setminus \mathcal{W}_{\mathbb{Q}}$ is finite and the assertion of the theorem holds for sufficiently large p. □

The following corollary is an immediate consequence of Theorem 7.9.5.

Corollary 7.9.6. *Let K be a totally real extension of $\mathbb{Q}$ or a degree-2 totally complex extension of a totally real number field such that $K/\mathbb{Q}$ is Galois and not cyclic. Then for any $\varepsilon > 0$ there exists a set $\bar{\mathcal{W}}_K$ of primes of K such that $\bar{\mathcal{S}}_{\mathbb{Q}}$, the set of rational primes below $\bar{\mathcal{W}}_K$, is of density greater than $1-\varepsilon$ and $\mathbb{Z}$ is definable over $O_{K,\bar{\mathcal{W}}_K}$.*

Proof. The only thing we need to observe here is that in a Galois but not cyclic extension of number fields all primes split. (See Section 1 of Chapter III of [37].) □

We finish this section with a few remarks.

Remark 7.9.7. As in Corollary 7.3.2, the definability and undecidability results for rings of $\mathcal{W}$-integers proved above for totally real fields and their totally complex extensions of degree 2 cover all abelian extensions of $\mathbb{Q}$.

Remark 7.9.8. The undecidability of HTP over any ring is an interesting fact only if the ring itself is recursive. So we should say a few words about the recursiveness of some of the rings mentioned in the theorems above. As it happens, if $\mathcal{W}_K$, as constructed in Theorems 7.9.4 and 7.9.5, is as large as possible then it is recursive, implying that $O_{K,\mathcal{W}_K}$ and consequently $O_{K,\bar{\mathcal{W}}_K}$ are recursive by Proposition A.8.6. Indeed, by Lemmas B.4.7 and B.4.8, a prime of K splits in the extension EK/K, where E is as above, if and only if the prime below it in $\mathbb{Q}$ splits in the extension $E/\mathbb{Q}$. Thus the set of all primes of K not splitting in the extension EK/K is recursive, by Proposition A.8.7 and Corollary A.8.2, and so is the set of all the primes of K lying above primes of $\mathbb{Q}$ splitting in the extension $K/\mathbb{Q}$. Therefore the intersection of these sets is also recursive. It remains to decide how to remove a factor of the largest relative degree from each complete set of conjugates over $\mathbb{Q}$ in this intersection. Using the presentation of primes selected in Section A.8, this can be done by selecting primes $\mathfrak{p}$ corresponding to the pair $(p, \alpha(\mathfrak{p}))$, where $\alpha(\mathfrak{p})$ has the lowest possible sequence number (under some fixed ordering of K) among all $\alpha(\mathfrak{p})$, $\mathfrak{p}$ being a factor of p of the highest relative degree over $\mathbb{Q}$.

Remark 7.9.9. The density results above were stated in terms of the Dirichlet density. They can also be restated in terms of the natural density, producing somewhat stronger assertions. (See Definition B.5.9 and Proposition B.5.10 for the definition of natural density and its relation to Dirichlet density.) The only modification which would be required is the substitution of the natural density version of the Chebotarev density theorem for the usual Chebotarev density theorem. (See Theorem 1 of [88]).

7.10 Big picture for number fields revisited

We are now ready to reconsider the big picture for number fields. First we review our notation:

- $\mathcal{S}$ is a finite set of rational primes;
- $\mathcal{W}$ is an infinite set of rational primes;
- $\mathcal{V}$ is a set of rational primes with a finite complement in the set of all primes $\mathcal{P}(\mathbb{Q})$;
- K is a number field;
- $\mathcal{S}_K$ is a finite set of K-primes;
- $\mathcal{W}_K$ is an infinite set of K-primes;
- $\mathcal{V}_K$ is a set of K-primes with a finite complement in the set of all K-primes $\mathcal{P}(K)$.

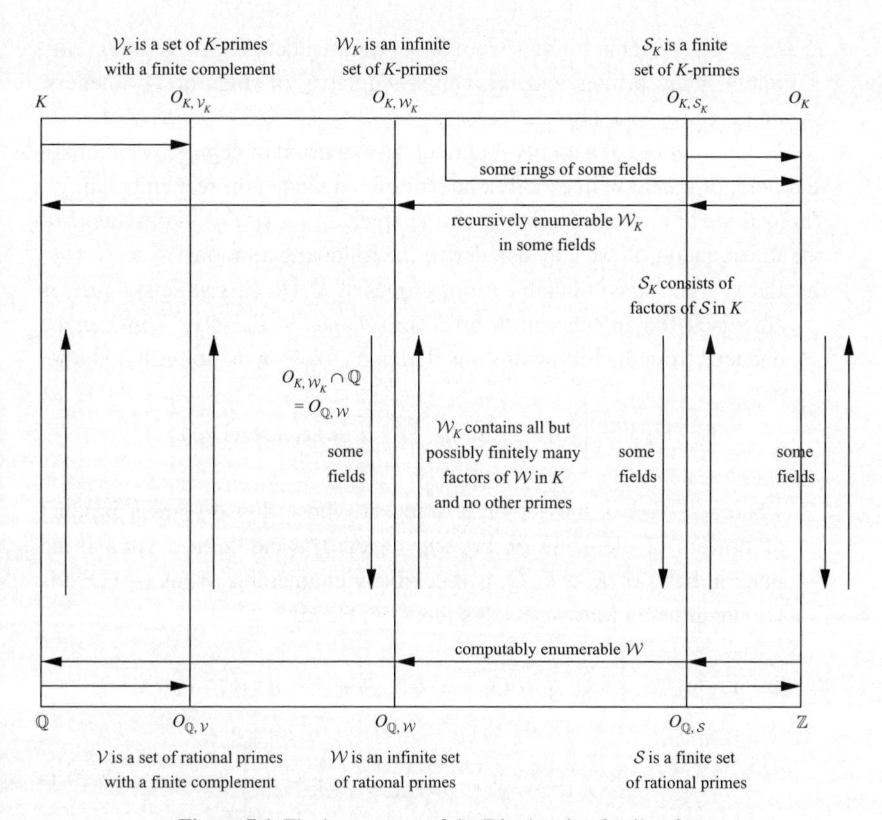

Figure 7.1 The known part of the Diophantine family of $\mathbb{Z}$.

Next consider Figure 7.1.fig 1 As before, arrows indicate Diophantine generation. The diagram represents the relations described in the following list.

- Let $\mathcal{U}$ be a collection of primes of $\mathbb{Q}$ and let $\bar{\mathcal{U}}_K$ be all the factors of $\mathcal{U}$ in K. Let $\mathcal{U}_K \subseteq \bar{\mathcal{U}}_K$ and assume that $\bar{\mathcal{U}}_K \setminus \mathcal{U}_K$ is a finite set. Then $O_{K,\mathcal{U}_K} \leq_{Dioph} O_{\mathbb{Q},\mathcal{U}}$. In the diagram this relation corresponds to the vertical arrows from $\mathbb{Q}$ to K, $O_{\mathbb{Q},\mathcal{V}}$ to $O_{K,\mathcal{V}_K}$, $O_{\mathbb{Q},\mathcal{W}}$ to $O_{K,\mathcal{W}_K}$, $O_{\mathbb{Q},\mathcal{S}}$ to $O_{K,\mathcal{S}_K}$, and $\mathbb{Z}$ to O_K, with the assumption that $\mathcal{W}_K$, $\mathcal{S}_K$ contain all but possibly finitely many K-factors of primes in $\mathcal{W}$ and $\mathcal{V}$ respectively and no other primes. These relations are consequences of the transitivity of Diophantine generation (see Theorem 2.1.15), the Diophantine generation of integral closure (see Proposition 2.2.1), the Diophantine regularity of rings of $\mathcal{W}$-integers (see Proposition 2.2.4), and the fact that integrality at finitely many primes has a Diophantine definition over any ring of $\mathcal{W}$-integers of a number field (see Theorem 4.2.4).

- $K \equiv_{Dioph} O_{K,\mathcal{V}_K}$. This follows from the existential definability of integrality at finitely many primes and the Dioph-regularity of rings of $\mathcal{W}$-integers. Similarly, $O_K \leq_{Dioph} O_{K,\mathcal{S}_K}$.
- Let K be a subfield of a totally real field, an extension of degree 2 of a totally real field, or a field with exactly one pair of conjugate non-real embeddings. Then, as we have shown earlier in this chapter, $\mathbb{Z} \leq_{Dioph} O_K$. From this Diophantine generation we can also derive the following relations:
  (a) Let $\mathcal{W}_K$ be any computable set of primes of K. (Recursive sets of primes are discussed in Section A.8.) Then $O_{K,\mathcal{W}_K} \leq_{Dioph} O_K$. This can be deduced from the following considerations. By Proposition A.8.4, the set

$$C_{\mathbb{Q}}(O_K) = \left\{(a_1, \ldots, a_n) \in \mathbb{Z}^n \,\middle|\, \sum_{i=1}^{n} a_i\omega_i \in O(\mathcal{W}_K)\right\},$$

  where $O(\mathcal{W}_K)$ is the set of K-integers whose divisors are a product of powers of elements of $\mathcal{W}_K$, $n = [K : \mathbb{Q}]$, and $\{\omega_1, \ldots, \omega_n\}$ is an integral basis of K over $\mathbb{Q}$, is recursively enumerable. Thus $C_{\mathbb{Q}}(O_K)$ is Diophantine by Matiyasevich's theorem. Hence

$$\begin{aligned} O_{K,\mathcal{W}_K} = \Bigg\{ \frac{x}{y} \Big| x, y \in O_K,\ y \neq 0,\ \exists a_1, \ldots, a_n \in C_{\mathbb{Q}}(O_K), \\ \exists z \in O_K \setminus \{0\},\ x\left(\sum_{i=1}^{n} a_i\omega_i\right) = yz \Bigg\}. \end{aligned} \tag{7.10.1}$$

  Since $C_{\mathbb{Q}}(O_K)$ is Diophantine over $\mathbb{Z}$ and $\mathbb{Z} \leq_{Dioph} O_K$, the set defined in (7.10.1) is Diophantine over O_K.
  (b) $O_{\mathbb{Q},\mathcal{S}} \equiv_{Dioph} O_{K,\mathcal{S}_K}$. Indeed,

$$O_{K,\mathcal{S}_K} \equiv_{Dioph} O_K \equiv_{Dioph} \mathbb{Z} \equiv_{Dioph} O_{\mathbb{Q},\mathcal{S}},$$

  where the first and last equivalences follow from the existential definability of the integrality at finitely many primes and the Diophantine generation of the rings of $\mathcal{W}$-integers with recursive denominator sets, as described above, over the rings of integers. The middle equivalence expresses a combination of the Diophantine definability of integers over O_K and the Diophantine generation of integral closure.
- Let K be a subfield of a totally real field or a totally complex extension of degree 2 of a totally real field.
  (a) For any $\varepsilon > 0$ there exists a set $\mathcal{W}_K$ of primes of K such that the Dirichlet density of $\mathcal{W}_K$ is greater than $1 - \varepsilon$ and $\mathbb{Q} \cap O_{K,\mathcal{W}_K} \leq_{Dioph} O_{K,\mathcal{W}_K}$. (See Theorem 7.9.1.)

(b) For any $\varepsilon > 0$ there exists a set $\mathcal{W}_K$ of primes of K such that the Dirichlet density of $\mathcal{W}_K$ is greater than $1 - [K : \mathbb{Q}]^{-1} - \varepsilon$ and $O_K \leq_{Dioph} O_{K,\mathcal{W}_K}$. (See Theorem 7.9.4.)

7.11 Further results

The results discussed in this chapter leave unanswered the big questions posed at the beginning of this book. In particular, we still do not know the Diophantine status of rings of integers in general or that of any number field, including $\mathbb{Q}$. The issue of $\mathbb{Q}$ and some of its "large" subrings will come up again in later chapters of this book where we discuss Mazur's conjectures and the results of Poonen, though this will be in the context of constructing a more general Diophantine model of $\mathbb{Z}$, as opposed to a Diophantine definition of integers. Unfortunately, up to the time of the writing of this book the general problem of defining integrality at infinitely many primes remains intractable. There are, however, a few more things to be said about defining $\mathbb{Z}$ over the rings of algebraic numbers before we completely abandon this subject. Recently, in [73], Bjorn Poonen proved the following theorem.

Theorem 7.11.1. *Let M/K be a number field extension. Suppose that there exists an elliptic curve of rank 1 defined over K such that this curve retains rank 1 over M. Then $O_K \leq_{Dioph} O_K$.*

The proof is partly based on the "weak vertical method" and on bound equations of the type described in Chapter 5. This result has provided a new avenue for our investigation, part of which will need to concentrate on determining exactly for which pairs of number fields such curves exist. Further extensions of this method were provided by Cornelissen, Pheidas, and Zahidi in [8] and Poonen and the present author independently in [72] and [111] respectively. Cornelissen, Pheidas, and Zahidi showed that the conditions in Theorem 7.11.1 can be replaced by the following requirements: the existence of a rank-1 elliptic curve over M and of an abelian variety over K whose positive rank over K is the same as its rank over M. Poonen and the present author, however, showed that the requirement that the elliptic curve in Theorem 7.11.1 has rank 1 can be eliminated. In [111] it was shown also that this elliptic curve method can be adjusted for "big rings" in the same way as for the norm method described in this book.

In [107] an attempt was made by the present author to distill the sources of difficulty in giving a Diophantine definition of $\mathbb{Z}$ over the rings of integers of an arbitrary number field. It turns out that these difficulties are traceable to the lack of good bounds on archimedean valuations over non-real number fields. The reader should be reminded that for a real number field the relationship "$x \leq y$" is Diophantine via the use of quadratic forms. (See Lemma 5.1.1.) However, we currently have no means of coding a relationship "$|x| \leq |y|$" over non-real number fields. The bounds from Chapter 5 are in some senses too rough for such a relationship and a better way is needed to enforce the bounds.

8

Diophantine undecidability of function fields

Fields of positive characteristic do not contain integers, and therefore constructing Diophantine definitions of integers to establish Diophantine undecidability, as we have done for some number rings, is not an option here. However, function fields over finite fields of constants do possess Diophantine models of integers, a fact which will allow us to show that the analog of Hilbert's Tenth Problem is undecidable over these fields. It will take us some time to arrive at the desired results and we will start with a seemingly unrelated point.

Before proceeding with our investigation we should note that the main ideas presented in this chapter are due to Cornelissen, Eisenträger, Pheidas, Videla, Zahidi, and the present author, and can be found in [6], [22], [67], [66], [69], [98], [102], and [117].

8.1 Defining multiplication through localized divisibility

This section contains some technical definability results which will allow us to make a transition from characteristic 0 to positive characteristic. The original idea underlying this method belongs to Denef (see [17]) and Lipshitz (see [48]–[50]). It was developed further by Pheidas in [66]. We reproduce Pheidas's results below.

We start with fixing notation and a definition.

Notation 8.1.1. In this section p will denote a fixed rational prime.

Definition 8.1.2. For $x, y \in \mathbb{N}$, define $x \mid_p y$ to mean

$$\exists f \in \mathbb{N} : y = xp^f.$$

Lemma 8.1.3. *Let n, m be positive integers and let s be a natural number. Then $m = p^s n$ if and only if*

$$n \,|_p\, m, \tag{8.1.1}$$

$$(n+1) \,|_p\, (m + p^s), \tag{8.1.2}$$

$$(n+2) \,|_p\, (m + 2p^s). \tag{8.1.3}$$

Proof. One direction of the lemma is obvious. If $m = p^s n$ then of course $n \,|_p\, m$ and $m + p^s = (n+1)p^s$, $m + 2p^s = (n+2)p^s$ so that $(n+1) \,|_p\, (m + p^s)$ and $(n+2) \,|_p\, (m + 2p^s)$.

Suppose now that $n \,|_p\, m$, $(n+1) \,|_p\, (m + p^s)$, and $(n+2) \,|_p\, (m + 2p^s)$. Then for some $k, l \in \mathbb{N}$, $m/n = p^k$ and $(m + p^s)/(n+1) = p^l$. First assume that $n \not\equiv 0 \bmod p$ and note the following:

$$\frac{m}{n} - \frac{m + p^s}{n+1} = \frac{mn + m - mn - p^s n}{n(n+1)} = \frac{p^k - p^s}{n+1} = p^k - p^l.$$

Now we consider three cases: $k = s$; $k > l$; $l > k$. In the first case we also conclude that $l = k = s$ and we are done. In the second case, we must also have $k > s$ and, further, since $p^k - p^s > p^k - p^l$ we have to conclude that $s < l$. At the same time,

$$\mathrm{ord}_p(p^k - p^l) = l > s \geq \mathrm{ord}_p \frac{p^k - p^s}{n+1} = \mathrm{ord}_p(p^k - p^l),$$

and so we have a contradiction.

In the third case, $(1 - p^{s-k})/(n+1) = 1 - p^{l-k}$ and

$$n = \frac{p^{s-k} - 1}{p^{l-k} - 1} - 1 = \frac{p^{s-k} - p^{l-k}}{p^{l-k} - 1} \equiv 0 \bmod p.$$

Thus, the condition $n \not\equiv 0 \bmod p$, together with conditions (8.1.1), and (8.1.2), implies that $m = np^s$. However, we can rewrite conditions (8.1.2) and (8.1.3) as

$$n' \,|_p\, m', \tag{8.1.4}$$

$$(n'+1) \,|_p\, (m' + p^s), \tag{8.1.5}$$

where $n' = n + 1$ and $m' = m + p^s$. Thus, if n is divisible by p we have that $n' = n + 1 \not\equiv 0 \bmod p$ and, from conditions (8.1.2) and (8.1.3) (or alternatively (8.1.4) and (8.1.5)), we conclude that $m' = n'p^s \Leftrightarrow m + p^s = (n+1)p^s \Leftrightarrow m = np^s$. □

Notation 8.1.4. Denote the system

$$\big(n \,|_p\, m\big) \wedge \big((n+1) \,|_p\, (m + u)\big) \wedge \big((n+2) \,|_p\, (m + 2u)\big)$$

by $PDIV(n, m, u)$. Then Lemma 8.1.3 can be restated as follows. *For* $n, m, s \in \mathbb{N}, n > 0, m > 0$, *we have that* $PDIV(n, m, p^s) \Leftrightarrow m = np^s$.

Below we give an easy but important corollary of Lemma 8.1.3, whose proof we leave to the reader.

Corollary 8.1.5. *Let* $u_1, u_2 \in \mathbb{N} \setminus \{0\}$ *with*

$$1 \mid_p u_1. \tag{8.1.6}$$

Then

$$PDIV(u_1, u_2, u_1) \quad \Leftrightarrow \quad \exists s \in \mathbb{N}, u_2 = u_1^2 = p^{2s}. \tag{8.1.7}$$

We also leave the proof of the following lemma to the reader.

Lemma 8.1.6.

1. *Let* $l, m, r \in \mathbb{N}$. *Then* $(p^l - 1) \mid (p^m - 1)$ *if and only if* $l | m$.
2. $\dfrac{p^{lr} - 1}{p^l - 1} \equiv r \bmod (p^l - 1)$.

The next lemma shows us how to compute squares using pth powers.

Lemma 8.1.7. *Let* $m, n \in \mathbb{Z}_{>0}$. *Then* $m = n^2$ *if and only if there exists* $r, s \in \mathbb{Z}_{>0}$ *such that the following conditions are satisfied:*

$$n < p^s - 1; \tag{8.1.8}$$

$$m < p^s - 1; \tag{8.1.9}$$

$$(p^{2s} - 1) \mid (p^{2r} - 1); \tag{8.1.10}$$

$$\frac{p^{2r} - 1}{p^{2s} - 1} \equiv n \bmod (p^{2s} - 1); \tag{8.1.11}$$

$$\frac{(p^{2r} - 1)^2}{(p^{2s} - 1)^2} \equiv m \bmod (p^{2s} - 1). \tag{8.1.12}$$

Proof. Suppose that $m = n^2$. Pick an s such that $m < p^s - 1$. Then inequalities (8.1.8) and (8.1.9) are satisfied. Now let $r = sn$ to satisfy conditions (8.1.10)–(8.1.12) via Lemma 8.1.6.

Suppose now that (8.1.8)–(8.1.12) are satisfied for some positive integers r and s. First of all, we note that (8.1.8) implies that $n^2 < (p^s - 1)(p^s + 1) =$

$p^{2s} - 1$. Also we obviously have that $m < p^{2s} - 1$. Next, (8.1.10) implies by Lemma 8.1.6 that $r = sk$ for some $k \in \mathbb{N}$. Therefore

$$n \equiv \frac{p^{2r} - 1}{p^{2s} - 1} \equiv k \bmod (p^{2s} - 1),$$

while

$$m \equiv \frac{(p^{2r} - 1)^2}{(p^s - 1)^2} \equiv k^2 \bmod (p^{2s} - 1).$$

Thus

$$n^2 \equiv k^2 \equiv m \bmod (p^{2s} - 1).$$

Since m, n^2 are positive integers and are less than $p^{2s} - 1$, the last congruence implies $m = n^2$.

Our next step is to show that conditions (8.1.8)–(8.1.12) can be rewritten using variables ranging over non-negative integers, integer constants, and the operations "+" and "$|_p$." In the listing below we provide the "translation" of each expression in the language of "+" and "$|_p$." This translation provides the proof of the lemma. □

Lemma 8.1.8. *Given $m, n \in \mathbb{Z}_{>0}$ there exist $r, s \in \mathbb{Z}_{>0}$ satisfying conditions (8.1.8)–(8.1.12) if and only if there exist $x, y, z, z_2, w, v, v_2, w_2, v_4, w_4, t, u_1, u_2 \in \mathbb{Z}_{>0}$ satisfying the conditions (8.1.13)–(8.1.25) set out below:*

$$(1 \,|_p\, w) \wedge (1 \,|_p\, v), \tag{8.1.13}$$
translation, $\exists s \in \mathbb{N}, w = p^s$ and $\exists r \in \mathbb{N}, v = p^r$;

$$PDIV(w, w_2, w), \tag{8.1.14}$$
translation, $w_2 = p^{2s}$, by Corollary 8.1.5;

$$PDIV(w_2, w_4, w_2), \tag{8.1.15}$$
translation, $w_4 = p^{4s}$, by Corollary 8.1.5;

$$PDIV(v, v_2, v), \tag{8.1.16}$$
translation, $v_2 = p^{2r}$, by Corollary 8.1.5;

$$PDIV(v_2, v_4, v_2), \tag{8.1.17}$$
translation, $v_4 = p^{4r}$, by Corollary 8.1.5;

$$PDIV(z_2, u_2, w_4), \tag{8.1.18}$$
translation, $u_2 = z_2 w_4 = z_2 p^{4s}$, by Lemma 8.1.3;

$$PDIV(z_2, u_1, w_2), \tag{8.1.19}$$

translation, $u_1 = z_2 w_2 = z_2 p^{2s}$, *by Lemma 8.1.3;*

$$v_4 - 2v_2 + 1 = u_2 - 2u_1 + z_2, \tag{8.1.20}$$

translation, $(p^{2r} - 1)^2 = z_2(p^{2s} - 1)^2$;

$$n + x = w, \tag{8.1.21}$$

translation, $n < p^s - 1$, *since* $x \in \mathbb{Z}_{>0}$;

$$m + y = w, \tag{8.1.22}$$

translation, $m < p^s - 1$, since $y \in \mathbb{Z}_{>0}$;

$$PDIV(z, v_2 - 1 + z, w_2), \tag{8.1.23}$$

translation, $p^{2r} - 1 + z = zp^{2s}$, $z(p^{2s} - 1) = p^{2r} - 1$, *by Lemma 8.1.3;*

$$PDIV(u, z - n + u, w_2), \tag{8.1.24}$$

translation, $z - n + u = up^{2s}$, $z - n = u(p^{2s} - 1)$, *by Lemma 8.1.3;*

$$PDIV(t, z_2 - m + t, w_2), \tag{8.1.25}$$

translation, $z_2 - m + t = tw$, $z_2 - m = t(p^{2s} - 1)$, *by Lemma 8.1.3;*

The last condition completes the "translation" of Lemma 8.1.7 into the language of "+" and "$|_p$."

We are now ready for the main theorem of this section.

Theorem 8.1.9. *There exist linear polynomials*

$$L_1(u, z, w, x_1, \ldots, x_k),\ H_1(u, z, w, x_1, \ldots, x_k), \ldots, L_r(u, z, w, x_1, \ldots, x_k),$$
$$H_r(u, z, w, x_1, \ldots, x_k),\ T_1(u, z, w, x_1, \ldots, x_k), \ldots, T_m(u, z, w, x_1, \ldots, x_k)$$

with coefficients in $\mathbb{Z}$ *such that, for any* $z, u, w \in \mathbb{Z}_{>0}$, $\exists x_1, \ldots, x_k \in \mathbb{Z}_{>0}$ *with*

$$\begin{cases} \bigwedge_{i=1}^{r} \big(H_i(u, z, w, x_1, \ldots, x_k) \,|_p\, L_i(u, z, w, x_1, \ldots, x_k)\big) \\ \bigwedge_{i=1}^{m} (T_i(u, z, w, x_1, \ldots, x_k) = 0) \end{cases}$$

if and only if $w = uz$.

Proof. It is enough to note that $2uz = (u + z)^2 - u^2 - z^2$. □

8.2 pth power equations over function fields I: Overview and preliminary results

The usefulness of Theorem 8.1.9 derives from its relationship to pth power equations in the characteristic $p > 0$, where equations of the form $y = x^{p^k}$ play the same fundamental role as played by Pell equations and other norm equations over number fields. We devote several sections below to the study of ways in which these equations can be rewritten as polynomial equations over function fields over finite fields of constants. These equations together with Theorem 8.1.9 will constitute a key step in the construction of a Diophantine model of $\mathbb{Z}$ over function fields.

The main result concerning the definability of pth powers over global fields of positive characteristic p is the following theorem.

Theorem 8.2.1. *Let M be a function field over a finite field of constants of characteristic $p > 0$. Then the set $P(M) = \{(x, y) \in M^2 | \exists s \in \mathbb{N},\ y = x^{p^s}\}$ is Diophantine over M.*

The proof of this theorem is provided by the material in Sections 8.2–8.4. We start with an overview of the proof, preliminary results, and notation.

8.2.2. Overview of the proof of the main theorem 8.2.1 The proof can be divided into three parts corresponding to Sections 8.2–8.4 respectively. In the present section we lay the groundwork with some technical results. More specifically we prove that, in some finite constant extension K of the given function field, there exists a special element t such that its divisor is a ratio of two primes, each of degree p^h for some natural number h. The existence of this element also implies the existence of a rational subfield F of K such that K is of degree p^h and separable over F. Further, one can arrange for the constant field of K to be sufficiently large so that it contains a "sufficiently large" number of constants c such that the divisor of $t + c$ in K is also a ratio of two primes, each of degree p^h. The proof of the existence of t and sufficiently many constants c (together with the exact definition of "sufficiently many") is in Theorem 8.2.3. The existence of a rational subfield as described above plays a crucial technical role in the proof of Theorem 8.2.1 by helping to establish sufficient conditions for an element of a field to be a pth power in K. These conditions can be summarized as follows.

- Since the field of constants is perfect and in a rational function field every degree-0 divisor is principal, an element of the rational subfield is a pth power if and only if its divisor is a pth power.

- An element of K is a pth power if and only if all the coefficients of its minimal polynomial over F are pth powers. (See Lemma 8.2.4 below.)
- If the values of a polynomial at sufficiently many constants are pth powers then the coefficients of the polynomial are pth powers. (See Lemma 8.2.5.)
- If $x \in K$ and $a \in F$ then $\mathbf{N}_{F(x)/F}(a - x)$ is the value of the minimal polynomial of x over F at a. (See Lemma 8.2.5.)
- Let F' be a subfield of K containing F but not equal to F. Let $\mathfrak{P}$ be a prime of K lying above a non-splitting prime of F, and let $\mathfrak{P}'$ be the F'-prime below $\mathfrak{P}$. Then $\mathbf{N}_{F'/F}(\mathfrak{P}')$ is a pth power of some prime in F. Thus if an element $v \in K \setminus F$ has poles and zeros of order divisible by p at all the primes except possibly at primes not splitting in the extension K/F, then the divisor of $\mathbf{N}_{F(v)/F}(v)$ is a pth power of a divisor in F. (This argument will be used in the proof of Lemma 8.3.5.)

In Section 8.3 we will prove that the set $\{x \in K : \exists s \in \mathbb{N}, x = t^{p^s}\}$ is Diophantine over K. This is Proposition 8.3.8. In Section 8.4, using pth powers of t, we first consider pth powers of elements of K that have simple zeros and poles only. (This is done in the proof of Proposition 8.3.8.) Finally, we examine the pth powers of arbitrary elements. Here, owing to some technical complications, we will consider separately the case of even and odd characteristics. (See Propositions 8.4.8 and 8.4.10.)

We now proceed with the technical preliminaries, starting with the existence theorem discussed above.

Theorem 8.2.3. *Let M be a function field over a finite field of constants C_M of characteristic $p > 0$. Then for any $l \in \mathbb{N}$ and for any sufficiently large positive integer h, a finite constant extension K of M contains a non-constant element t and constants $c_0 = 0, c_1, \ldots, c_l$ such that for all $i = 0, \ldots, l$ the divisor of $t + c_i$ in K is of the form $\mathfrak{p}_i/\mathfrak{q}$, where $\mathfrak{p}_i$, $\mathfrak{q}$ are primes of K of degree p^h, and for any $s \in \mathbb{N}, i \neq m$, we have $c_i^{p^s} \neq c_m$. Further, let $e(K, t)$ be the number of primes ramifying in the extension $K/C_K(t)$, where C_K is the constant field of K. Then for any $r \in \mathbb{N}$ we can arrange for l to be greater than $2e(K, t) + r$.*

Proof. Let z be a non-constant element of M which is not a pth power. (Such an element exists by the Strong Approximation Theorem.) Then by Lemma B.1.32 the extension $M/C(z)$ is finite and separable, and therefore is simple. Thus, for some $\alpha \in M$ we have that $M = C_M(z, \alpha)$. Let M_G be the Galois closure of M over $C_M(z)$. Let C be the constant field of M_G. Then $M_G/C(z)$

and, by assumption, $M_G/C(z,\alpha)$ are Galois extensions and all three fields have the same field of constants. Let $E = C(z,\alpha)$.

The following diagram describes the extensions involved.

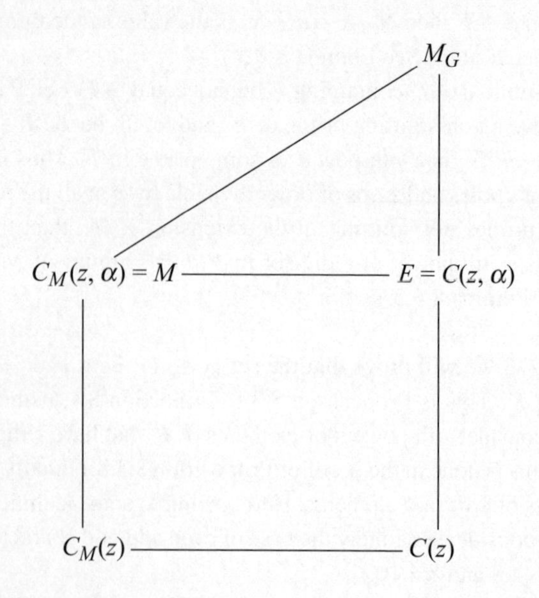

Fix a positive integer h. Let $|C| = p^r$. Then $C(z)$ has exactly

$$\frac{p^{rp^h} - p^{rp^{h-1}}}{p^h} \longrightarrow \infty \text{ as } h \to \infty$$

irreducible polynomials of degree p^h. Indeed, $p^{rp^h} - p^{rp^{h-1}}$ is the number of elements of the algebraic closure of C of degree p^h over C. Each element has exactly p^h conjugates over C. Let h_E be the class number of E. Then for any sufficiently large h, it is the case that $C(z)$ will contain at least $h_E + 2$ primes of degree p^h.

Next consider the Galois extension $M_G/C(z)$. Let $\mathfrak{t}$ be a prime of $C(z)$ of degree p^h. Assume that $\mathfrak{t}$ is unramified and splits completely in the extension $M_G/C(z)$. (Such a prime exists for all sufficiently large h by Corollary B.4.28 applied to $\sigma = \text{id}$.) Then, we claim, it splits completely in E and its factors in E are all of degree p^h. Indeed, assume that

$$\mathfrak{t} = \prod_{i=1}^{[E:C(z)]} \mathfrak{T}_i$$

is the factorization of $\mathfrak{t}$ in E. For each i, the relative degree of $\mathfrak{T}_i$ over $\mathfrak{t}$ is equal to 1. This fact together with the fact that there is no constant field extension from $C(z)$ to E implies that the $C(z)$-degree of $\mathfrak{t}$ must be the same as the E-degree of $\mathfrak{T}_i$. Thus, for sufficiently large h we know that E has at least $h_E + 2$ primes of degree p^h. Let $\mathfrak{b}_1, \ldots, \mathfrak{b}_{h_E+2}$ be these primes. Next, consider the following $h_E + 1$ divisors of E of degree 0: $\mathfrak{b}_2/\mathfrak{b}_1, \ldots, \mathfrak{b}_{h_E+2}/\mathfrak{b}_1$. At least two of these divisors belong to the same divisor class and so, for some $1 < i \neq j \leq h_E + 2$, $\mathfrak{b}_i/\mathfrak{b}_j$ is a principal divisor. Thus there exists $t \in E$ such that its divisor is of the form $\mathfrak{p}/\mathfrak{q}$, where $\mathfrak{p}$, $\mathfrak{q}$ are primes of E of degree p^h. Note that t is of order 1 at a prime of E and therefore is not a pth power in E. Hence, the extension $E/C(t)$ is finite and separable by Lemma B.1.32. Let $\mathfrak{P}$ be a prime of $C(t)$ corresponding to the zero of t. As we have established above, this prime remain tos prime in E. Thus $\mathfrak{P}$ is not ramified. Further, since the constant fields of $C(t)$ and E are the same, $\mathfrak{P}$ is of degree 1 in $C(t)$, and $\mathfrak{p}$ is of degree p^h in E, we must conclude that the relative degree of $\mathfrak{p}$ over $\mathfrak{P}$ is p^h. Therefore, by Proposition B.1.11 we have that $[E : C(t)] = p^h$.

Next, from Lemma B.4.23 and Corollaries B.4.24 and B.4.28, we conclude that for all sufficiently large k there are primes of $C(t)$ of degree k which remain prime in the extension $E/C(t)$. Let $k_1, \ldots, k_l$ be integers large enough in the sense above and let $\mathfrak{T}_1, \ldots, \mathfrak{T}_l$ be primes, not splitting in the extension $E/C(t)$, of degrees $k_1, \ldots, k_l$ respectively. Assume additionally that for all $i \neq j$ we have $(k_i, k_j) = 1$, $(k_i, p) = 1$, and $\text{ord}_{\mathfrak{T}_i} t = 0$. Let C_K/C be the extension of degree $k_1 \cdots k_l$. Let $K = C_K E$. Let $C_0 = C$ and let C_i/C be the extension of degree $\prod_{j=1}^{i} k_j$. Let $P_i(t) \in C[t]$ be an irreducible polynomial of degree k_i. We claim that the following assertions are true.

- $[C_i E : C_i(t)] = p^h$ for all $i = 0, \ldots, l$.
- In the extension $C_i(t)/C(t)$, the primes $\mathfrak{T}_1, \ldots, \mathfrak{T}_i$ split completely into degree-1 factors and $\mathfrak{T}_{i+1}, \ldots, \mathfrak{T}_l$ remain prime.
- Any factor of $\mathfrak{T}_j$, $j = 1, \ldots, n$, remains prime in the extension $C_i E/C_i(t)$.
- Let $c_i \in C_i$ be a root of $P_i(t)$. Then for $j = 1, \ldots, i - 1$ we have that c_i, c_j are not conjugates over $\mathbb{F}_p$, the field of p elements, and therefore for any non-negative integer u we have that $c_i^{p^u} \neq c_j$.

We will proceed by induction. Assume that the above statements are for the extension $C_i E/E$ for some i with $0 \leq i < l$ and consider the extension $C_{i+1}E/E$. Let $\mathfrak{t}_{i,j}$ be a factor of $\mathfrak{T}_j$ in $C_i(t)$. Let $\mathfrak{t}_{i+1,j}$ be a factor of $\mathfrak{t}_{i,j}$ in $C_{i+1}(t)$. Let $\mathfrak{U}_j$ be the prime above $\mathfrak{T}_j$ in E and let $\mathfrak{u}_{i,j}$, $\mathfrak{u}_{i+1,j}$ be factors of $\mathfrak{t}_{i,j}$ and $\mathfrak{t}_{i+1,j}$ in $C_i E$ and $C_{i+1} E$ respectively. The following diagram describes the extensions that we will consider for the inductive step.

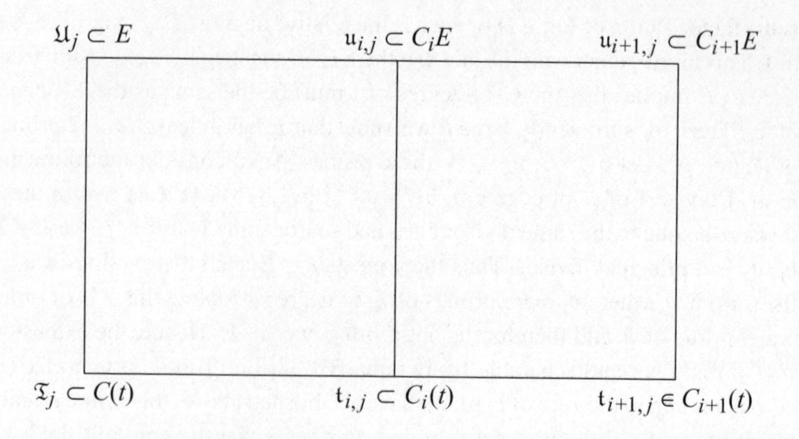

First of all, by Lemma B.3.4 we observe that $[C_iE : C_i(t)] = p^h$. Next consider $\mathfrak{t}_{i,j}, 0 \leq j \leq i$. By induction, $\mathfrak{t}_{i,j}$ is of degree 1 and therefore by Lemma B.4.16 we have that $\mathfrak{t}_{i+1,j}$ will be the only factor of $\mathfrak{t}_{i,j}$ in $C_{i+1}(t)$. However, the residue field of $\mathfrak{u}_{i,j}$ is of degree p^h over C_i, since by induction $\mathfrak{u}_{i,j}$ is the only factor of $\mathfrak{t}_{i,j}$ in C_iE. Since $[C_{i+1}E : C_iE] = [C_{i+1} : C_i] = k_{i+1}$ and $(k_{i+1}, p) = 1$, by Lemma B.4.22, we have that $\mathfrak{u}_{i+1,j}$ is the only factor of $\mathfrak{u}_{i,j}$ in $C_{i+1}E$ and therefore the only factor of $\mathfrak{t}_{i+1,j}$ in $C_{i+1}E$.

Next we note that by our induction hypothesis, $\mathfrak{t}_{i,i+1}$ is the only factor of $\mathfrak{T}_{i+1}$ in $C_i(t)$. By Lemma B.4.21 the degree of $\mathfrak{t}_{i,i+1}$ in $C_i(t)$ is the same as the degree of $\mathfrak{T}_{i+1}$ in $C(t)$, i.e. the degree is equal to k_{i+1}. Since a finite field has exactly one extension of every degree, C_{i+1} is the residue field of $\mathfrak{t}_{i,i+1}$. Thus, by Lemmas B.4.15 and B.4.26, in the extension $C_{i+1}(t)/C_i(t)$ we have $\mathfrak{t}_{i,i+1}$ splitting completely into degree-1 factors, and none of these factors splits in the extension $C_{i+1}E/C_{i+1}(t)$.

Now consider $\mathfrak{t}_{i,j}, \mathfrak{u}_{i,j}$, for $j > i + 1$. Their residue fields are of degrees k_j and k_jp^h over C_i respectively. Since by assumption $(k_{i+1}, k_jp^h) = 1$ for $i + 1 < j$, we can use Lemma B.4.22 again to conclude that the primes will remain prime in the extensions $C_{i+1}(t)/C_i(t)$ and $C_{i+1}E/C_iE$ respectively.

Finally we note that for all i we have $c_i \in C_i$, since a finite field has a unique extension of every degree and $k_i \mid [C_i : C]$. Further, suppose that c_i, c_j are conjugate over $\mathbb{F}_p$. Then for some $\sigma \in \mathrm{Gal}(C(c_i, c_j)/\mathbb{F}_p)$ we have that $\sigma(c_i) = c_j$. Thus $\sigma(P_i) = \sigma(P_j)$, which is impossible due to the difference in degrees. This concludes the proof of the above assertions.

We now consider the divisor of t in K. In E, the divisor of t is $\mathfrak{p}/\mathfrak{q}$, where $\mathfrak{p}, \mathfrak{q}$ are primes of degree p^h. Therefore, as in the discussion above, neither $\mathfrak{p}$ nor $\mathfrak{q}$ will split in the extension K/E. Similarly, the degree-1 primes below $\mathfrak{p}, \mathfrak{q}$ in $C(t)$ will remain prime and of degree 1 in the extension $C_K(t)/C(t)$. Further, consider the divisor of $t - c_i$. This element has a unique degree-1 pole at $\mathfrak{q}$. Since $(t - c_i) \mid P_i(t)$, it must have a unique zero at a degree-1 factor $\mathfrak{t}_{l,i}$ of $\mathfrak{T}_i$.

Finally we note the following. By Proposition B.4.33 we have that $e(K, t) \leq \deg \mathfrak{E}$, where $\mathfrak{E} = \prod_{\mathfrak{e} \in \mathcal{E}} \mathfrak{e}$ and $\mathcal{E}$ is the set of all the primes of E ramified in the extension $E/C(t)$. Thus, we set $l > \deg \mathfrak{E} + r$ to satisfy the last requirement of the theorem. □

Lemma 8.2.4. *Let F/G be a finite separable extension of fields of positive characteristic p. Let $\alpha \in F$ be such that, for some positive integer a, all the coefficients of its monic irreducible polynomial over G are p^ath powers in G. Then α is a p^ath power in F.*

Proof. Let $a_0^{p^a} + a_1^{p^a} + \cdots + a_{m-1}^{p^a} T^{m-1} + T^m$ be the monic irreducible polynomial of α over G. Let β be the element of the algebraic closure of F such that $\beta^{p^a} = \alpha$. Then β is of degree at most m over G. At the same time, $G(\alpha) \subseteq G(\beta)$. Therefore $G(\alpha) = G(\beta)$. □

Lemma 8.2.5. *Let F/G be a finite separable extension of fields of positive characteristic p. Let $[F : G] = n$. Let a be a positive integer. Let $x \in F$ be such that $F = G(x)$ and for distinct $a_0, \ldots, a_n \in G$ we have that $\mathbf{N}_{F/G}(a_i^{p^a} - x) = y_i^{p^a}$. Then x is a p^ath power in F.*

Proof. Let $H(T) = A_0 + A_1 T + \cdots + A_{n-1} T^{n-1} + T^n$ be the monic irreducible polynomial of x over G. Then for $i = 0, \ldots, n$ we have that $H(a_i^{p^a}) = y_i^{p^a}$. Further, we have the following linear system of equations:

$$\begin{pmatrix} 1 & a_0^{p^a} & \cdots & a_0^{p^a(n-1)} & a_0^{p^a n} \\ \vdots & \vdots & \vdots & \vdots & \vdots \\ 1 & a_n^{p^a} & \cdots & a_n^{p^a(n-1)} & a_n^{p^a n} \end{pmatrix} \begin{pmatrix} A_0 \\ \vdots \\ 1 \end{pmatrix} = \begin{pmatrix} y_0^{p^a} \\ \vdots \\ y_n^{p^a} \end{pmatrix}$$

Using Cramer's rule to solve the system, it is not hard to conclude that for $i = 0, \ldots, n$ it is the case that A_i is a p^ath power in G. Then, by Lemma 8.2.4, x is a p^ath power in F. □

Lemma 8.2.6. *Let F be a function field. Let $w \in F$, let $\mathfrak{a}_1, \ldots, \mathfrak{a}_r$ be primes of K and let $a_1, \ldots, a_{r+1}$ be a set of distinct constants of F. Then the set $\{w + a_1, \ldots, w + a_{r+1}\}$ contains at least one element of F having no zero at any of the primes $\mathfrak{a}_1, \ldots, \mathfrak{a}_r$.*

Proof. The lemma follows from the fact that each prime $\mathfrak{a}_i$ can be a zero of at most one element of the set $\{w + a_1, \ldots, w + a_{r+1}\}$. □

Lemma 8.2.7. *Let w be a non-constant element of a function field K and let $b, c \in C$, the constant field of K. Then all the zeros of $(w+c)/(w+b)$ are zeros of $w+c$ and all the poles of $(w+c)/(w+b)$ are zeros of $w+b$. Further, the height of $(w+c)/(w+b)$ is equal to the height of w. (Here by height we mean the degree of the zero or pole divisor of an algebraic function. See Definition B.1.25.)*

Proof. Let $\mathfrak{p}$ be a prime of K. Then on the one hand $\mathfrak{p}$ is a pole of w if and only if $\mathfrak{p}$ is also a pole of $w+c$ and a pole of $w+b$. Moreover, the order of the pole at all the three functions will be the same. On the other hand, any zero of $(w+c)/(w+b)$ will come from zeros of $w+c$ or poles of $w+b$. So let $\mathfrak{p}$ be a pole of $w+b$. Then $\text{ord}_{\mathfrak{p}}(w+c) = \text{ord}_{\mathfrak{p}}(w+b)$ and therefore $\text{ord}_{\mathfrak{p}}(w+c)/(w+b) = 0$. A similar argument shows that $(w+c)/(w+b)$ is a unit at any valuation which is a pole of $w+c$. Consequently, all the zeros of $(w+c)/(w+b)$ are zeros of $w+c$ and all the poles of $(w+c)/(w+b)$ are zeros of $w+b$.

Finally, note that

$$\frac{w+c}{w+b} = 1 + \frac{c-b}{w+b}.$$

Let $H_K((w+c)/(w+b))$ denote the K-height of $(w+c)/(w+b)$. Then by Remark B.1.26 we have the following equalities:

$$H_K\left(\frac{w+c}{w+b}\right) = H_K\left(1 + \frac{c-b}{w+b}\right) = H_K\left(\frac{c-b}{w+b}\right) = H_K(w+b) = H_K(w).$$

The last follows from the fact, mentioned above, that the pole divisors of $w+b$ and w are the same. □

We leave the proof of the following lemma to the reader.

Lemma 8.2.8. *Let K be a function field. Let $m > 1$ be an integer. Let $x \in K$. Then for any prime $\mathfrak{p}$ of K we have that $\mathfrak{p}$ is a pole of $x^m - x$ if and only if it is a pole of x. If $\text{ord}_{\mathfrak{p}} x < 0$ then $\text{ord}_{\mathfrak{p}}(x^m - x) \equiv 0 \bmod m$.*

In the remainder of this section and in Sections 8.3 and 8.4 we will use the following notation and assumptions.

8.2.9. Notation and assumptions

- Let M denote a function field over a finite field of constants of characteristic $p > 0$.

- Let $a = 1$ if $p > 2$ and let $a = 2$ if $p = 2$.
- Let K denote a finite separable constant field extension of M.
- Let C_K denote the finite constant field of K.
- There exist $c_0 = 0, \ldots, c_l \in C_K \setminus \{\pm 1\}$ such that, for some $t \in K$ and for all $i = 0, \ldots, l$, the divisor of $t + c_i$ is of the form $\mathfrak{p}_i/\mathfrak{q}$, where $\mathfrak{p}_i$, $\mathfrak{q}$ are primes of K.
- Let $C(K) = \{c_0, \ldots, c_l\}$.
- r_i will denote the smallest positive integer such that $c_i^{p^{r_i}} = c_i$.
- For any $0 < j \leq r_i, m \neq i$, we have that $c_i^{p^j} \neq c_m$.
- Let $d_{ij} = c_i^{p^j}, j \in \mathbb{N}$.
- $[K : C_K(t)] = p^h = n$ for some $h \in \mathbb{N}$.
- Let $\mathcal{E}(K, t)$ be the set of all primes of K ramifying in the extension $K/C_K(t)$ together with the primes $\mathfrak{p}$ and $\mathfrak{q}$.
- Let $e(K, t) = |\mathcal{E}(K, t)|$ denote the number of primes ramifying in the extension $K/C_K(t)$.
- $l > (p^h + 2e(K, t)) + 2$.
- For any $w \in K$, let

$$C_w = \left\{c \in C(K) : (\forall j \in \mathbb{N})(\forall \mathfrak{p} \in \mathcal{E}(K, t))\left(\text{ord}_{\mathfrak{p}}\left(w - c^{p^j}\right) \leq 0\right)\right\}.$$

- For any $b \in \mathbb{N}$, $w \in K$, let $B(b, w, u, v)$ denote the system of equations

$$y - t = u^{p^b} - u, \tag{8.2.1}$$

$$y^{-1} - t^{-1} = v^{p^b} - v. \tag{8.2.2}$$

- For $i, k \in \{1, \ldots, l\}$, $j_i \in \{1, \ldots, r_i\}$, $j_k \in \{1, \ldots, r_k\}$, $w, u_{i,j_i,k,j_k}, v_{i,j_i,k,j_k} \in K$, let $C(i, k, j_i, j_k, w, u_{i,j_i,k,j_k}, v_{i,j_i,k,j_k})$ denote the system of equations

$$w_{i,j_i,k,j_k} = \frac{w - d_{i,j_i}}{w - d_{k,j_k}}, \tag{8.2.3}$$

$$t_{i,k} = \frac{t - c_i}{t - c_k}, \tag{8.2.4}$$

$$w_{i,j_i,k,j_k} - t_{i,k} = u^p_{i,j_i,k,j_k} - u_{i,j_i,k,j_k}, \tag{8.2.5}$$

$$\frac{1}{w_{i,j_i,k,j_k}} - \frac{1}{t_{i,k}} = v^p_{i,j_i,k,j_k} - v_{i,j_i,k,j_k}. \tag{8.2.6}$$

- For $s \in \mathbb{N}, i, k \in \{1, \ldots, l\}$, $j_i \in \{1, \ldots, r_i\}$, $j_k \in \{1, \ldots, r_k\}$, $e = -1, 1$, $m = 0, 1$, and $u, v, \mu_{i,j_i,k,j_k,e,m}, \sigma_{i,j_i,k,j_k,e}, \nu_{i,j_i,e} \in K$, let

$$D\big(s, i, j_i, k, e, m, j_k, u, v, \mu_{i,j_i,k,j_k,e,m}, \sigma_{i,j_i,k,j_k,e}, \nu_{i,j_i,e}\big)$$

be the following system of equations:

$$u_{i,k} = \frac{u + c_i}{u + c_k}, \tag{8.2.7}$$

$$v_{i,j_i,k,j_k} = \frac{v + d_{i,j_i}}{v + d_{k,j_k}}, \tag{8.2.8}$$

$$v^{2e}_{i,j_i,k,j_k} t^{mp^{as}} - u^{2e}_{i,k} t^m = \mu^{p^{as}}_{i,j_i,k,j_k,e,m} - \mu_{i,j_i,k,j_k,e,m}, \tag{8.2.9}$$

$$v^e_{i,j_i,k,j_k} - u^e_{i,k} = \sigma^{p^a}_{i,j_i,k,j_k,e} - \sigma_{i,j_i,k,j_k,e}. \tag{8.2.10}$$

- Let $j, r, s \in \mathbb{N}$, and $u, \tilde{u}, v, \tilde{v}, x, y \in K$. Let $E(u, \tilde{u}, v, \tilde{v}, x, y, j, r, s)$ denote the following system of equations:

$$v = u^{p^r}, \tag{8.2.11}$$

$$\tilde{v} = \tilde{u}^{p^j}, \tag{8.2.12}$$

$$u = \frac{x^p + t}{x^p - t}, \tag{8.2.13}$$

$$\tilde{u} = \frac{x^p + t^{-1}}{x^p - t^{-1}}, \tag{8.2.14}$$

$$v = \frac{y^p + t^{p^s}}{y^p - t^{p^s}}, \tag{8.2.15}$$

$$\tilde{v} = \frac{y^p + t^{-p^s}}{y^p - t^{-p^s}}. \tag{8.2.16}$$

- Let $j, r, s \in \mathbb{N}$ and $u, \tilde{u}, v, \tilde{v}, x, y \in K$, and let $E2(u, \tilde{u}, v, \tilde{v}, x, y, j, r, s)$ denote the following system of equations:

$$v = u^{2^r}, \tag{8.2.17}$$

$$\tilde{v} = \tilde{u}^{2^j}, \tag{8.2.18}$$

$$u = \frac{x^2 + t^2 + t}{x^2 + t}, \tag{8.2.19}$$

$$\tilde{u} = \frac{x^2 + t^{-2} + t^{-1}}{x^2 + t^{-1}}, \tag{8.2.20}$$

$$v = \frac{y^2 + t^{2^{s+1}} + t^{2^s}}{y^2 + t^{2^s}}, \tag{8.2.21}$$

$$\tilde{v} = \frac{y^2 + t^{-2^{s+1}} + t^{-2^s}}{y^2 + t^{-2^s}}. \tag{8.2.22}$$

As will become clear in the following section, it will be necessary to be able to assume that the "important" variables do not have poles or zeros at primes ramifying in the extension $K/C(t)$. The next lemma assures us that, under our assumptions, we have enough constants to replace a variable by its sum with a constant from a fixed set, if this is needed to ensure that the non-ramification condition is satisfied.

Lemma 8.2.10. *For any $u, w \in K$ we have that $|C_w|$ and $|C_w \cap C_u|$ contain more than $n+2$ elements.*

Proof. Consider the following table:

$$\begin{bmatrix} w - c_1 & w - c_1^p & \cdots & w - c_1^{p^j} & \cdots \\ \vdots & \vdots & \vdots & \vdots & \vdots \\ w - c_l & w - c_l^p & \cdots & w - c_l^{p^j} & \cdots \end{bmatrix}.$$

Observe that by assumption on the elements of $C(K)$ no two rows share an element and the difference between any two elements of the table is constant. Thus by Lemma 8.2.6 elements of $l - e(K,t)$ rows have no zero at any element of $\mathcal{E}(K,t)$ and consequently $|C_w| \geq n + 2 + e(k,t)$. Next consider the table

$$\begin{bmatrix} u - b_1 & u - b_1^p & \cdots & u - b_1^{p^j} & \cdots \\ \vdots & \vdots & \vdots & \vdots & \vdots \\ u - b_{|C_w|} & u - c_{|C_w|}^p & \cdots & u - c_{|C_w|}^{p^j} & \cdots \end{bmatrix},$$

where $b_i \in C_w$. By an analogous argument, at least $|C_w| - e(K,t)$ rows of this table contain no element with a zero at any valuation of $\mathcal{E}(K,t)$. Thus at least $|C_w| - 2e(K,t) = n + 2$ elements are contained in $C_w \cap C_u$. □

The final lemma of this section is a technical result which will help us to eliminate a case in a later section.

Lemma 8.2.11. *Let $\sigma, \mu \in K$. Assume that none of the primes that are poles of σ or μ ramify in the extension $K/C_K(t)$. Further, assume the following equality is true:*

$$t(\sigma^{p^a} - \sigma) = \mu^{p^a} - \mu. \tag{8.2.23}$$

Then $\sigma^{p^a} - \sigma = \mu^{p^a} - \mu = 0$.

Proof. Let $\mathfrak{A}$, $\mathfrak{B}$ be integral divisors of K, relatively prime to each other and to $\mathfrak{p}$ and $\mathfrak{q}$, such that the divisor of σ is of the form $(\mathfrak{A}/\mathfrak{B})\, \mathfrak{p}^i \mathfrak{q}^k$, where i, k are integers. Then it is not hard to see that for some integral divisor $\mathfrak{C}$, relatively prime to $\mathfrak{B}$, $\mathfrak{p}$, and $\mathfrak{q}$, and for some integers j, m the divisor of μ is of the form

$(\mathfrak{C}/\mathfrak{B})\, \mathfrak{p}^j \mathfrak{q}^m$. Indeed, let $\mathfrak{t}$ be a pole of μ such that $\mathfrak{t} \neq \mathfrak{p}$ and $\mathfrak{t} \neq \mathfrak{q}$. Then

$$\begin{aligned} 0 > p^a \mathrm{ord}_{\mathfrak{t}}\mu &= \mathrm{ord}_{\mathfrak{t}}(\mu^{p^a} - \mu) = \mathrm{ord}_{\mathfrak{t}}(t(\sigma^{p^a} - \sigma)) = \mathrm{ord}_{\mathfrak{t}}(\sigma^{p^a} - \sigma) \\ &= p^a \mathrm{ord}_{\mathfrak{t}}\sigma. \end{aligned}$$

Conversely, let $\mathfrak{t}$ be a pole of σ such that $\mathfrak{t} \neq \mathfrak{p}$ and $\mathfrak{t} \neq \mathfrak{q}$. Then

$$\begin{aligned} 0 > p^a \mathrm{ord}_{\mathfrak{t}}\sigma &= \mathrm{ord}_{\mathfrak{t}}(\sigma^{p^a} - \sigma) = \mathrm{ord}_{\mathfrak{t}}(t(\sigma^{p^a} - \sigma)) = \mathrm{ord}_{\mathfrak{t}}(\mu^{p^a} - \mu) \\ &= p^a \mathrm{ord}_{\mathfrak{t}}\mu. \end{aligned}$$

By the Strong Approximation Theorem B.2.1, there exists $b \in K$ such that the divisor of b is of the form $\mathfrak{B}\mathfrak{D}/\mathfrak{q}^c$, where $\mathfrak{D}$ is an integral divisor relatively prime to $\mathfrak{A}$, $\mathfrak{C}$, $\mathfrak{p}$, $\mathfrak{q}$ and c is a natural number. Then $b\sigma = s_1 t^i$ and $b\mu = s_2 t^j$, where s_1, s_2 are integral over $C_K[t]$ and have zero divisors relatively prime to $\mathfrak{p}$ and $\mathfrak{B}$. Indeed, consider the divisors of $b\sigma$:

$$\frac{\mathfrak{B}\mathfrak{D}}{\mathfrak{q}^s}\frac{\mathfrak{A}}{\mathfrak{B}}\mathfrak{p}^i \mathfrak{q}^k = \mathfrak{D}\mathfrak{A}\mathfrak{p}^i \mathfrak{q}^{k-c} = \mathfrak{D}\mathfrak{A}\mathfrak{q}^{k-c+i}\frac{\mathfrak{p}^i}{\mathfrak{q}^i}.$$

Thus the divisor of s_1 is of the form $\mathfrak{D}\mathfrak{A}\mathfrak{q}^{k-c+i}$ and therefore $\mathfrak{q}$ is the only pole of s_1, making it integral over $C_K[t]$. Further, by construction $\mathfrak{A}$ and $\mathfrak{D}$ are integral divisors relatively prime to $\mathfrak{p}$ and $\mathfrak{B}$. A similar argument applies to s_2.

Multiplying through by b^{p^a} we will obtain the following equation:

$$t(s_1^{p^a} t^{ip^a} - b^{p^a-1} s_1 t^i) = s_2^{p^a} t^{jp^a} - b^{p^a-1} s_2 t^j. \tag{8.2.24}$$

Suppose that $i < 0$. Then the left-hand side of (8.2.24) would have a pole of order $ip^a + 1$ at $\mathfrak{p}$. This would imply that $j < 0$ and that the right-hand side has a pole of order jp^a at $\mathfrak{p}$. Thus, we can assume that i, j are both non-negative. We can now rewrite (8.2.24) in the form

$$s_1^{p^a} t^{ip^a+1} - s_2^{p^a} t^{jp^a} = b^{p^a-1}(s_1 t^{i+1} - s_2 t^j). \tag{8.2.25}$$

Let $\mathfrak{t}$ be any prime factor of $\mathfrak{B}$ in K. Then $\mathfrak{t}$ does not ramify in the extension $K/C_K(t)$ and since $p^a > 2$ we know that $\mathrm{ord}_{\mathfrak{t}}(s_1^{p^a} t^{ip^a+1} - s_2^{p^a} t^{jp^a}) \geq 2$. Further, since t is not a pth power in K the global derivation with respect to t is defined in Definition B.9.2, and by Corollary B.9.7 we also have

$$\mathrm{ord}_{\mathfrak{t}} \frac{d(s_1^{p^a} t^{ip^a+1} - s_2^{p} t^{jp^a})}{dt} > 0.$$

Finally,

$$\mathrm{ord}_{\mathfrak{t}} \frac{d(s_1^{p^a} t^{ip^a+1} - s_2^{p^a} t^{jp^a})}{dt} = \mathrm{ord}_{\mathfrak{t}}(s_1^{p^a} t^{ip^a}).$$

Therefore, since by assumption $\mathfrak{t}$ is not a zero of t, we have that s_1 has a zero at $\mathfrak{t}$. This, however, is impossible. Consequently $\mathfrak{B}$ is a trivial divisor, and in

(8.2.23) all the functions are integral over $C_K[t]$, i.e they can have poles at $\mathfrak{q}$ only. Assuming that μ is not a constant and thus has a pole at $\mathfrak{q}$, we note that the left-hand side has a pole at $\mathfrak{q}$ of order equivalent to 1 modulo p, while the right-hand side has a pole at $\mathfrak{q}$ of order equivalent to 0 modulo p. Thus μ is a constant. But the only way in which the product of t and a function integral over $C_K[t]$ can be a constant is for that function to be equal to zero. □

In the following sections we prove a sequence of lemmas which will describe the equations constituting a Diophantine definition of pth powers in K. Occasionally we will have to deal separately with the cases $p > 2$ and $p = 2$. We start with pth powers of t.

8.3 pth power equations over function fields II: pth powers of a special element

We start this section by observing in the first two lemmas that in the equations below, under some circumstances, we can replace the "most important" variable by its pth root.

Lemma 8.3.1. *Suppose that $B(b, w, u, v)$ holds for some $w, u, v \in K$ and that $\tilde{w}^{p^b} = w$. Then, for some $\tilde{u}, \tilde{v} \in K$, $B(b, \tilde{w}, \tilde{u}, \tilde{v})$ holds.*

Proof. Set $\tilde{u} = u - \tilde{w}$, $\tilde{v} = v - \tilde{w}^{-1}$ and observe that the following equations hold:

$$\tilde{w} - t = (u - \tilde{w})^{p^b} - (u - \tilde{w}),$$
$$\tilde{w}^{-1} - t^{-1} = (v - \tilde{w}^{-1})^{p^b} - (v - \tilde{w}^{-1}).$$

□

Lemma 8.3.2. *Suppose that $w, u_{i,j_i,k,j_k}, v_{i,j_i,k,j_k}$, for $i, k = 0, \ldots, l$, $j_i = 1, \ldots, r_i$, $j_k = 1, \ldots, r_k$, are elements of $K \setminus C(t)$ such that*

$$\left(\bigwedge_{i=0}^{l} \bigvee_{j_i \in \{1,\ldots,r_i\}} \bigwedge_{k=0,k\neq i}^{l} \bigvee_{j_k \in \{1,\ldots,r_k\}} C\big(i, k, j_i, j_k, w, u_{i,j_i,k,j_k}, v_{i,j_i,k,j_k}\big)\right)$$

holds. Suppose that $w = \tilde{w}^p$ for some $\tilde{w} \in K$. Then K contains elements $\tilde{u}_{i,j_i,k,j_k}, \tilde{v}_{i,j_i,k,j_k}$, for $i, k = 0, \ldots, l$, $j_i = 1, \ldots, r_i$, $j_k = 1, \ldots, r_k$, such that

$$\left(\bigwedge_{i=0}^{l} \bigvee_{j_i \in \{1,\ldots,r_i\}} \bigwedge_{k=0,k\neq i}^{l} \bigvee_{j_k \in \{1,\ldots,r_k\}} C\big(i, k, j_i, j_k, \tilde{w}, \tilde{u}_{i,j_i,k,j_k}, \tilde{v}_{i,j_i,k,j_k}\big)\right)$$

holds.

Proof. Observe the following:

$$w_{i,j_i,k,j_k} = \frac{w - d_{i,j_i}}{w - d_{k,j_k}} = \frac{w - c_i^{p^{j_i}}}{w - c_k^{p^{j_k}}} = \left(\frac{\tilde{w} - c_i^{p^{m_i}}}{\tilde{w} - c_k^{p^{m_k}}}\right)^p = (\tilde{w}_{i,m_i,k,m_k})^p,$$

where $m_i = j_i - 1, m_k = j_k - 1$ if $j_k, j_i > 1$ and $m_i = r_i, m_k = r_k$ if $j_k = 1, j_i = 1$. Note that since for all k we have that j_k can take any value $1, \ldots, r_k$, the same will be true of m_k. Thus equations (8.2.5) and (8.2.6) can be rewritten in the following manner:

$$\tilde{w}_{i,m_i,k,m_k} - t_{i,k} = \left(u^p_{i,j_i,k,j_k} - \tilde{w}^p_{i,m_i,k,m_k}\right) - \left(u_{i,j_i,k,j_k} - \tilde{w}_{i,m_i,k,m_k}\right), \tag{8.3.1}$$

$$\frac{1}{\tilde{w}_{i,m_i,k,m_k}} - \frac{1}{t_{i,k}} = \left(v^p_{i,j_i,k,j_k} - \frac{1}{\tilde{w}^p_{i,m_i,k,m_k}}\right) - \left(v_{i,j_i,k,j_k} - \frac{1}{\tilde{w}_{i,m_i,k,m_k}}\right), \tag{8.3.2}$$

□

where $1 \leq m_i \leq r_i, 1 \leq m_k \leq r_k$.

The next lemma and its corollary treat the case of a rational function.

Lemma 8.3.3. *Let $u, v \in K$, let $y \in C(t)$, and assume that y does not have zeros or poles at any valuation of K ramifying in the extension $K/C_K(t)$ and that y is not a p^ath power. Assume further that $B(a, y, u, v)$ holds. Then $y = t$.*

Proof. First of all, note that the poles of $v^{p^a} - v$ and $u^{p^a} - u$ in K are of orders divisible by p^a by Lemma 8.2.8. Since the zero and the pole of t are of orders equal to ± 1, we must conclude from (8.2.1) and (**??**) that the divisor of y is of the form $\mathfrak{U}^{p^a}\mathfrak{p}^{b_1}\mathfrak{q}^{b_2}$. Indeed, let $\mathfrak{t}$ be a prime which is not equal to $\mathfrak{p}$ or $\mathfrak{q}$. Without loss of generality assume that $\mathfrak{t}$ is a pole of y. Then, since $\text{ord}_{\mathfrak{t}} t = 0$,

$$0 > \text{ord}_{\mathfrak{t}} y = \text{ord}_{\mathfrak{t}}(t - y) = \text{ord}_{\mathfrak{t}}\left(u^{p^a} - u\right) \equiv 0 \bmod p^a.$$

Now consider the order at $\mathfrak{q}$. We have

$$\text{ord}_{\mathfrak{q}}(y - t) = \text{ord}_{\mathfrak{q}}\left(u^{p^a} - u\right).$$

Therefore either $\text{ord}_{\mathfrak{q}} y < -1$ and $\text{ord}_{\mathfrak{q}} y \equiv 0 \bmod p^a$ or $\text{ord}_{\mathfrak{q}} y = \text{ord}_{\mathfrak{q}} t = -1$. Similarly, either $\text{ord}_{\mathfrak{p}} y > 1$ and $\text{ord}_{\mathfrak{p}} y \equiv 0 \bmod p^a$ or $\text{ord}_{\mathfrak{p}} y = \text{ord}_{\mathfrak{p}} t = 1$. Further, since the divisor of y must be of degree 0, the orders at $\mathfrak{p}$ and $\mathfrak{q}$ must be equivalent modulo p^a. If $\text{ord}_{\mathfrak{q}} y \equiv \text{ord}_{\mathfrak{p}} y \equiv 0 \bmod p^a$ then, taking into account the fact that no prime which is a pole or zero of y ramifies in the extension

$K/C_K(t)$, we can conclude that the divisor of y in the rational field is also a p^ath power of another divisor. Thus, since in the rational field every degree-0 divisor is principal and the field of constants is perfect, y is a p^ath power. Therefore we must conclude that $|\mathrm{ord}_{\mathfrak{q}} y| = |\mathrm{ord}_{\mathfrak{p}} y| = 1$. Then we can deduce, using an argument similar to the one above, that yt^{-1} is a p^ath power in the rational field. Thus (8.2.1) can be rewritten as

$$t(f-1)^{p^a} = u^{p^a} - u, \tag{8.3.3}$$

where $f \in C_K(t)$. Since $f - 1$ is a rational function in t, we can rewrite (8.3.3) further as

$$t\left(f_1^{p^a}/f_2^{p^a}\right) = u^{p^a} - u, \tag{8.3.4}$$

where f_1, f_2 are relatively prime polynomials in t over C and f_2 is monic. From this equation it is clear that any valuation that is a pole of u is either a pole of t or a zero of f_2. Further, the absolute value of the order of any pole of u at any valuation which is a zero of f_2 must be the same as the order of f_2 at this valuation. Therefore $s = f_2 u$ will have poles only at the valuations which are poles of t. Thus we can rewrite (8.3.4) in the form

$$-tf_1^{p^a} + s^{p^a} = sf_2^{p^a-1}.$$

Let $\mathfrak{c}$ be a zero of f_2. Then, since f_2 is a polynomial in t, $\mathfrak{c}$ is not a pole of t. Since $p^a - 1 \geq 2$ and s is integral over $C_K[t]$, we have that $\mathrm{ord}_{\mathfrak{c}}(s^{p^a} - tf_1^{p^a}) \geq 2$.

Now observe that by Proposition B.9.3

$$d\left(-tf_1^{p^a} + s^{p^a}\right)/dt = -f_1^{p^a}$$

and, since by assumption f_2 does not have any zeros at valuations ramifying in the extension $K/C_K(t)$, by Corollary B.9.7 we have

$$\mathrm{ord}_{\mathfrak{c}}\left(-f_1^{p^a}\right) = \mathrm{ord}_{\mathfrak{c}}\left(d\left(-tf_1^{p^a} + s^{p^a}\right)/dt\right) > 0.$$

Thus f_1 has a zero at $\mathfrak{c}$. But f_1 and f_2 are supposed to be relatively prime polynomials. Hence, f_2 does not have any zeros, and thus is equal to 1. Therefore y is a polynomial in t. Similarly, we can show that $1/y$ is a polynomial in $1/t$. Hence y is a power of t and, more specifically, unless $y = t$ we must have that y is a power of t divisible by p^a. If $y = t$ we are done. Otherwise, we have shown that y is a p^ath power of another rational function in t over C_K, contradicting our assumptions. □

Corollary 8.3.4. Let u, v, y be as in Lemma 8.3.3 with the exception that we will now allow y to be a p^ath power in K. Then for some non-negative integer s we have that $y = t^{p^{as}}$.

Proof. First of all, observe the following. Either y is a constant or, for some non-negative integer k, it is the case that $y = w^{p^{ak}}$, where $w \in C_K(t)$, w is not a p^ath power of any element and, by Lemma 8.3.1 and induction, there exist $\bar{u}, \bar{v} \in K$ such that $B(a, w, \bar{u}, \bar{v})$ holds. Now, it is not hard to see that y cannot be a constant. Indeed, if $y \in C_K$ then $\text{ord}_{\mathfrak{q}}(t - y) = -1$ while, as discussed above, the degrees of all the poles of the right-hand side quantities in (8.2.1) are divisible by p^a. Thus there exist $\bar{u}, \bar{v} \in K$ such that $B(a, w, \bar{u}, \bar{v})$ holds. Note further that, since $K/C(t)$ is separable, $w \in C(t)$. Hence we can apply Lemma 8.3.3 to $w, t, \bar{u}, \bar{v}$ in place of y, t, u, v to conclude that $w = t$. But in this case $y = w^{p^{ak}} = t^{p^{ak}}$ and so the corollary holds. □

We now proceed to eliminate the assumption that the function in question is rational.

Lemma 8.3.5. *Let w be an element of $K \setminus C_K(t)$. Let $u, v, u_{i,j_i,k,j_k}, v_{i,j_i,k,j_k}$, for $i, k = 0, \ldots, l, j_i = 1, \ldots, r_i, j_k = 1, \ldots, r_k$, be elements of K satisfying*

$$\bigwedge_{i=0}^{l} \bigvee_{j_i \in \{1,\ldots,r_i\}} \bigwedge_{k=0,k\neq i}^{l} \bigvee_{j_k \in \{1,\ldots,r_k\}} C(i, k, j_i, j_k, w, u_{i,j_i,k,j_k}, v_{i,j_i,k,j_k}). \quad (8.3.5)$$

Then w is a p*th power of some element of K.*

Proof. By Corollary 8.2.10, we can choose distinct natural numbers

$$i, k_1, \ldots, k_{n+1} \in \{1, \ldots, l\}$$

such that $\{c_i, c_{k_1}, \ldots, c_{k_{n+1}}\} \subset C_w$. Fix the indices $i, k_1, \ldots, k_{n+1}$. By the assumption of the lemma, (8.2.3)–(8.2.6) hold for the quadruples

$$(i, j_i, k_1, j_1), \quad \ldots, \quad (i, j_i, k_{n+1}, j_{n+1})$$

for some index j_i with $1 \leq j_i \leq r_i$ and indices $j_{k_1}, \ldots, j_{k_{n+1}}$ with $1 \leq j_{k_m} \leq r_{k_m}$. Next consider $w_{i,j_i,k_r,j_{k_r}} = (w - d_{i,j_i})/(w - d_{k_m,j_{k_m}})$ for $m = 1, \ldots, n+1$. Note that neither the numerator nor the denominator of this fraction has a zero at a valuation ramifying in the extension $K/C_K(t)$. Thus by Lemma 8.2.7, for all $m = 1, \ldots, n+1$, we have that $w_{i,j_i,k_m,j_{k_m}}$ has no zeros or poles at any valuation ramifying in the extension $K/C_K(t)$.

By assumption $w \notin C_K(t)$. This implies that for all $m = 1, \ldots, n+1$ we also have that $w_{i,j_i,k_m,j_{k_m}} \notin C_K(t)$. Further, by an argument similar to that used in the proof of Lemma 8.3.3, for all $m = 1, \ldots, n+1$ equations (8.2.5) and (8.2.6) imply that the divisor of $w_{i,j_i,k_m,j_{k_m}}$ is of the form $\mathfrak{A}^p \mathfrak{p}_{k_m}^{b_1} \mathfrak{p}_i^{b_2}$, where b_1 is either -1 or 0 and b_2 is either 1 or 0.

Let $K_w = C_K(w, t)$ and note that for all $m = 1, \ldots, n+1$ we have that $w_{i,j_i,m,j_m} \in K_w$ and $[K_w : C_K(t)] = p^{\bar{h}}$, where $0 < \bar{h} \le h$. (The left-hand inequality is strict owing to our assumption that $w \notin C_K(t)$). Further, since $w_{i,j_i,k_m,j_{k_m}}$ does not have any zeros or poles ramifying in the extension $K/C_K(t)$, the divisor of $w_{i,j_i,k_m,j_{k_m}}$ will be of the form $\mathfrak{A}^p_{K_w}\mathfrak{P}^{b_1}_{i,w}\mathfrak{P}^{b_2}_{k_m,w}$ in K_w, where $\mathfrak{A}_{K_w}$ is the K_w-divisor below the divisor $\mathfrak{A}$ and for all s we have that $\mathfrak{P}_{s,w}$ denotes the prime below $\mathfrak{p}_s$ in $C_K(t, w)$. Next we note that, on the one hand, by Proposition B.4.2 the divisor of

$$\mathbf{N}_{K_w/C_K(t)}\big(w_{i,j_i,k_m,j_{k_m}}\big)$$

is equal to the norm of the divisor of $w_{i,j_i,k_m,j_{k_m}}$. On the other hand,

$$\mathbf{N}_{K_w/C_K(t)}\mathfrak{P}_{s,w} = \mathfrak{P}_s^{f(\mathfrak{P}_{s,w}/\mathfrak{P}_s)} = \mathfrak{P}_s^{p^{\bar{h}}}.$$

Thus the divisor of the norm of $w_{i,j_i,k_m,j_{k_m}}$ in $C_K(t)$ is a pth power of some other divisor of $C_K(t)$. Since in $C_K(t)$ every degree-0 divisor is principal, we must conclude that, for all $m = 1, \ldots, n+1$, the $K/C_K(t)$-norm of $w_{i,j_i,k_m,j_{k_m}}$ is a pth power of some element of $C_K(t)$. Further,

$$\begin{aligned}\frac{1}{w_{i,j_i,k_m,j_{k_m}}} &= \frac{w - d_{k_m,j_{k_m}}}{w - d_{i,j_i}} = 1 + \frac{d_{i,j_i} - d_{k_m,j_{k_m}}}{w - d_{i,j_i}} \\ &= \big(d_{i,j_i} - d_{k_m,j_{k_m}}\big)\left(\frac{1}{d_{i,j_i} - d_{k_m,j_{k_m}}} - \frac{1}{d_{i,j_i} - w}\right).\end{aligned}$$

Thus we can conclude that for $m = 1, \ldots, n+1$ it is the case that

$$\mathbf{N}_{K_w/C_K(t)}\left(\frac{1}{d_{i,j_i} - d_{k_m,j_{k_m}}} - \frac{1}{d_{i,j_i} - w}\right)$$

is a pth power. Then, by Lemma 8.2.5, taking into account our assumption that for all natural numbers s and for $m \ne j$ we have that $c_m^{p^s} \ne c_j$, we can conclude that $w - d_{i,j_i}$ is a pth power in K. Consequently, w is a pth power in K. □

Corollary 8.3.6. *Let $w, u, v, u_{i,j_i,k,j_k}, v_{i,j_i,k,j_k}$, for $i, k = 0, \ldots, l, j_i = 1, \ldots, r_i, j_k = 1, \ldots, r_k$, be as above. Then $w \in C_K(t)$.*

Proof. Unless w is a constant and consequently is in $C_K \subset C_K(t)$, for some natural number m there exists $\bar{w}$ such that $w = \bar{w}^{p^m}$ and $\bar{w}$ is not a pth power. By Lemma 8.3.2 and induction, there exist $\bar{u}, \bar{v}, \bar{u}_{i,j_i,k,j_k}, \bar{v}_{i,j_i,k,j_k}$, for, $i, k = 0, \ldots, l, j_i = 1, \ldots, r_i, j_k = 1, \ldots, r_k$ such that

$$\bigwedge_{i=0}^{l} \bigvee_{j_i \in \{1,\ldots,r_i\}} \bigwedge_{k=0,k\ne i}^{l} \bigvee_{j_k \in \{1,\ldots,r_k\}} C\big(i, k, j_i, j_k, \bar{w}, \bar{u}_{i,j_i,k,j_k}, \bar{v}_{i,j_i,k,j_k}\big).$$

Since $\bar{w}$ is not a pth power, we conclude that $\bar{w} \in C_K(t)$. Thus for some s we have that $w^{p^s} \in C_K(t)$. But the extension $K/C_K(t)$ is separable. Thus $w \in C_K(t)$. □

Corollary 8.3.7. *Let* $w, u, v, u_{i,j_i,k,j_k}, v_{i,j_i,k,j_k}$, *for* $i, k = 0, \ldots, l, j_i = 1, \ldots, r_i, j_k = 1, \ldots, r_k$, *be elements of* K *satisfying*

$$\left(\bigwedge_{i=0}^{l} \bigvee_{j_i \in \{1,\ldots,r_i\}} \bigwedge_{k=0,k\neq i}^{l} \bigvee_{j_k \in \{1,\ldots,r_k\}} C\big(i, k, j_i, j_k, w, u_{i,j_i,k,j_k}, v_{i,j_i,k,j_k}\big)\right)$$
$$\bigwedge B(a, w, u, v)$$

Then for some $s \in \mathbb{N}$ *we have that* $w = t^{p^{as}}$.

Proof. From Corollary 8.3.6 we conclude that $w \in C_K(t)$. Therefore we can apply Corollary 8.3.4 to conclude that $w = t^{p^{as}}$ for some non-negative integer s. □

Finally we prove the main result of this section.

Proposition 8.3.8. *The set* $\{w \in K \mid \exists s \in \mathbb{N} : w = t^{p^s}\}$ *is Diophantine over* K.

Proof. First of all observe that, for any $x \in K$ and any $s \in \mathbb{N}$,

$$x^{p^{as}} - x = \left(x^{p^{a(s-1)}} + x^{p^{a(s-2)}} + \cdots + x\right)^{p^a} - \left(x^{p^{a(s-1)}} + x^{p^{a(s-2)}} + \cdots + x\right), \tag{8.3.6}$$

$$x^{p^{as}} - x = \left(x^{p^{as-1}} + x^{p^{as-2}} + \cdots + x\right)^{p} - \left(x^{p^{as-1}} + x^{p^{as-2}} + \cdots + x\right). \tag{8.3.7}$$

Next we want to show that, assuming $w = t^{p^{as}}$, it follows that (8.3.5) is true over K. In view of equality (8.3.6), it is enough to show that, for some $1 \leq j_i \leq r_i, 1 \leq j_k \leq r_k, w_{i,j_i,k,j_k} = (t_{i,k})^{p^{as}}$. Choose $j_i \equiv as \bmod r_i$. (Such a j_i exists since the set of all possible values of j_i contains a representative of every class modulo r_i.) Then for some integer m we have that $c_i^{p^{as}} = \left(c_i^{p^{j_i}}\right)^{p^{mr_i}} = c_i^{p^{j_i}}$. Similarly, choose $j_k \equiv as \bmod r_k$ so that $c_k^{p^{as}} = c_k^{p^{j_k}}$. Now we conclude that the set $\{w \in K | \exists s \in \mathbb{N} : w = t^{p^s}\}$ is Diophantine over K for $p > 2$. Finally, in the case $p = 2$ and $a = 2$ we note that there exists a non-negative r such that $w = t^{2^r}$ if and only if either there exists a non-negative s such that $w = t^{4^s}$ or $w = (t^{4s})^2$. □

8.4 *p*th power equations over function fields III: *p*th powers of arbitrary functions

In this section we will use pth powers of t to construct pth powers of arbitrary elements. We will start with a lemma which is a slightly different version of an idea that we have used already in Section 8.3.

Lemma 8.4.1. *Let m be a positive integer. Let $v \in K$ and assume that for some distinct $a_0 = 0, a_1, \ldots, a_n \in C_K$, the divisor of $v + a_0, \ldots, v + a_n$ is a p^mth power of some other divisor of K. Then, assuming that for all i we have that $v + a_i$ does not have any zeros or poles at any prime ramifying in the extension $K/C_K(t)$, it is the case that v is a p^mth power in K.*

Proof. First assume that $v \in C_K(t)$. Since $v + a_i$ does not have any zeros or poles at primes ramifying in the extension $K/C_K(t)$, the divisor of $v + a_i$ in $C_K(t)$ is a p^mth power of another $C_K(t)$ divisor. Since in $C_K(t)$ every degree-0 divisor is principal and the constant field is perfect, v is a pth power in $C_K(t)$ and therefore in K. Next assume that $v \notin C_K(t)$. Note that no zero or pole of $v + a_i$ is at any valuation ramifying in the extension $K/C_K(t, v)$. Hence in $C_K(t, v)$ the divisor of $v + a_i$ is also a p^mth power of another divisor. Finally note that $\mathbf{N}_{C_K(t,v)/C_K(t)}(v + a_i)$ will be a p^mth power in $C_K(t)$ and apply Lemma 8.2.5. □

Before we produce a Diophantine definition of the pth powers of arbitrary elements of K, we will carry out the construction for elements with simple zeros and poles. The next two lemmas deal with the construction of such elements. We will use a global derivation with respect to t to verify that an element has simple poles at certain valuations. A discussion of global and local derivations and their relationship to the order of zeros can be found in appendix section B.9.

Lemma 8.4.2. *Let $p > 2$. Let $x \in K$. Let $u = (x^p + t)/(x^p - t)$. Let $a \in C_K, a \neq \pm 1$. Then all zeros and poles of $u + a$ are simple except possibly for zeros or poles at $\mathfrak{p}$, $\mathfrak{q}$ or primes ramifying in the extension $K/C_K(t)$.*

Proof. It is enough to show that the proposition holds for u. The argument for u^{-1} follows by symmetry. First of all we note that the global derivation with respect to t is defined over K and that the derivation follows the usual rules by Definition B.9.2 and Proposition B.9.4. So consider

$$\frac{d(u + a)}{dt} = \frac{2x^p}{(x^p - t)^2}.$$

If $\mathfrak{t}$ is a prime of K that does not ramify in the extension $K/C_K(t)$ and is not a pole or zero of t then, by Corollary B.9.7, we have that

$$\text{ord}_{\mathfrak{t}}(u+a) = \text{ord}_{\mathfrak{a}} \frac{(1+a)x^p + (1-a)t}{x^p - t} > 1$$

if and only if $\mathfrak{t}$ is a common zero of $u+a$ and $d(u+a)/dt$. If

$$\text{ord}_{\mathfrak{t}} \frac{2x^p}{(x^p - t)^2} > 0$$

then $\mathfrak{t}$ is either a zero of x or a pole of $x^p - t$. Any zero of x which is not a zero of t is not a zero of $u+a$ for $a \neq 1$. Further, any pole of x is also not a zero of $u+a$. Thus all zeros of $u+a$ at primes not ramifying in the extension $K/C_K(t)$ and different from $\mathfrak{p}$ and $\mathfrak{q}$ are simple. Next we note that poles of $u+a$ are zeros of u^{-1}. Further,

$$\frac{du^{-1}}{dt} = \frac{-2x^p}{(x^p+t)^2},$$

and by a similar argument u^{-1} and du^{-1}/dt do not have any common zeros at any primes that do not ramify in the extension $K/C_K(t)$ and are not a pole or a zero of t. □

We leave to the reader the proof of the next lemma, which deals with the case $p = 2$.

Lemma 8.4.3. *Let $p = 2$. Let $x \in K$. Let $u = (x^2 + x + t)/(x^2 + t)$. Let $a \in C_K$, $a \neq 1$. Then all zeros and poles of $u + a$ are simple except possibly for zeros or poles at $\mathfrak{p}$, $\mathfrak{q}$ or at primes ramifying in the extension $K/C_K(t)$.*

Our next task is to note that the equations we are going to be using for this case (the case of a function with simple zeros or poles) can be reproduced if we take the pth root of the "main" variable, just as in the case of Lemma 8.3.2. The proof is also analogous to the proof of Lemma 8.3.2 and so we omit it. □

Lemma 8.4.4. *Let $s \in \mathbb{N}$, $s > 0$. Let $x, v \in K \setminus \{0\}$ and assume that for some $\tilde{v} \in K$ we have that $\tilde{v}^{p^a} = v$. Let $u = (x^p + t)/(x^p - t)$ if $p > 2$ and let $u = (x^2 + x + t)/(x^2 + t)$ if $p = 2$. Further, assume that*

$$\exists \mu_{i,j_i,k,j_k,e,m}, \sigma_{i,j_i,k,j_k,e}, \nu_{i,j_i,e} \in K$$

$$\forall i \, \exists j_i \, \forall (k \neq i) \, \exists j_k \, \forall m \, \forall e : \tag{8.4.1}$$

$$D\big(s, i, j_i, k, j_k, m, e, u, v, \mu_{i,j_i,k,j_k,e,m}, \sigma_{i,j_i,k,j_k,e}, \nu_{i,j_i,e}\big)$$

holds. Then

$$\exists \tilde{\mu}_{i,j_i,k,j_k,e,m}, \tilde{\sigma}_{i,j_i,k,j_k,e}, \tilde{v}_{i,j_i,e} \in K$$
$$\forall i \ \exists j_i \ \forall (k \neq i) \ \exists j_k \ \forall m \ \forall e : \tag{8.4.2}$$
$$D\big(s-1, i, j_i, k, j_k, m, e, u, \tilde{v}, \tilde{\mu}_{i,j_i,k,j_k,e,m}, \tilde{\sigma}_{i,j_i,k,j_k,e}, \tilde{v}_{i,j_i,e}\big)$$

holds.

Lemma 8.4.5. *Let $s \in \mathbb{N}$, $x, v \in K \setminus \{0\}$. Let $u = (x^p + t)/(x^p - t)$, if $p > 2$ and let $u = (x^2 + x + t)/(x^2 + t)$ if $p = 2$. Further, assume that* (8.4.1) *holds. Then, for some natural number d, we have that $v = u^{p^d}$.*

Proof. First of all, we claim that for all i, k it is the case that $u_{i,k}$ has no multiple zeros or poles except possibly at the primes ramifying in $K/C_K(t)$, or at $\mathfrak{p}$ or $\mathfrak{q}$. Indeed, by Lemma 8.2.7, all the poles of $u_{i,k}$ are zeros of $u + c_k$ and all the zeros of $u_{i,k}$ are zeros of $u + c_i$. However, by Lemma 8.4.2 and by the assumption on c_i and c_k (see Notation and Assumptions 8.2.9), all the zeros of $u + c_k$ and $u + c_i$ are simple, except possibly for zeros at $\mathfrak{p}$, $\mathfrak{q}$, or primes ramifying in the extension $K/C_K(t)$. For future use, we also note that u is not a pth power in K, assuming $x \neq 0$. (This can be established by computing the derivative of u, which is not 0 if x is not 0.)

We will show that if $s > 0$ then v is a p^ath power in K, and if $s = 0$ then $u = v$. This assertion together with Lemma 8.4.4 will produce the desired conclusion.

Note that by Lemma 8.2.10, we can choose distinct natural numbers $i, k_1, \dots, k_{n+1} \in \{0, \dots, l\}$ such that $\{c_i, c_{k_1}, \dots, c_{k_{n+1}}\} \subset C_v \cap C_u$ and, for all $1 \leq j_i \leq r_i$, $1 \leq j_{k_f} \leq r_{k_f}$, with $f = 1, \dots, n+1$, we have that $u_{i,k_f,1}$ and $v_{i,j_i,k_f,j_{k_f},1}$ have no zeros or poles at the primes of K ramifying in the extension $K/C(t)$, or at $\mathfrak{p}$ or $\mathfrak{q}$. Note also that for the indices thus selected, all the poles and zeros of $u_{i,k_f,1}$ are simple. Thus we can pick natural numbers $i, k_1, \dots, k_{n+1}, j_i, j_{k_1}, \dots, j_{k_{n+1}}$ such that equations (8.2.7)–(8.2.10) are satisfied for these values of the indices, and $u_{i,k_1}, v_{i,j_i,k_1,j_{k_1}}, \dots, u_{i,k_{n+1}}, v_{i,j_i,k_{n+1},j_{k_{n+1}}}$ have no poles or zeros at primes ramifying in the extension $K/C(t)$, or at $\mathfrak{p}$ or $\mathfrak{q}$.

Now assume that $s > 0$ and let f range over the set $\{1, \dots, n+1\}$. First let $e = \pm 1$, while $m = 0$, and consider the two versions of (8.2.9) with these values of e and m:

$$v^2_{i,j_i,k_f,j_{k_f}} - u^2_{i,k_f} = \mu^{p^a}_{i,j_i,k_f,j_{k_f},1,0} - \mu_{i,j_i,k_f,j_{k_f},1,0}, \tag{8.4.3}$$

$$v^{-2}_{i,j_i,k_f,j_{k_f}} - u^{-2}_{i,k_f} = \mu^{p^a}_{i,j_i,k_f,j_{k_f},-1,0} - \mu_{i,j_i,k_f,j_{k_f},-1,0}. \tag{8.4.4}$$

By an argument similar to that used in the proof of Lemma 8.3.3, either for all $f = 1, \ldots, n+1$ the divisor of $v_{i,j_i,k_f,j_{k_f}}$ in K is a p^ath power of another divisor or for some f and some prime $\mathfrak{t}$ not ramifying in $K/C(t)$ and not equal to $\mathfrak{p}$ or $\mathfrak{q}$ we have that $\text{ord}_{\mathfrak{t}} v_{i,j_i,k_f,j_{k_f}} = \pm 1$.

In the first case, given the assumption that the $v_{i,j_i,k_f,j_{k_f}}$ do not have poles or zeros at ramifying primes and Lemma 8.4.1, we have that v is a p^ath power in K. So suppose that the second alternative holds. In this case, without loss of generality, assume that $\mathfrak{t}$ is a pole of $v_{i,j_i,k_f,j_{k_f}}$ for some f. Next, consider the equations

$$v^2_{i,j_i,k_f,j_{k_f}} t^{p^{as}} - u^2_{i,k_f} t = \mu^{p^a}_{i,j_i,k_f,j_{k_f},1,1} - \mu_{i,j_i,k_f,j_{k_f},1,1} \tag{8.4.5}$$

$$v^2_{i,j_i,k_f,j_{k_f}} - u^2_{i,k_f} = \mu^{p^a}_{i,j_i,k_f,j_{k_f},0,1} - \mu_{i,j_i,k_f,j_{k_f},0,1} \tag{8.4.6}$$

obtained from (8.2.9) by first making $e = 1, m = 1$ and then $e = 1, m = 0$. (If $\mathfrak{t}$ were a zero of $v_{i,j_i,k_f,j_{k_f}}$ then we would set e equal to -1 in both equations.) Since t does not have a pole or zero at $\mathfrak{t}$ and $p^a > 2$, we must conclude that

$$\text{ord}_{\mathfrak{t}}\big(v^2_{i,j_i,k_f,j_{k_f}} t^{p^{as}} - u^2_{i,k_f} t\big) = \text{ord}_{\mathfrak{t}}\big(\mu^{p^a}_{i,j_i,k_f,j_{k_f},1,1} - \mu_{i,j_i,k_f,j_{k_f},1,1}\big) \geq 0$$

and

$$\text{ord}_{\mathfrak{t}}\big(v^2_{i,j_i,k_f,j_{k_f}} - u^2_{i,k_f}\big) = \text{ord}_{\mathfrak{t}}\big(\mu^{p^a}_{i,j_i,k_f,j_{k_f},0,1} - \mu_{i,j_i,k_f,j_{k_f},0,1}\big) \geq 0.$$

Thus

$$\begin{aligned}
&\text{ord}_{\mathfrak{t}} v^2_{i,j_i,k_f,j_{k_f}} \big(t^{p^{as}} - t\big) \\
&\quad = \text{ord}_{\mathfrak{t}}\big(\mu^{p^a}_{i,j_i,k_f,j_{k_f},1,1} - \mu_{i,j_i,k_f,j_{k_f},1,1} \\
&\qquad - t\mu^{p^a}_{i,j_i,k_f,j_{k_f},0,1} + t\mu_{i,j_i,k_f,j_{k_f},0,1}\big) \geq 0.
\end{aligned}$$

Finally, we deduce that $\text{ord}_{\mathfrak{t}}(t^{p^{as}} - t) \geq 2\,|\text{ord}_{\mathfrak{t}}\, v|$. But in $C_K(t)$ all the zeros of $t^{p^{as}} - t$ are simple. Thus this function can have multiple zeros only at primes ramifying in the extension $K/C_K(t)$. By assumption $\mathfrak{t}$ is not one of these primes and thus we have a contradiction, unless v is a p^ath power.

Suppose now that $s = 0$. Let $i, k_1, \ldots, k_{n+1}$ be selected as above. Then from (8.4.5) and (8.4.6) we obtain, for $k_f \in \{k_1, \ldots, k_{n+1}\}$,

$$\mu^{p^a}_{i,j_i,k_f,j_{k_f},1,1} - \mu_{i,j_i,k_f,j_{k_f},1} = t\big(\mu^{p^a}_{i,j_i,k_f,j_{k_f},0,1} - \mu_{i,j_i,k_f,j_{k_f},0,1}\big).$$

Note here that all the poles of $\mu_{i,j_i,k_f,j_{k_f},1,1}$ and $\mu_{i,j_i,k_f,j_{k_f},0,1}$ are poles of u_{i,k_f}, $v_{i,j_i,k_f,j_{k_f}}$, or t and thus are not at any valuation ramifying in the extension $K/C_K(t)$. By Lemma 8.2.11 we can then conclude that, for all $k_f \in \{k_1, \ldots, k_{n+1}\}$,

$$v^2_{i,j_i,k_f,j_{k_f}} - u^2_{i,k_f} = 0.$$

Thus $v_{i,j,k_f,j_{k_f}} = \pm u_{i,k_f}$. Since all the poles of u_{i,k_f} are simple, (8.2.10) with $e = 1$ rules out the negative option. Therefore

$$v_{i,j,k_f,j_{k_f}} = u_{i,k_f}. \tag{8.4.7}$$

Rewriting (8.4.7) we obtain

$$\frac{d_{i,j} - d_{k_f,j_{k_f}}}{v + d_{k_f,j_{k_f}}} = \frac{c_i - c_{k_f}}{u + c_{k_f}}$$

or

$$v = au + b, \tag{8.4.8}$$

where a, b are constants. However, unless $b = 0$, we have a contradiction with (8.2.10) for $e = -1$ because, unless $b = 0$, we have that v^{-1} and u^{-1} have different simple poles. Finally, if $a \neq 1$ then we have a contradiction for (8.2.10) for $e = 1$ again, because the difference, unless it is 0 (and therefore $a = 1$), will have simple poles. □

The following corollary completes the construction for the "simple pole and zero" case.

Corollary 8.4.6. *Let $x \in K$ and let $u = (x^p + t)/(x^p - t)$ if $p > 2$ and let $u = (x^2 + x + t)/(x^2 + t)$ if $p = 2$. Then the set $\{w \in K \mid \exists s \in \mathbb{N} : w = u^{p^s}\}$ is Diophantine over K.*

Proof. Given Lemma 8.4.5, as in Proposition 8.3.8 it is enough to show that if $w = u^{p^{as}}$ for some natural number s then (8.4.1) can be satisfied in the remaining variables over K. This assertion can be shown to be true in the same way as the analogous statement in Proposition 8.3.8. □

Remark 8.4.7. The reader should note that in Lemmas 8.4.2–8.4.5 and in Corollary 8.4.6 we can systematically replace t by t^{-1} without changing the conclusions.

We are now ready for the last sequence of propositions before the main theorem. We will have to separate the case $p = 2$ again. We start with the case $p > 2$.

Proposition 8.4.8. *Let $p > 2$. Let $x, y \in K$. Then there exist $v, \tilde{v}, u, \tilde{u}, v_1, \tilde{v}_1, u_1, \tilde{u}_1 \in K$ and $s, r, j, r_1, j_1 \in \mathbb{N}$ such that*

$$\begin{cases} E(u, \tilde{u}, v, \tilde{v}, x, y, j, r, s), \\ E(u_1, \tilde{u}_1, v_1, \tilde{v}_1, x+1, y+1, j_1, r_1, s) \end{cases} \tag{8.4.9}$$

hold if and only if $y = x^{p^s}$. (Each expression in (8.4.9) corresponds to a system of equations, as defined in Notation and Assumptions 8.2.9.)

Proof. Suppose that (8.4.9) is satisfied over K. Then, using the fact that $E(u, \tilde{u}, v, \tilde{v}, x, y, j, r, s)$ holds, from (8.2.11), (8.2.13), and (8.2.15), we obtain

$$\frac{x^{p^{r+1}} - t^{p^r}}{x^{p^{r+1}} + t^{p^r}} = \frac{y^p - t^{p^s}}{y^p + t^{p^s}},$$

and so

$$y = x^{p^r} t^{p^{s-1} - p^{r-1}}.$$

Similarly, from (8.2.12), (8.2.14), and (8.2.16),

$$y = x^{p^j} t^{-p^{s-1} + p^{j-1}}.$$

Thus, $x^{p^j - p^r} = t^{2p^{s-1} - p^{j-1} - p^{r-1}}$. From $E(u_1, \tilde{u}_1, v_1, \tilde{v}_1, x+1, y+1, j_1, r_1, s)$ we similarly conclude that

$$(y+1) = (x+1)^{p^{r_1}} t^{p^{s-1} - p^{r_1 - 1}},$$

and so

$$(x+1)^{p^{j_1} - p^{r_1}} = t^{2p^{s-1} - p^{j_1 - 1} - p^{r_1 - 1}}.$$

If on the one hand x is a constant then $s = r = s = j_1 = r_1$ and $y = x^{p^s}$. Suppose, however, that x is not a constant. If $2p^{s-1} - p^{j-1} - p^{r-1} > 0$ then x has a zero at $\mathfrak{p}$ and a pole at $\mathfrak{q}$. Further, we also conclude that $x + 1$ has a pole at $\mathfrak{q}$, $2p^{s-1} - p^{j_1 - 1} - p^{r_1 - 1} > 0$, and $x + 1$ has a zero at $\mathfrak{p}$, which is impossible. We can similarly rule out the case $2p^{s-1} - p^{j-1} - p^{r-1} < 0$. Thus $2p^{s-1} - p^{j-1} - p^{r-1} = 0$, $s = r = s = j_1 = r_1$, and $y = x^{p^s}$. On the other hand, if $y = x^{p^s}$ and then we set $s = r = s = j_1 = r_1$ then we can certainly find $v, \tilde{v}, u, \tilde{u}, v_1, \tilde{v}_1, u_1, \tilde{u}_1 \in K$ to satisfy (8.4.9). □

The following propositions treat the characteristic-2 case.

Lemma 8.4.9. *Let* $p = 2$. *Then for* $x, y = \tilde{y}^2 \in K$, $j, r, s \in \mathbb{N} \setminus \{0\}$, *and* $u, \tilde{u} \in K$ *there exist* $v, \tilde{v} \in K$ *such that*

$$E2(u, \tilde{u}, v, \tilde{v}, x, y, j, r, s) \tag{8.4.10}$$

holds if and only if there exist $v_1, \tilde{v}_1 \in K$ *such that*

$$E2(u_1, \tilde{u}_1, v_1, \tilde{v}_1, x, \tilde{y}, j-1, r-1, s-1) \tag{8.4.11}$$

holds.

Proof. Suppose that, on the one hand, for some $x, y = \tilde{y}^2 \in K, j, r, s \in \mathbb{N} \setminus \{0\}$, and $u, \tilde{u} \in K$, we have that (8.4.10) holds. Then from (8.2.21) we derive

$$v = \frac{\tilde{y}^4 + t^{2^{s+1}} + t^{2^s}}{\tilde{y}^4 + t^{2^s}} = \left(\frac{\tilde{y}^2 + t^{2^s} + t^{2^{s-1}}}{\tilde{y}^2 + t^{2^{s-1}}}\right)^2. \tag{8.4.12}$$

Thus, if we set

$$v_1 = \frac{\tilde{y}^2 + t^{2^s} + t^{2^{s-1}}}{\tilde{y}^2 + t^{2^{s-1}}}, \tag{8.4.13}$$

we conclude that $v_1 = u^{2^{j-1}}$. Similarly, if we set

$$\tilde{v}_1 = \frac{\tilde{y}^2 + t^{-2^s} + t^{-2^{s-1}}}{\tilde{y}^2 + t^{-2^{s-1}}} \tag{8.4.14}$$

then $\tilde{v}_1 = \tilde{u}^{2^{r-1}}$. Thus (8.4.11) holds. On the other hand, it is clear that if for some $x, y = \tilde{y}^2 \in K, j, r, s \in \mathbb{N} \setminus \{0\}$, and $u, \tilde{u} \in K$ there exist $v_1, \tilde{v}_1 \in K$ such that (8.4.11) holds, then by setting $v = v_1^2, \tilde{v} = \tilde{v}_1^2$ we will ensure that (8.4.10) holds. □

Proposition 8.4.10. *Let $p = 2$. Then for $x, y \in K, s \in \mathbb{N}$ there exist $j, r \in \mathbb{N}$, and $u, \tilde{u}, v, \tilde{v} \in K$ such that (8.4.10) holds if and only if $y = x^{2^s}$.*

Proof. Since $E2(u, \tilde{u}, v, \tilde{v}, x, y, j, r, s)$ holds, from (8.2.17), (8.2.19), and (8.2.21) we conclude that

$$y^2 = (x^{2^{r+1}} t^{2^{s+1}} + t^{2^r + 2^{s+1}} + t^{2^r + 2^{s+1}}) t^{-2^{r+1}}. \tag{8.4.15}$$

Similarly, from the same three equations we conclude that

$$y^2 = (x^{2^{j+1}} t^{-2^{s+1}} + t^{-2^j - 2^{s+1}} + t^{-2^j - 2^{s+1}}) t^{2^{j+1}}. \tag{8.4.16}$$

Thus to show that (8.4.10) implies $y = x^{2^s}$ we simply need to show that $s = r$ or $s = j$.

By Lemma 8.4.9 it is enough to consider two cases: the case where one of r, j, s is equal to zero and the case where y is not a square. First suppose that $s = 0$. Then

$$v = \frac{y^2 + t^2 + t}{y^2 + t}$$

and v is not a square, since

$$\frac{dv}{dt} = \frac{t^2}{y^4 + t^2} \neq 0.$$

But $v = u^{2^r}$ and therefore $r = 0 = s$.

Suppose now that $r = 0$. Then $v = u$ and therefore v is not a square by an argument similar to the one above. However, if $s > 0$ then v is a square. Thus to avoid contradiction we must conclude that $s = 0$.

Finally, if $j = 0$ then $\tilde{v} = \tilde{u}$ is not a square and $s = 0$ again. Thus we have reduced the problem to the case where y is not a square and r, s, j are positive. In this case, from (8.4.15), taking the square root of both sides of the equation we obtain

$$y = \left(x^{2^r} t^{2^s} + t^{2^{r-1}+2^s} + t^{2^r+2^{s-1}}\right)t^{-2^r}, \tag{8.4.17}$$

implying that, unless either $r = 1, s > 1$ or $s = 1, r > 1$, we have that y is a square.

Similarly, from (8.4.16), taking the square root we obtain

$$y = \left(x^{2^j} t^{-2^s} + t^{-2^{j-1}-2^s} + t^{-2^j-2^{s-1}}\right)t^{2^j}, \tag{8.4.18}$$

implying that, unless either $j = 1, s > 1$ or $s = 1, j > 1$, y is again a square.

Thus we have $s = 1, r > 1, j > 1$, or $r = j = 1, s > 1$, or y is a square. First, suppose that $s = 1$. Then eliminating y from (8.4.17) and (8.4.18) and substituting 1 for s yields

$$\left(x^{2^r} t^2 + t^{2^{r-1}+2} + t^{2^r+1}\right)t^{-2^r} = \left(x^{2^j} t^{-2} + t^{-2^{j-1}-2} + t^{-2^j-1}\right)t^{2^j},$$
$$x^{2^r} t^{2-2^r} + t^{-2^{r-1}+2} + t = x^{2^j} t^{2^j-2} + t^{2^{j-1}-2} + t^{-1}.$$

Then $t + 1/t$ is a square in K. This is impossible, since this element is not a square in $C_K(t)$ and the extension $K/C_K(t)$ is separable.

Suppose now that $r = j = 1, s > 1$. Eliminating y from (8.4.17) and (8.4.18) and substituting 1 for j and r produces

$$t^{2^s-2}x^2 + t^{2-2^s}x^2 = t^{2^s-1} + t^{2^{s-1}} + t^{1-2^s} + t^{-2^{s-1}}. \tag{8.4.19}$$

This equation implies that $t^{2^s-1} + t^{1-2^s}$ is a square in K. But

$$\text{ord}_{\mathfrak{p}}\left(t^{2^s-1} + t^{1-2^s}\right) = \text{ord}_{\mathfrak{p}} t^{1-2^s} = 1 - 2^s$$

is odd and $\mathfrak{p}$ is not ramified in the extension $K/C_K(t)$. Thus we have a contradiction again.

Hence, if y is not a square, then $r = s = j = 0$. Thus we have shown that (8.4.10) implies $y = x^{2^s}$. Conversely, if $y = x^{2^s}$ then we can set $j = r = s$ and (8.4.10) will be satisfied. □

8.4.11. The proof of Theorem 8.2.1 completed To complete the proof of Theorem 8.2.1 we note the following. Let M be any function field over a finite field of constants. Let K be a finite constant extension of M satisfying the conditions listed in 8.2.9. Such an extension exists by Theorem 8.2.3. Then the set $P(K)$ is Diophantine over K. However, $P(M) = P(K) \cap K^2$ and therefore, by the "going up and then down" method, $P(M)$ is Diophantine over M.

We end this section with two corollaries that we will use in Chapter 10, on the Diophantine classes of function fields.

Lemma 8.4.12. *Let $t \in K \setminus \{0\}$. Let a be a fixed natural number different from zero. Let $\mathcal{W}$ be any set of primes of K. Then the set*

$$\left\{(x, y) \in O_{K,\mathcal{W}^2} | \exists r \in \mathbb{N} : x = t^{p^r} \wedge y = t^{p^{ar}}\right\}$$

is Diophantine over $O_{K,\mathcal{W}}$.

Proof. We will proceed by induction. Suppose that we can write down a set of polynomial equations specifying that $x = t^{p^r}$ and $y_{a-1} = t^{p^{(a-1)r}}$. Next consider the following set of equations:

$$w_a = t(ty_{a-1} + 1), \tag{8.4.20}$$

$$\exists s \in \mathbb{N}, \quad z_a = w_a^{p^s}, \tag{8.4.21}$$

$$\text{ord}_{\mathfrak{p}} z_a = \text{ord}_{\mathfrak{p}} x, \tag{8.4.22}$$

$$y_a = \frac{1}{x}\left(\frac{z_a}{x} - 1\right). \tag{8.4.23}$$

It is clear, given our inductive assumptions, that equations (8.4.20)–(8.4.23) will have solutions in K if and only if $y_a = t^{p^{ar}}$. Furthermore, by the discussion above and Theorem 4.2.4, equations (8.4.21) and (8.4.22) can be rewritten in a polynomial form. □

Corollary 8.4.13. *Let $x, t \in K$. Then the set*

$$\left\{(x, y, w, t) \in K^4 \mid \exists r \in \mathbb{N} : y = x^{p^r}, w = t^{p^r}\right\}$$

is Diophantine over K. (The proof is similar to the proof of Lemma 8.4.12.)

8.5 Diophantine model of $\mathbb{Z}$ over function fields over finite fields of constants

Using pth power equations and the fact that we can assert integrality at finitely many primes using polynomial equations, we now show the Diophantine undecidability of algebraic function fields over finite fields of constants, by constructing a Diophantine model of $\mathbb{Z}$ over such fields. The construction we present here is based on the results by Cornelissen and Zahidi from [6]. We also remind the reader that a definition of Diophantine models and their relation to Diophantine undecidability can be found in Definition 3.4.3 and Proposition 3.4.4. We start by introducing the notation for this section.

Notation 8.5.1.

- Let M be a function field over a finite field of constants of characteristic $p > 0$.
- Let $P(M) = \{(x, y) : \exists s \in \mathbb{N},\ y = x^{p^s}\}$.
- Let $\mathfrak{p}$ be any prime of M.
- Let $INT(\mathfrak{p}) = \{x \in M : \text{ord}_{\mathfrak{p}}x \geq 0\}$.
- Let $DIV(\mathfrak{p}) = \{(x, y) \in K^2 : \text{ord}_{\mathfrak{p}}x\,|_p\text{ord}_{\mathfrak{p}}y\}$
- Let $MULT(M) = \{(x, y, z) \in M^3 : (\text{ord}_{\mathfrak{p}}x)(\text{ord}_{\mathfrak{p}}y) = \text{ord}_{\mathfrak{p}}z\}$.
- Let $ADD(M) = \{(x, y, z) \in M^3 : \text{ord}_{\mathfrak{p}}x + \text{ord}_{\mathfrak{p}}y = \text{ord}_{\mathfrak{p}}z\}$.

Proposition 8.5.2. *$DIV(\mathfrak{p})$ is Diophantine over M.*

Proof. For any $(x, y) \in M^2$ we have $(x, y) \in DIV(\mathfrak{p})$ if and only if there exists $z \in M$ such that $(x, z) \in P(M)$ and $\{y/z,\ z/y\} \subset INT(\mathfrak{p})$. Now $P(M)$ is Diophantine over M by Theorem 8.2.1. $INT(\mathfrak{p})$ is Diophantine over M by Theorem 4.3.4. Thus the proposition holds. □

Proposition 8.5.3. *$MULT(M)$ and $ADD(M)$ are Diophantine over M.*

Proof. Note that $(x, y, z) \in ADD(M)$ if and only if $\{xy/z,\ z/xy\} \subset INT(\mathfrak{p})$. Thus $ADD(M)$ is Diophantine over M. Now by Theorem 8.1.9 the fact that $ADD(M)$, $INT(\mathfrak{p})$, and $DIV(\mathfrak{p})$ are Diophantine over M implies that

$$MULT^+(M) = MULT(M) \cap INT(\mathfrak{p})^3$$

is Diophantine over M. We need to take care of elements which have negative orders at $\mathfrak{p}$. Observe that for any $(x, y, z) \in M^3$, we have that $(x, y, z) \in MULT(M)$ if and only if one of the following statements is true: $(x, y, z) \in MULT^+(M)$; $(1/x,\ y,\ 1/z) \in MULT^+(M)$; $(x,\ 1/y,\ 1/z) \in MULT^+(M)$; $(1/x,\ 1/y,\ z) \in MULT^+(M)$. □

Corollary 8.5.4. *Let $A \subset \mathbb{Z}^k$ be a Diophantine subset of $\mathbb{Z}^k$. Then $\{(z_1, \ldots, z_k) \in M^k : (\text{ord}_{\mathfrak{p}}z_1, \ldots, \text{ord}_{\mathfrak{p}}z_k) \in A\}$ is a Diophantine subset of M^k.*

Proof. The inductive argument necessary to prove this corollary is similar to the argument used to prove Proposition 3.4.7 and Lemma 3.2.2. □

Theorem 8.5.5. *M has a Diophantine model of* $\mathbb{Z}$.

Proof. First of all we observe that by Corollary A.7.8 it is the case that M is recursive. Therefore, by Proposition 3.4.7 it is enough to construct a map from $\mathbb{Z}$ into M making the image of the graph of multiplication and addition Diophantine. Consider a map $\phi : \mathbb{Z} \to M$ defined by $\phi(k) = t^{p^k}$ if $k \geq 0$ and $\phi(k) = t^{-p^{-k}}$ otherwise. From Section 8.3, we know that $\phi(\mathbb{Z})$ is a Diophantine subset of M. Next consider a subset of $\mathbb{Z}$ consisting of the pairs of integers $A = \{(k, p^k) : k \geq 0\} \cup \{(k, -p^{-k}) : k < 0\}$. By Corollary A.1.6, A is recursive and therefore it is r.e. by Lemma A.2.2. Thus by Theorem 1.2.2 we have that A is Diophantine. Hence by Corollary 8.5.4 the set

$$M(A) = \{(u, v) \in M^2 : (\text{ord}_{\mathfrak{p}} u, \text{ord}_{\mathfrak{p}} v) \in A\}$$

is Diophantine over M. Next consider the following sets:

$$B_\times = \{(x, y, z) \in \phi(\mathbb{Z})^3 : \exists x_1, y_1, z_1, \{(x_1, x), (y_1, y), (z_1, z)\} \subset M(A), \\ (x_1, y_1, z_1) \in MULT(M)\},$$

and

$$B_+ = \{(x, y, z) \in \phi(\mathbb{Z})^3 : \exists x_1, y_1, z_1, \{(x_1, x), (y_1, y), (z_1, z)\} \subset M(A), \\ (x_1, y_1, z_1) \in ADD(M)\}.$$

Observe that the sets $B_\times$ and B_+ are both Diophantine subsets of M^3, $B_\times$ being the ϕ-image of the graph of multiplication while B_+ is the ϕ-image of the graph of addition. □

9

Bounds for function fields

In this chapter we will discuss some bound equations specialized for function fields. These bounds will be used in the next chapter in our discussion of Diophantine classes of function fields. Some methods used below should be familiar to the reader from Chapter 5.

9.1 Height bounds

In this section we will consider how to obtain information about the height of a function, given information on the height of a polynomial evaluated at this function. We also compare the height of the coordinates of a field element with respect to a chosen basis and the height of the element itself. (The reader is reminded that the definition of the height of a function field element can be found in B.1.25.)

Lemma 9.1.1. *Let K be a function field and let $F(T) \in K[T]$ be a polynomial of degree greater than or equal to* 1. *Let $H_K(x)$ denote the height of x in K. Then there exists a positive constant C_F, depending on $F(T)$ only, such that for all $x \in K$ we have that $H_K(x) \leq C_F \cdot (H_K(F(x))$.*

Proof. Since the case where the degree of $F(T)$ is equal to 1 is obvious, we will assume that the degree of $F(T)$ is greater than 1. Let

$$F(T) = A_0 + A_1T + \cdots + A_nT^n, \qquad n > 1.$$

Let C_{1F} be the maximum of the heights of the coefficients of $F(T)$. Let $\mathfrak{p}$ be a pole of x such that either $\mathfrak{p}$ is not a zero or a pole of any coefficient of F or $|\text{ord}_{\mathfrak{p}}\, x| > 2C_{1F}$. Then $|\text{ord}_{\mathfrak{p}}\, F(x)| \geq |\text{ord}_{\mathfrak{p}}\, x|$, where $\mathfrak{p}$ is the pole of both x

and $F(x)$. Indeed, in this case, for any $i = 1, \ldots, n-1$ we have that

$$|\mathrm{ord}_{\mathfrak{p}} A_n x^n| > |\mathrm{ord}_{\mathfrak{p}} A_{n-i} x^{n-i}|.$$

This is true because

$$|\mathrm{ord}_{\mathfrak{p}} A_n x^n| \geq n\,|\mathrm{ord}_{\mathfrak{p}} x| - |\mathrm{ord}_{\mathfrak{p}} A_n| \geq n\,|\mathrm{ord}_{\mathfrak{p}} x| - C_{1F},$$

while

$$|\mathrm{ord}_{\mathfrak{p}} A_{n-i} x^{n-i}| \leq (n-i)\,|\mathrm{ord}_{\mathfrak{p}} x| + |\mathrm{ord}_{\mathfrak{p}} A_{n-i}| \leq (n-i)\,|\mathrm{ord}_{\mathfrak{p}} x| + C_{1F}.$$

Thus

$$|\mathrm{ord}_{\mathfrak{p}} A_n x^n| - |\mathrm{ord}_{\mathfrak{p}} A_{n-i} x^{n-i}| \geq |\mathrm{ord}_{\mathfrak{p}} x| - 2C_{1F} > 0.$$

Therefore

$$|\mathrm{ord}_{\mathfrak{p}} F(x)| \geq |n\,\mathrm{ord}_{\mathfrak{p}} x| - C_{1F} \geq |\mathrm{ord}_{\mathfrak{p}} x|.$$

Let $\mathcal{P}$ be the set of primes of K satisfying the conditions described above. That is, if $\mathfrak{p} \in \mathcal{P}$ then $\mathfrak{p}$ is a pole of x such that either it is not a zero or a pole of any coefficient of $F(T)$ or the absolute value of the order of the pole is greater than $2C_{1F}$. Note that the number of primes of K which are not in $\mathcal{P}$ but are poles of x is less or equal to $2(n+1)C_{1F}$ and their degrees are also bounded by C_{1F}. Then we have the following inequality:

$$\begin{aligned}
H_K(x) &= \left|\sum_{\mathfrak{p} \text{ is a pole of } x,} \deg \mathfrak{p}\,\mathrm{ord}_{\mathfrak{p}} x\right| \\
&= \left|\sum_{\mathfrak{p}\in\mathcal{P}} \deg \mathfrak{p}\,\mathrm{ord}_{\mathfrak{p}} x\right| + \left|\sum_{\mathfrak{p} \text{ is a pole of } x,\, \mathfrak{p}\notin\mathcal{P}} \deg \mathfrak{p}\,\mathrm{ord}_{\mathfrak{p}} x\right| \\
&\leq \left|\sum_{\mathfrak{p}\in\mathcal{P}} \deg \mathfrak{p}\,\mathrm{ord}_{\mathfrak{p}} F(x)\right| + \sum_{\mathfrak{p} \text{ is a pole of } x,\, \mathfrak{p}\notin\mathcal{P}} 2(C_{1F})^2 \\
&\leq H_K(F(x)) + 4(n+1)(C_{1F})^3 \\
&\leq 4(n+1)(C_{1F})^3 H_K(F(x)) = C_F H_K(F(x)).
\end{aligned}$$

□

Lemma 9.1.2. *Let $C_1(t)/C_2(t)$ be a constant field extension of rational function fields over finite fields of constants. Let $r = [C_1(t) : C_2(t)]$. Let $w \in C_1(t)$ be of $C_1(t)$-height h. Then $w = f/g$, where $g \in C_2(t)$, $H_{C_1(t)}(f) \leq rh$, and $H_{C_2(t)}(g) = H_{C_1(t)}(g) \leq rh$.*

Proof. First of all, we note that since the constant extension is separable, by Lemma B.4.21, the $C_1(t)$-height and $C_2(t)$-height of an element of the smaller field will be the same. Next write $w = u/v$, where u and v are relatively prime polynomials in t. Then

$$\max\left(H_{C_1(t)}(u), H_{C_1(t)}(v)\right) \leq H_{C_1(t)}(w) = h.$$

Let $v = v_1, \ldots, v_r$ be all the conjugates of v over $C_2(t)$. Then, on the one hand, $\prod_{i=1}^{r} v_i \in C_2(t)$ and for all i we have that $H_{C_1(t)}(v) = H_{C_1(t)}(v_i)$. On the other hand, $H_{C_1(t)}(uv_2 \cdots v_r) \leq rh$ and $H_{C_1(t)}(vv_2 \cdots v_r) \leq rh$. □

The following lemma is quite similar to Corollary 5.2.2. We leave the proof to the reader.

Lemma 9.1.3. *Let K/L be a finite separable extension of function fields. Let K_N be the Galois closure of K over L. Let $h \in K$ and let $h = h_1, \ldots, h_k \in K_N$ be all the conjugates of h over L. Let $\Omega = \{\omega_1, \ldots, \omega_k\}$ be a basis of K over L and assume that $h = \sum_{i=0}^{k-1} f_i\omega_i$, where $f_i \in L$ for $i = 0, \ldots, k-1$. Then $H_{K_N}(f_i) \leq kaH_{K_N}(h)$, where $a = a(\Omega)$ is a positive constant depending on the basis elements only.*

9.2 Using pth powers to bound the height

Lemma 9.2.1. *Let $\mathcal{W}$ be a collection of primes of a function field K such that in some finite extension M of K only finitely many primes of $\mathcal{W}$ have relative-degree-1 factors. Let $F(T)$ be a polynomial over K as described in Lemma B.4.19. Let $t \in K$ be such that all the poles of t are among the primes of $\mathcal{W}$. Let $x \in K$. Let k, m be arbitrary non-negative integers. Then the following statements are true.*

1. *$\left(t^{p^k} - t\right)^{p^m}/F(x) \in O_{K,\mathcal{W}}$ implies that all the poles of $\left(t^{p^k} - t\right)/F(x)$ are poles of t and all the zeros of $F(x)$ are among the zeros of $t^{p^k} - t$.*
2. *For every $x \in K$ there exist $k, m \in \mathbb{N}$ such that $\left(t^{p^k} - t\right)^{p^m}/F(x) \in O_{K,\mathcal{W}}$.*
3. *If $\left(t^{p^k} - t\right)^{p^m}/F(x) \in O_{K,\mathcal{W}}$ then $H_K(x) \leq C_F p^{(k+m)} H_K(t)$.*

Proof.

1. Any pole of $\left(t^{p^k} - t\right)^{p^m}/F(x)$ is either a pole of t or a zero of $F(x)$. But by Lemma B.4.18 none of the zeros of $F(x)$ is in $\mathcal{W}$, while the quotient $\left(t^{p^k} - t\right)^{p^m}/F(x) \in O_{K,\mathcal{W}}$ and thus must have poles at primes of $\mathcal{W}$ only. Therefore all the poles of the quotient must come from poles of t.

2. If $\mathfrak{p}$ is a prime of K which is not a pole of t then for some $k \in \mathbb{N}$ it is the case that $\mathfrak{p}$ is a zero of $t^{p^k} - t$.
3. By Lemma 9.1.1,

$$H_K(x) \leq C_F H_K(F(x)) \leq C_F H_K\left(t^{p^k} - t\right)^{p^m} = C_F p^{(m+k)} H_K(t).$$

□

Corollary 9.2.2. *Let K be a function field over a finite field of constants of characteristic $p > 0$. Let $\mathcal{W}_K$ be a set of primes of K without relative-degree-1 factors in some finite separable extension of K. Let $a, b \in \mathbb{N}$ and assume that $t \in O_{K,\mathcal{W}_K}$. Then there exists a set $A \subset O_{K,\mathcal{W}} \times O_{K,\mathcal{W}}$ such that the following statements are true.*

1. *A is Diophantine over $O_{K,\mathcal{W}_K}$.*
2. *$(x, y) \in A \Rightarrow \exists r \in \mathbb{N}, y = t^{p^r}$ and $H_K(y) > aH_K(x) + b$.*
3. *For every $x \in O_{K,\mathcal{W}}$ there exists $y \in O_{K,\mathcal{W}}$ such that $(x, y) \in A$.*

Proof. Let $\mathfrak{p} \in \mathcal{W}_K$ be a pole of t. Let $s \in \mathbb{N}$ be such that $p^s > C_F H_K(t)$, where $F(T)$ is as defined in Lemma B.4.19 and C_F is as defined in Lemma 9.1.1. Next let A be a set of pairs $(x, y) \in O^2_{K,\mathcal{W}}$ such that there exist $k, m \in \mathbb{Z}_{>0}$, $z \in O_{K,\mathcal{W}}$ satisfying the following equations:

$$z = \frac{\left(t^{p^k} - t\right)^{p^m}}{F\left(x^{a+b}\right)}, \tag{9.2.1}$$

$$y = t^{p^{k+m+s}}. \tag{9.2.2}$$

Suppose that (9.2.1) and (9.2.2) are satisfied over $O_{K,\mathcal{W}}$. Then

$$H_K\left(F\left(x^{a+b}\right)\right) \leq H_K\left(\left(t^{p^k} - t\right)^{p^m}\right) = H_K\left(t^{p^{k+m}}\right)$$

and, by Lemma 9.1.1,

$$\begin{aligned} aH_K(x) + b &\leq (a + b)H_K(x) = H_K\left(x^{(a+b)}\right) \\ &\leq C_F H_K\left(t^{p^{k+m}}\right) = C_F p^{(k+m)} H_K(t) < p^{(k+m)} p^s < H_K(y). \end{aligned}$$

Finally we note that A is Diophantine over $O_{K,\mathcal{W}_K}$ by Lemma 8.4.13. □

10

Diophantine classes over function fields

Having resolved the issues of Diophantine decidability over global function fields, we turn our attention to Diophantine definability over these fields. Our goal for this chapter is to produce vertical and horizontal definability results for "large" rings of functions, as we did for "large" number rings. The original results discussed in this chapter can be found in [93], [100] and [103].

We start with a function field version of the weak vertical method.

10.1 The weak vertical method revisited

In this section we revisit the weak vertical method and adjust it for function fields. As will be seen below, very little "adjusting" will be required.

Theorem 10.1.1. The weak vertical method for function fields *Let K/L be a finite separable extension of function fields over finite fields of constants and let K_N be the normal closure of K over L. Let $\{\omega_1 = 1, \ldots, \omega_k\}$ be a basis of K over L. Let $z \in K$. Further, let $\mathcal{V}$ be a finite set of primes of K satisfying the following conditions.*

1. *Each prime of $\mathcal{V}$ is unramified over L and is the only K-factor of the prime below it in L.*
2. *z is integral at all the primes of $\mathcal{V}$.*
3. *For each $\mathfrak{p} \in \mathcal{V}$ there exists $b(\mathfrak{p}) \in C_L$ (the constant field of L) such that*

$$z - b(\mathfrak{p}) \equiv 0 \bmod \mathfrak{p}.$$

4. *$\{\omega_1, \ldots, \omega_k\}$ is a local integral basis with respect to every prime of $\mathcal{V}$.*
5. *$|\mathcal{V}| > ka(\Omega)H_{K_N}(z)$, where $a(\Omega)$ is a constant defined as in Lemma 9.1.3.*

Then $z \in L$.

Proof. Write $z = \sum_{i=1}^{k} f_i\omega_i$, where for all $i = 1, \ldots, k$ we have that $f_i \in L$. By assumption, $\{\omega_1, \ldots, \omega_k\}$ is a local integral basis for all the primes in $\mathcal{V}$. By the strong approximation theorem B.2.1 there exists $f \in L$ such that, for every $\mathfrak{p} \in \mathcal{V}$, we have that $f \equiv b(\mathfrak{p}) \bmod \mathfrak{p}$. Thus $z - f$ will be zero modulo every prime of $\mathcal{V}$. At the same time, $z - f = (f_1 - f) + f_2\omega_2 + \cdots + f_k\omega_k$. By Lemma B.4.12 and an argument similar to the one used in Corollary 6.1.5, for $i = 2, \ldots, k$ and for all $\mathfrak{p} \in \mathcal{V}$, we have that $\text{ord}_{\mathfrak{p}} f_i > 0$. Furthermore, by Lemma 9.1.3 we have that

$$H_K(f_i) \leq H_{K_N}(f_i) \leq akH_{K_N}(z) < |\mathcal{V}|.$$

Thus, unless $f_2 = \cdots = f_k = 0$, we have a contradiction. But if $f_2 = \cdots = f_k = 0$ then $z \in L$. □

Using the weak vertical method we will prove the following theorem concerning vertical Diophantine definability.

Theorem 10.1.2. *Let K/L be a finite separable extension of global fields. Then the following statements are true.*

1. *For any $\varepsilon > 0$ there exists a set $\mathcal{W}_K$ of primes of K of density greater than $1 - \varepsilon$ such that $L \cap O_{K,\mathcal{W}_K} \leq_{Dioph} O_{K,\mathcal{W}_K}$.*
2. *For any $\varepsilon > 0$ there exists a set $\mathcal{W}_L$ of primes of L of density greater than $1 - \varepsilon$ such that $O_{L,\mathcal{W}_L}$ has a Diophantine definition in its integral closure in K.*

The proof of the theorem is contained in Sections 10.2–10.4. The overall plan for the proof is very similar to the plan we used for number fields. Essentially, as over number fields, it is enough to take care of the cyclic case. The only difference here is that we will have two kinds of cyclic extensions: non-constant cyclic extensions (with no change in the constant field) and constant extensions (which will automatically be cyclic).

10.2 The weak vertical method applied to non-constant cyclic extensions

In this section we will consider a cyclic extension of global function fields over the same field of constants. We will start by fixing the notation for this section.

Notation 10.2.1.

- Let G denote an algebraic function field of positive characteristic $p > 0$.
- Let $\mathcal{P}(G)$ be the set of all primes of G.

- Let $\mathcal{W}_G \subset \mathcal{P}(G)$ be such that, in some finite separable extension of G, all but possibly finitely many primes of $\mathcal{W}_G$ have no factors of relative degree 1.
- Let F be a subfield of G such that
  (a) both G and F have the same field constant C, which is of size p^r for some positive integer r;
  (b) G/F is a cyclic extension.
- Let t denote a non-constant element of F such that t is not a pth power and $t \in O_{G,\mathcal{W}_G}$.
- Let $\mathcal{Z}(t)$ be the set of all G-primes occurring in the divisor of t.
- Let $d = [G : C(t)]$.
- Let $\Omega = \{1, \alpha, \ldots, \alpha^{m-1}\}$ be a basis of G over F.
- Let $q \neq p$ be a rational prime and assume that $q > d, (q, r) = 1, (q, m) = 1$.
- Let β be an element of the algebraic closure of C such that $[C(\beta) : C] = q$.
- Let $n(\alpha)$ be the constant defined in Lemma B.4.34. This constant depends on α and K only.
- Let $a(\Omega)$ be as in Lemma 9.1.3. This constant depends on the power basis of α only.
- Let g_F, g_G be the genuses of F and G respectively.
- Let b be a fixed positive integer such that

$$b > 2\log_p 2(m + 4g_G + 3mg_F + 1 + 2dm).$$

Next we prove an easy lemma which makes the construction work.

Lemma 10.2.2. *Let $f \in C[t]$. Let l be a positive integer such that $l \equiv 0 \bmod r$. Then*

$$\frac{f^{p^l} - f}{t^{p^l} - t} \in C[t].$$

Proof. Let $f(t) = \sum_{i=0}^{k} a_i t^i$, $a_i \in C$. Then, since $l \equiv 0 \bmod r \Rightarrow a_i^{p^l} = a_i$, we have

$$f^{p^l} - f = \sum_{i=0}^{k} a_i^{p^l} t^{ip^l} - \sum_{i=0}^{k} a_i t^i = \sum_{i=0}^{k} a_i \left(t^{ip^l} - t^i\right)$$
$$= \left(t^{p^l} - t\right) \sum_{i=0}^{k} a_i \frac{\left(t^{p^l}\right)^i - t^i}{t^{p^l} - t}.$$

Thus the lemma holds. □

Lemma 10.2.3. *Suppose that the following equations and inequalities are satisfied for some $x \in G$ and some $u \in \mathbb{N} \setminus \{0\}$:*

$$x_j = tx^q + x^j, \qquad j = 0, 1, \tag{10.2.1}$$

$$H_G\big(t^{p^u}\big) > \max\big\{2mH_G(t)\big(ma(\Omega)\big(qH_G(x) + H_G(t)\big) + 2d + n(\alpha)\big), H_G\big(t^{p^b}\big)\big\} \tag{10.2.2}$$

$$l = (qu+1)r, \tag{10.2.3}$$

and either

$$\mathrm{ord}_{\mathfrak{P}} \frac{x_j^{p^l} - x_j}{t^{p^l} - t} \equiv 0 \bmod q, \tag{10.2.4}$$

or

$$\mathrm{ord}_{\mathfrak{P}} \frac{x_j^{p^l} - x_j}{t^{p^l} - t} \geq 0, \tag{10.2.5}$$

for all G-primes $\mathfrak{P}$ not splitting in the extension $G(\beta)/G$. Then $x \in F$.

Proof. Suppose that (10.2.1)–(10.2.5) are satisfied, as indicated in the statement of the lemma. Then from (10.2.1) we conclude that for all $\mathfrak{p} \in \mathcal{P}(G) \setminus \mathcal{Z}(t)$ such that $\mathrm{ord}_{\mathfrak{p}} x_j < 0$ we have that $\mathrm{ord}_{\mathfrak{p}} x_j \equiv 0 \bmod q$. Further, observe that in G all the zeros of $t^{p^l} - t$ are of order at most $d < q$ by Proposition B.1.11. Similarly, $|\mathcal{Z}(t)| \leq 2d$ by Lemma B.4.21, taking into account that in $C(t)$ both the zero and the pole divisors of t have degree 1.

Next consider a prime $\mathfrak{p} \in \mathcal{P}(G) \setminus \mathcal{Z}(t)$ such that it does not split in the extension $G(\beta)/G$ and is a zero of $t^{p^l} - t$. If $\mathfrak{p}$ is a pole of x_j or if $x_j^{p^l} - x_j$ is a unit at $\mathfrak{p}$ then

$$\mathrm{ord}_{\mathfrak{p}} \frac{x_j^{p^l} - x_j}{t^{p^l} - t} \equiv -\mathrm{ord}_{\mathfrak{p}}(t^{p^l} - t) \not\equiv 0 \bmod q.$$

Therefore from (10.2.4) and (10.2.5) we can conclude that $x_j^{p^l} - x_j$ has a zero at every $\mathfrak{p}$ such that $\mathfrak{p} \notin \mathcal{Z}(t)$, $\mathrm{ord}_{\mathfrak{p}}(t^{p^l} - t) > 0$, and $\mathfrak{p}$ does not split in the extension $G(\beta)/G$.

Next let C_l be the splitting field of the polynomial $X^{p^l} - X$. Note that by (10.2.3) we have that $l \equiv 0 \bmod r$ and therefore $C \subset C_l$. Let $\mathcal{Z}_l(t)$ be the set of C_lG primes lying above the primes of $\mathcal{Z}(t)$. Observe that $|\mathcal{Z}_l(t)| < 2d$ also, by Lemma B.4.21. Further, $(l, q) = 1$ and, since $p^u > p^b$ by (10.2.2), we also have that

$$l > u > \log_p \left(2 \log_p 2(m + 4g_G + 3mg_F + 1 + 2dm)\right).$$

Therefore the following statements are true.

1. There are more than $p^l/2m$ degree-1 primes of C_lF which do not split in the extension $C_l(\beta)G/C_lF$, by Proposition B.4.31.
2. For any prime $\mathfrak{p}_{C_lG}$ of C_lG, it is the case that $\mathfrak{p}_{C_lG}$ lies above a degree-1 C_lF prime which is not a pole of t if and only if $\text{ord}_{\mathfrak{p}_{C_lG}}(t^{p^l} - t) > 0$.
3. Let $\mathfrak{p}_{C_lG}$ be a prime of C_lG lying above a degree-1 C_lF prime not splitting in the extension $C_l(\beta)G/C_lF$. Let $\mathfrak{p}_G$ be the G-prime below $\mathfrak{p}_{C_lG}$. Then $\mathfrak{p}_G$ does not split in the extension $C(\beta)G/G$, by Proposition B.4.31.

Therefore in C_lG we have that $x_j^{p^l} - x_j$ has a zero at every prime $\mathfrak{p}_{C_lG}$ such that $\mathfrak{p}_{C_lG} \notin \mathcal{Z}_l(t)$, $\text{ord}_{\mathfrak{p}}(t^{p^l} - t) > 0$, and $\mathfrak{p}_{C_lG}$ lies above a degree-1 prime of C_lF that does not split in the extension $C_l(\beta)G/C_lF$. But this means that, for at least $p^l/2m - 2d$ distinct primes $\mathfrak{p}$ of C_lG lying above non-splitting primes of C_lF,

$$\text{ord}_{\mathfrak{p}}(x_j - a(\mathfrak{p})) > 0$$

for some constant $a(\mathfrak{p}) \in C_l \subset C_lF$. Now going back to (10.2.2) and (10.2.3) we observe that

$$\begin{aligned} p^l/2m > p^u/2m &> a(\Omega)m(qH_G(x) + H_G(t)) + 2d + n(\alpha) \\ &\geq a(\Omega)mH_G(x_j) + n(\alpha) + 2d, \end{aligned}$$

so that, by the weak vertical method for function fields (Theorem 10.1.1), we have that $x_j \in C_lF$. Since $x_j \in G$, we must conclude that $x_j \in C_lF \cap G = F$, since in the extension F/G the constant field remains the same. Finally, if $x^qt, x^qt + x \in F$ then $x \in F$. □

Theorem 10.2.4. $O_{G,\mathcal{W}_G} \cap F \leq_{Dioph} O_{G,\mathcal{W}_G}$.

Proof. First of all we observe the following. Let $x \in C_F[t]$. Then, by Corollary 9.2.2, for some u we have that (10.2.2) is satisfied. Now, by Lemma 10.2.2, it is the case that $(x_j^{p^l} - x_j)/(t^{p^l} - t)$ is a polynomial of degree $p^l(q \deg x + 1) - p^l = p^lq \deg x$. Therefore, (10.2.4) and (10.2.5) are satisfied. Next, let γ be a generator of F over $C_F(t)$. Then $x \in O_{G,\mathcal{W}_G} \cap F$ if and only if $x \in O_{G,\mathcal{W}_G}$ and $x = \sum_{i=0}^{d-1} (z_i/v_i)\gamma^i$, where $z_i, v_i \in C_F(t)$ and $v_i \neq 0$.

The only remaining task is to rewrite (10.2.1)–(10.2.5) in a Diophantine fashion. We can rewrite (10.2.2) using Corollary 9.2.2. Next, given x, by

Lemma 8.4.12 and Corollary 8.4.13 the set $\{(t^{p^u}, t^{p^l}, x_j^{p^l}), l = (qu+1)r, u \in \mathbb{N}\}$ is Diophantine over G. Finally, (10.2.4) and (10.2.5) can be transformed using Theorem 4.5.2. □

10.3 The weak vertical method applied to constant field extensions

In this section we will consider finite constant extensions of rational function fields. We will again start by describing the notation for the section.

Notation 10.3.1.

- Let C be a finite field of characteristic p.
- Let t be transcendental over C.
- Let $\mathcal{W}_{C(t)}$ be a set of primes of $C(t)$ such that in some finite separable extension M of $C(t)$ only finitely many primes of $\mathcal{W}_{C(t)}$ have relative-degree-1 factors. Assume also that the prime which is the pole of t is included in $\mathcal{W}_{C(t)}$.
- Let $q \neq p$ be a rational prime.
- Let $r = [C : \mathbb{F}_p]$.
- Let β be an element of the algebraic closure of C such that $[C(\beta) : C] = q$.
- Let $\mathfrak{p}_t$ be the zero of t in $C(t)$.

Lemma 10.3.2. *Suppose that the equations and inequalities below are satisfied over $O_{C(t),\mathcal{W}_{C(t)}}$ for some $x \in O_{C(t),\mathcal{W}_{C(t)}}$ and some $u \in \mathbb{N}$. Then $x \in O_{C(t),\mathcal{W}_{C(t)}} \cap \mathbb{F}_p(t)$,*

$$\text{ord}_{\mathfrak{p}_t} x \geq 0, \tag{10.3.1}$$

$$x_j = tx^q + x^j, \qquad j = 0, 1, \tag{10.3.2}$$

$$H_{C(t)}(t^{p^u}) > 2(r+1) \cdot H_{C(t)}(x_j), \tag{10.3.3}$$

$$l = 2qru + 1, \tag{10.3.4}$$

and either

$$\text{ord}_{\mathfrak{p}} \frac{x_j^{p^l} - x_j}{t^{p^l} - t} \equiv 0 \bmod q, \tag{10.3.5}$$

or

$$\text{ord}_{\mathfrak{p}} \frac{x_j^{p^l} - x_j}{t^{p^l} - t} \geq 0, \tag{10.3.6}$$

for all zeros $\mathfrak{p}$ of $t^{p^l} - t$ in $C[t]$ such that $\mathfrak{p}$ does not split in the extension $C(\beta, t)/C(t)$.

Proof. Suppose that the equations and inequalities above have a solution with $l \in \mathbb{N}$ and all the other variables in $O_{C(t),\mathcal{W}_{C(t)}}$. Then by Lemma 9.1.2 we have that $x_j = f/g$, where $g \in \mathbb{F}_p(t)$ and $H_{C(t)}(f) \leq rH_{C(t)}(x_j)$. From (10.3.5), (10.3.6), and Lemma B.4.32 we conclude in the same fashion as in Lemma 10.2.3 that $x_j^{p^l} - x_j$ has zeros at all the primes which are zeros of $t^{p^l} - t$. Thus

$$f^{p^l}g - fg^{p^l}$$

is divisible by $t^{p^l} - t$ in $C[t]$. Since $g \in \mathbb{F}_p[t]$, we have on the one hand that $g^{p^l} - g \equiv 0 \bmod (t^{p^l} - t)$ in $C[t]$, and thus $g(f^{p^l} - f) \equiv 0 \bmod (t^{p^l} - t)$. On the other hand, let f_0 be the polynomial obtained from f^{p^l} by replacing t^{p^l} by t. Clearly, f_0 is of the same degree as f, and $f_0 \equiv f^{p^l} \bmod (t^{p^l} - t)$ and thus $g(f_0 - f) \equiv 0 \bmod t^{p^{l}}$. Therefore, unless $f_0 = f$, the degree of fg is greater than or equal to p^l. But by (10.3.3) we know that $\deg fg < p^l$. Thus $f_0 = f$. This means that every coefficient of f raised to the power p^l is equal to itself. At the same time, since C is of degree r over $\mathbb{F}_p$, every coefficient of f raised to the power p^{qru} is equal to itself. Thus every coefficient of f raised to the pth power is equal to itself and therefore belongs to $\mathbb{F}_p$. □

We can now prove the theorem below, following the same plan as in the proof of Theorem 10.2.4.

Theorem 10.3.3. $\mathbb{F}_p(t) \cap O_{C(t),\mathcal{W}_{C(t)}} \leq_{Dioph} O_{C(t),\mathcal{W}_{C(t)}}$.

10.4 Vertical definability for large subrings of global function fields

In this section we put together Theorems 10.2.4 and 10.3.3 to obtain vertical definability for large subrings of function fields and so complete the proof of Theorem 10.1.2. As in the case of number fields, these vertical results will later lead to a way to define integrality at infinitely many primes in the function field case. We start as usual with a notation list for the section.

Notation 10.4.1.

- Let K/L be a separable extension of global function fields.
- Let $\mathcal{S}_K$ be a finite set of primes of K.

- Let C_K, C_L be the constant fields of K and L respectively.
- Let $t \in L$ be such that it is not a pth power in K.
- Let N be the Galois closure of K over $C_K(t)$ and let C_N be the constant field of N.
- Let $\mathcal{P}(K)$ be the set of all the primes of K.
- Let $\mathcal{W}_K \subset \mathcal{P}(K)$ be such that $t \in O_{K,\mathcal{W}_K} \cap O_{K,\mathcal{S}_K}$.
- Let $\mathcal{W}_N$ be the set of primes of N lying above $\mathcal{W}_K$. (We remind the reader that this assumption implies that $O_{N,\mathcal{W}_N}$ is the integral closure of $O_{K,\mathcal{W}_K}$ in N, by Proposition B.1.22.)
- Let $\mathcal{W}_{C_N(t)}$ be the set of primes of $C_N(t)$ all of whose N-factors are in $\mathcal{W}_N$.
- Let $\mathcal{W}_L$ be the set of primes of L all of whose N-factors are in $\mathcal{W}_N$.
- Assume that in some finite separable extension F of N all but finitely many primes of $\mathcal{W}_N$ have no factors of relative degree 1.
- Let $\{E_i, i = 1, \ldots, l\}$ be the set of fields satisfying the following conditions: $C_N(t) \subset E_i \subset N$; N/E_i is a cyclic extension.

We now prove a more general vertical definability result. A reader might recognize the similarities with the number field case.

Theorem 10.4.2. $O_{K,\mathcal{W}_K} \cap L \leq_{Dioph} O_{K,\mathcal{W}_K}$.

Proof. First of all we observe the following. Since $N/C_K(t)$ is Galois, we also have that $N/C_N(t)$ is Galois. Next let $\sigma \in \text{Gal}(N/C_N(t))$ and observe that the fixed field of σ is one of finitely many cyclic subextensions of N containing $C_N(t)$. Thus, if $x \in \bigcap_{i=1}^{l} E_i$ then we have that x is fixed by all the elements of the Galois group and therefore must be in $C_N(t)$. Further, since for all i it is the case that $C_N(t) \subset E_i$ and C_N is the constant field of N, it follows that the constant field of E_i is also C_N. Thus, by Theorem 10.2.4, for all $i = 1, \ldots, l$ we have that $O_{N,\mathcal{W}_N} \cap E_i \leq_{Dioph} N$.

Using the intersection property of Diophantine generation, we immediately conclude that $O_{N,\mathcal{W}_N} \cap C_N(t) \leq_{Dioph} O_{N,\mathcal{W}_N}$. Next, note that $O_{N,\mathcal{W}_N} \cap C_N(t) = O_{C(t),\mathcal{W}_{C_N(t)}}$. Furthermore, in the extension $FN/C(t)$ none of the primes of $\mathcal{W}_{C_N(t)}$ has a relative-degree-1 factor. Thus by Theorem 10.3.3 we have that $O_{\mathcal{W}_{C_N(t)}} \cap \mathbb{F}_p(t) \leq_{Dioph} O_{\mathcal{W}_{C_N(t)}}$. Using the transitivity of generation, we now conclude that

$$O_{N,\mathcal{W}_N} \cap \mathbb{F}_p(t) \leq_{Dioph} O_{N,\mathcal{W}_N}.$$

Since $t \in O_{K,\mathcal{W}_K} \subset O_{N,\mathcal{W}_N}$, using Diophantine generation of the fraction field and its extensions, we now have

$$L \leq_{Dioph} \mathbb{F}_p(t) \leq_{Dioph} O_{N,\mathcal{W}_N} \cap \mathbb{F}_p(t) \leq_{Dioph} O_{N,\mathcal{W}_N}.$$

Using the fact that $O_{N,\mathcal{W}_N} \leq_{Dioph} O_{N,\mathcal{W}_N}$ and the intersection property again, we get

$$L \cap O_{N,\mathcal{W}_N} \leq_{Dioph} O_{N,\mathcal{W}_N}.$$

Finally, by the Diophantine generation of integral closure,

$$L \cap O_{K,\mathcal{W}_K} = L \cap O_{N,\mathcal{W}_N} \leq_{Dioph} O_{K,\mathcal{W}_K}. \qquad \square$$

As before we can give an estimate of the "size" of the rings to which our results are applicable.

Theorem 10.4.3. *Let $G/\mathbb{F}_p(t)$ be a cyclic extension of prime degree q such that $(q, [N : \mathbb{F}_p(t)]) = 1$. Let $\mathcal{W}_L$ be the set of all L-primes not splitting in the extension LG/L. Let $\mathcal{W}_K$ be the set of all K-primes lying above the primes of $\mathcal{W}_L$. Then*

$$O_{L,\mathcal{W}_L} = O_{K,\mathcal{W}_K} \cap L \leq_{Dioph} O_{K,\mathcal{W}_K}$$

and the Dirichlet density of both prime sets is $1 - 1/q$.

Proof. Let $\mathcal{W}_N$ be the set of N-primes above the primes of $\mathcal{W}_L$. On the one hand, by Lemmas B.4.7 and B.4.8 we have that $\mathcal{W}_N$ and $\mathcal{W}_K$ consist of all the primes of N and K respectively not splitting in the extensions GN/N and KG/K, plus or minus a finite set of primes. Therefore we can apply Theorem 10.4.2 to conclude that $O_{K,\mathcal{W}_K} \cap L \leq_{Dioph} O_{K,\mathcal{W}_K}$. On the other hand, by the Chebotarev density theorem the density of K-primes and the density of L-primes not splitting in the extensions KG/K and LG/L is $1 - 1/q$. $\square$

Remark 10.4.4. Since q can be made arbitrarily large, Theorem 10.4.2 implies Theorem 10.1.2.

Finally we note that the vertical definability results certainly apply to rings of $\mathcal{S}$-integers since the conditions on the prime sets in Theorem 10.4.2 must hold up to a finite set of primes. Thus we also have the following result.

Theorem 10.4.5. $O_{K,\mathcal{S}_K} \cap L \leq_{Dioph} O_{K,\mathcal{S}_K}$.

10.5 Integrality at infinitely many primes over global function fields

In this section we convert our vertical definability results into horizontal definability results using the same methods as over the number fields. The main result which will be proved in this section is stated below.

Theorem 10.5.1. *Let K be a function field over a finite field of constants. Let $\mathcal{S}$ be a finite collection of primes of K. Then for any $\varepsilon > 0$ there exists a set of K-primes $\mathcal{W}$, containing $\mathcal{S}$ and of Dirichlet density greater than $1 - \varepsilon$, such that $O_{K,\mathcal{S}}$ has a Diophantine definition over $O_{K,\mathcal{W}}$.*

The proof of this theorem is derived from a technical result contained in Proposition B.5.6, which is the function field analog of Lemma B.5.5, and Proposition 10.5.3 below. Before we proceed, as usual we need to describe the objects under consideration.

Notation 10.5.2.

- Let K be a function field of characteristic $p > 0$ over a finite field of constants C.
- Let $\mathcal{S}_K$ be a finite collection of primes of K.
- Let $t \in K$ be such that t has poles at all the elements of $\mathcal{S}_K$, has no other poles, and is not a pth power. (Such a t exists by the strong approximation theorem B.2.1.)
- Let ε be a positive real number.
- Let s be a natural number not divisible by p and such that $s > 2/\varepsilon$.
- Let $E = C(t^s)$.
- Let M be a constant extension of E of prime degree $r > 2/\varepsilon$ and such that r does not divide $[K : C(t^s)]!$.
- Let K_G be the Galois closure of K over E.
- Let $\mathcal{V}_{K,1}$ be the set of primes of K splitting completely in the extension MK_G/K.
- Let $\mathcal{W}_{K,1}$ be the set of all the K-primes lying above E-primes splitting completely in the extension K/E.
- Let $\mathcal{Z}_{K,1} = \mathcal{W}_{K,1} \setminus \mathcal{V}_{K,1}$.
- Let $\mathcal{Z}_{E,1}$ be the set of E-primes below the primes $\mathcal{Z}_{K,1}$.
- Let $\mathcal{G}_{K,1}$ be a set of K-primes such that it contains exactly one prime above each prime in $\mathcal{Z}_{E,1}$.
- Let $\mathcal{V}_{K,2}$ be the set of all the primes of K of relative degrees greater than or equal to 2 over E.

- Let $\mathcal{V}_K = \mathcal{V}_{K,1} \cup \mathcal{G}_{K,1} \cup \mathcal{V}_{K,2}$.
- Let $\mathcal{W}_K = \mathcal{P}(K) \setminus \mathcal{V}_K$.
- Let $\tilde{\mathcal{W}}_K = \mathcal{W}_K \cup \mathcal{S}_K$.

The diagram below shows all the fields under consideration.

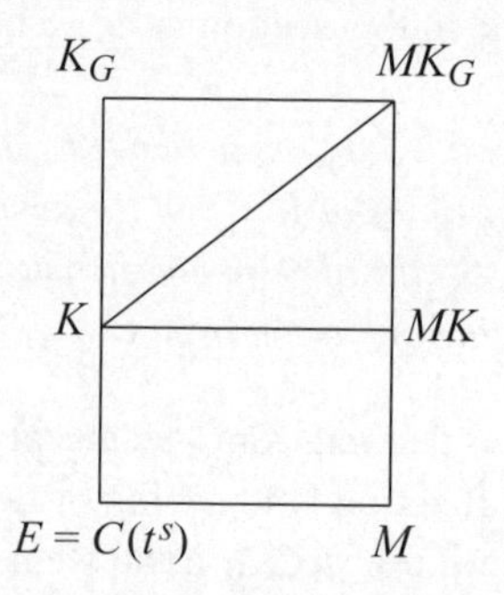

Proposition 10.5.3. *$O_{K,\mathcal{S}_K} \leq_{Dioph} O_{K,\tilde{\mathcal{W}}_K}$ and $\delta(\tilde{\mathcal{W}}_K) > 1 - \varepsilon$.*

Proof. First of all we observe the following. Since t is not a pth power in K and $s \not\equiv 0 \bmod p$, we have that t^s is not a pth power in K. Therefore the extension $K/C(t^s)$ is separable by Lemma B.1.32. Next, by Proposition B.5.6 and by construction, we know that $\tilde{\mathcal{W}}_K$ satisfies the requirements of Theorem 10.4.2. Further, again by Proposition B.5.6 and the construction of $\tilde{\mathcal{W}}_K$, $O_{K,\tilde{\mathcal{W}}_K} \cap E = C[t^s]$. Thus by Theorem 10.4.2 we have that $C[t^s]$ has a Diophantine definition over $O_{K,\tilde{\mathcal{W}}_K}$, or $C[t^s] \leq_{Dioph} O_{K,\mathcal{W}_K}$. At the same time, since $O_{K,\mathcal{S}_K}$ is the integral closure of $C[t^s]$ in K, by Proposition 2.2.1 we have that $O_{K,\mathcal{S}_K} \leq_{Dioph} C[t^s]$. Thus by the transitivity of Dioph-generation we have that $O_{K,\mathcal{S}_K} \leq_{Dioph} O_{K,\tilde{\mathcal{W}}_K}$. Finally, by Theorem B.5.6,

$$\delta(\tilde{\mathcal{W}}_K) = \delta(\mathcal{W}_K) > 1 - \frac{1}{[K:E]} - \frac{1}{[M:E]} > 1 - \varepsilon. \qquad \square$$

10.6 The big picture for function fields revisited

In this section we go back to the Diophantine family of a polynomial ring over a finite field of constants to review what we have learned about Diophantine generation and Diophantine classes within the family. With this plan in mind, consider Figure 10.1.fig 1 It illustrates the facts listed in Proposition 10.6.2 below, most of which we have already established. As before, the arrows signify Diophantine generation. In our cataloging of facts we will proceed from left to

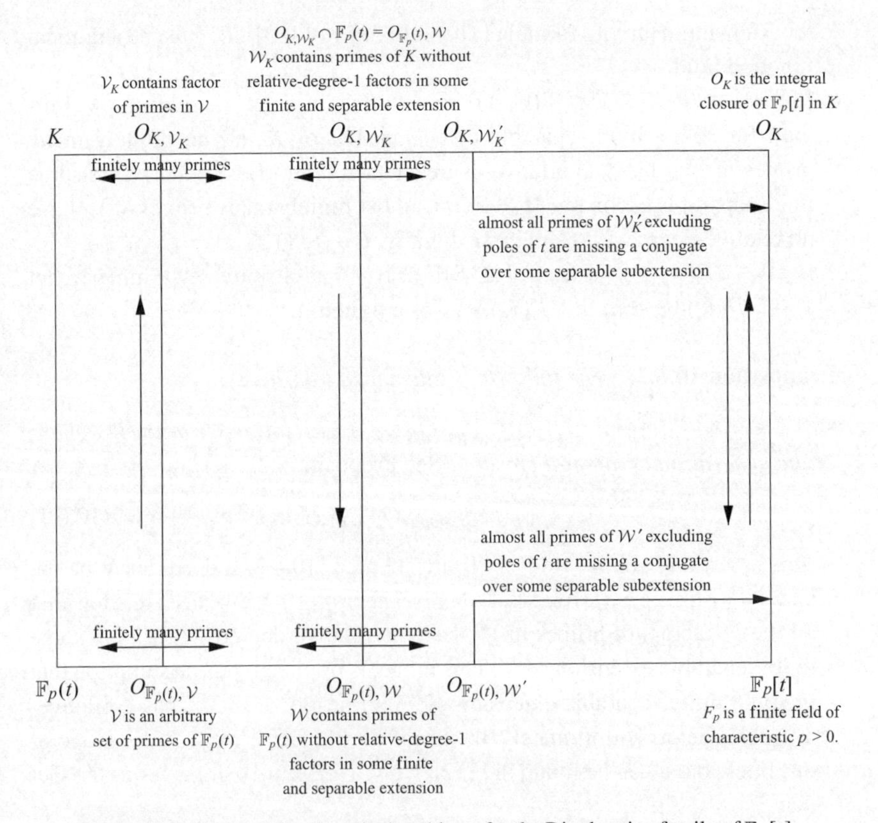

Figure 10.1 Horizontal and vertical problems for the Diophantine family of $\mathbb{F}_p[t]$ revisited.

right, starting with the lower level. First, however, we review the notation and assumptions.

10.6.1. Notation and assumptions

- Let $\mathbb{F}_p$ be a finite field of p elements.
- Let t be transcendental over $\mathbb{F}_p$.
- Let K be a finite separable extension of $\mathbb{F}_p(t)$.
- Let $\mathcal{P}(K)$, $\mathcal{P}(\mathbb{F}_p(t))$ be the sets of all the primes of K and $\mathbb{F}_p(t)$ respectively.
- Let $\mathcal{W}'_K \subset \mathcal{P}(K)$, $\mathcal{W}' \subset \mathcal{P}(\mathbb{F}_p(t))$ be prime sets containing all the poles of t in K and $\mathbb{F}_p(t)$ respectively. Assume further that, for some separable subextension U_K of K, every element of $\mathcal{W}'_K$ has a conjugate over U_K which is not in $\mathcal{W}'_K$. Similarly, for some separable subextension U of $\mathbb{F}_p(t)$, every element of $\mathcal{W}'$ has a conjugate over U which is not in $\mathcal{W}'$. We should note that by

the Riemann–Hurwitz formula (Theorem 3.6.1 of [33]), U must be a rational function field.

- Let $\mathcal{W}_K \subset \mathcal{P}(K)$, $\mathcal{W} \subset \mathcal{P}(\mathbb{F}_p(t))$ be such that $\mathcal{W}'_K \subset \mathcal{W}$ and $\mathcal{W}' \subset \mathcal{W}$. Further, for some finite separable extension M_K of K, all but finitely many primes of $\mathcal{W}_K$ have no relative-degree-1 factors in M_K. Similarly, for some finite separable extension M of $\mathbb{F}_p(t)$, all but finitely many primes of $\mathcal{W}$ have no relative-degree-1 factors in M. Finally, $O_{K,\mathcal{W}_K} \cap \mathbb{F}_p(t) = O_{\mathbb{F}_p(t),\mathcal{W}}$.
- Let $\mathcal{V} \subseteq \mathcal{P}(\mathbb{F}_p(t))$ be such that $\mathcal{P}(\mathbb{F}_p(t)) \setminus \mathcal{V}$ is a finite set. Similarly, let $\mathcal{V}_K \subseteq \mathcal{P}(K)$ be such that $\mathcal{P}(K) \setminus \mathcal{V}_K$ is a finite set.

Proposition 10.6.2. *The following statements are true.*

- *$O_K \leq_{Dioph} \mathbb{F}_p[t]$ (and $O_{K,\mathcal{V}_K} \leq_{Dioph} O_{\mathbb{F}_p(t),\mathcal{V}}$). This follows from the Diophantine generation of integral closure.* (See Proposition 2.2.1.)

$$\mathbb{F}_p[t] \leq_{Dioph} O_{\mathbb{F}_p(t),\mathcal{W}'}. \tag{10.6.1}$$

This assertion "almost" follows from Theorem 10.4.2. Indeed, let N be the Galois closure of $F_p(t)$ over U. To apply Theorem 10.4.2 we need to know that all the N-factors of primes in $\mathcal{W}'$ have no relative-degree-1 factors in some finite separable extension of N. This in fact is implied by the assumption that in some finite separable extension of $F_p(t)$ no prime in $\mathcal{W}'$ has a relative-degree-1 factor. Unfortunately, the proof of this fact is outside the scope of this book, but it can be found in [112]. Now, Theorem 10.4.2 gives us the fact that

$$O_{F_p(t),\mathcal{W}'} \cap U \leq_{Dioph} O_{F_p(t),\mathcal{W}'}.$$

Let $\mathcal{W}_U$ be the set of all primes of U with all their $F_p(t)$-factors in $\mathcal{W}'$. Then by assumption $\mathcal{W}_U$ is a finite set and $O_{F_p(t),\mathcal{W}'} \cap U = O_{U,\mathcal{W}_U}$. Let $O_{F_p(t),\mathcal{S}'}$ be the integral closure of $O_{U,\mathcal{W}_U}$ in $F_p(t)$. Then $\mathcal{S}'$ is a finite set and contains the pole of t. Thus using the Diophantine generation of integral closure and the fact that we can define integrality at finitely many primes, we can conclude that (10.6.1) holds.

- *The Diophantine class of $O_{F_p(t),\mathcal{W}}$ (or $O_{K,\mathcal{W}_K}$) does not change if we add or remove finitely many primes from $\mathcal{W}$ (or $\mathcal{W}_K$).*

This assertion is new and requires an (easy) argument. It is enough to carry out the argument for the case where we remove or add one prime. We will describe the argument for $\mathcal{W}$. An identical argument will also work for $\mathcal{W}_K$. So let $\mathfrak{q} \in \mathcal{W}$ and let $\mathcal{U} = \mathcal{W} \setminus \{\mathfrak{q}\}$. Since we know how to define integrality at finitely many primes in any global function field (see Theorem 4.3.4), and holomorphy rings of global function fields are Dioph-regular (see Note 2.2.5),

$O_{F_p(t),\mathcal{U}} \leq_{Dioph} O_{F_p(t),\mathcal{W}}$. Thus, the only assertion that requires proof is the assertion that

$$O_{F_p(t),\mathcal{W}} \leq_{Dioph} O_{F_p(t),\mathcal{U}}. \tag{10.6.2}$$

By the strong approximation theorem B.2.1 there exists an element $x \in O_{F_p(t),\mathcal{U}}$ such that x has only one zero at $\mathfrak{q}$. Let $y \in O_{F_p(t),\mathcal{W}}$. Then, for some $k \in \mathbb{N}$, we have that $z = x^{p^k} y \in O_{F_p(t),\mathcal{U}}$ and $y = z/x^{p^k}$. Thus

$$O_{F_p(t),\mathcal{W}} = \left\{ \frac{z}{w}, z, w \in O_{F_p(t),\mathcal{U}} \wedge \exists k \in \mathbb{N}, w = x^{p^k} \right\}.$$

Since by Theorem 8.2.1 we can rewrite $\exists k \in \mathbb{N}, w = x^{p^k}$ in a Diophantine fashion, we can conclude that in fact (10.6.2) holds.

- $F_p[t] \leq_{Dioph} O_K$.

This follows from Theorem 10.4.5.

- $O_K \leq_{Dioph} O_{K,\mathcal{W}'_K}$.

Here we can use an argument similar to the argument used to show (10.6.1).

- $O_{F_p(t),\mathcal{W}} \leq_{Dioph} O_{K,\mathcal{W}_K}$.

This assertion follows by Theorem 10.4.2 again with the same caveat as for (10.6.1).

11

Mazur's conjectures and their consequences

In this chapter we explore two conjectures due to Barry Mazur. These conjectures, which are a part of a series of conjectures made by Mazur concerning topology of rational points, have had a very important influence on the development of the subject. The conjectures first appeared in [55], and later in [56], [57], and [58]. They were explored further by among others, Colliot-Thélène, Skorobogatov, and Swinnerton-Dyer in [4], Cornelissen and Zahidi in [6], Pheidas in [70], and the present author in [108]. Perhaps the most spectacular result which has come out of attempts to prove or disprove the conjectures is a theorem of Poonen, which will be discussed in detail in the next chapter. Unfortunately, up to the time of writing, the conjectures are still unresolved.

11.1 The two conjectures

The first conjecture that we are going to discuss states the following.

Conjecture 11.1.1. *Let V be any variety over $\mathbb{Q}$. Then the topological closure of $V(\mathbb{Q})$ in $V(\mathbb{R})$ possesses at most a finite number of connected components.* (Conjecture 2 of [58].)

Remark 11.1.2. Let W be an algebraic set defined over a number field. Then $W = V_1 \cup \cdots \cup V_k$, $k \in \mathbb{N}$, where V_i is a variety and $\bar{W} = \bar{V}_1 \cup \cdots \cup \bar{V}_k$, and where $\bar{W}, \bar{V}_1, \ldots, \bar{V}_k$ denote the topological closures of $W, V_1, \ldots, V_k$ respectively in $\mathbb{R}$ if K is a real field and $\mathbb{C}$ otherwise. Further, if $n_W, n_1, \ldots, n_k$ are the numbers of connected components of $\bar{W}, \bar{V}_1, \ldots, \bar{V}_k$ respectively and $n_i < \infty$ for all $i = 1, \ldots, k$ then $n_W \leq n_1 + \cdots + n_k$. Thus, without changing its scope we can apply Conjecture 11.1.1 to algebraic sets instead of varieties.

This conjecture has an implication (the second Mazur conjecture) whose importance should be clear to a reader of this book.

Conjecture 11.1.3. *There is no Diophantine definition of $\mathbb{Z}$ over $\mathbb{Q}$.*

Instead of proving this implication right away, we will restate the first conjecture for a wider variety of objects and consider its implication in an extended setting.

11.2 A ring version of Mazur's first conjecture

Notation 11.2.1.

- For a number field K, let $\mathcal{P}(K)$ denote the set of all finite primes of K.
- Let $V \subset \mathbb{C}^n$ be an affine algebraic set defined over a field K. Let $A \subseteq K$. Then let $V(A) = \{\bar{x} = (x_1, \ldots, x_n) \in V \cap A^n\}$.

Question 11.2.2. *Let K be a number field and let $\mathcal{W}_K$ be a set of primes of K. Let V be any affine algebraic set defined over K. Let $\overline{V(O_{K,\mathcal{W}_K})}$ be the topological closure of $V(O_{K,\mathcal{W}_K})$ in $\mathbb{R}$ if $K \subset \mathbb{R}$ or in $\mathbb{C}$ otherwise. Then how many connected components does $\overline{V(O_{K,\mathcal{W}_K})}$ have?*

We start with the following simple observations.

Proposition 11.2.3. *Let T_1, T_2 be topological spaces. Consider $T = T_1 \times T_2$ under the product topology. Let $\pi : T \to T_1$ be a projection. Let $S \subset T$ be such that the topological closure $\overline{\pi(S)}$ of $\pi(S)$ has infinitely many components. Then the topological closure $\bar{S}$ of S also has infinitely many components.*

Proof. First of all, observe that $\pi(\bar{S}) \subseteq \overline{\pi(S)}$, since a projection maps limit points to limit points. Thus $\bar{S} \subseteq \pi^{-1}(\overline{\pi(S)})$. By assumption, $\overline{\pi(S)} = \bigcup_{i \in I} C_i$, where I is infinite and the C_i are closed and pairwise disjoint. Further, infinitely many C_i will contain points of $\pi(S)$. Indeed, suppose not. Let $C_1, \ldots, C_l$ be all the C_i's containing points of $\pi(S)$. Let $C_j \not\subset C_1 \cup C_2 \cup \cdots \cup C_l$. Let $a \in C_j$. Then $a \notin \pi(S) \subset C_1 \cup \cdots \cup C_l$, but every neighborhood of a has a point of C_i for some $i = 1, \ldots, l$. Therefore, for some $i = 1, \ldots, l$ we have that a is a limit point of C_i. Indeed, suppose not. Then for each i there is a neighborhood of a where C_i has no points. The intersection of all such neighborhoods is a neighborhood where $C_1 \cup C_2 \cup \cdots \cup C_l$ has no points, and we have a contradiction. Since the C_i are closed, $a \in C_i$ for some $i = 1, \ldots, l$. Thus we have another contradiction. Consequently $\bar{S} = \bigcup_{i\in I}(\bar{S} \cap \pi^{-1}(C_i))$, where infinitely many elements of the union are non-empty and all elements are closed. Hence $\bar{S}$ is a union of infinitely many pairwise-disjoint closed sets. □

From this observation we immediately derive several easy but useful corollaries concerning connected components.

Corollary 11.2.4. *Suppose that for some ring R contained in a number field and for some affine algebraic set V defined over the fraction field of R we have that $\overline{V(R)}$ has infinitely many connected components. Assume also that R has a Diophantine definition over a ring $\tilde{R} \supset R$, where the fraction field of $\tilde{R}$ is a number field K. Then for some affine algebraic set W defined over K, we have that $\overline{W(\tilde{R})}$ has infinitely many connected components.*

Proof. Let V be an algebraic set with, as described in the statement of the proposition, infinitely many components of $\overline{V(R)}$. Let $g(t, \bar{y})$ be a Diophantine definition of R over $\tilde{R}$. Let $\{f_i(\bar{x}), \bar{x} = (x_1, \ldots, x_n), i = 1, \ldots, m\}$ be polynomials defining V. Then consider the following system:

$$\begin{cases} g(x_i, \bar{y}_i) = 0, & i = 1, \ldots, n, \\ f_j(\bar{x}) \ \ = 0, & j = 1, \ldots, m. \end{cases} \tag{11.2.1}$$

Let W be the algebraic set defined by this system in K. Note that the projection of $W(\tilde{R})$ on the $\bar{x}$-coordinates is precisely $V(R)$ and therefore the topological closure of $W(\tilde{R})$ in $\mathbb{R}$ or $\mathbb{C}$ will have infinitely many connected components. □

Corollary 11.2.5. *Let $\mathcal{W}, \mathcal{S}$ be finite sets of primes of $\mathbb{Q}$, with $\mathcal{S} = \mathcal{P}(\mathbb{Q}) \setminus \mathcal{W}$. Suppose that Conjecture 11.1.1 holds over $\mathbb{Q}$. Let V be any variety defined over $\mathbb{Q}$. Then the real topological closure of $V(O_{\mathbb{Q},\mathcal{W}})$ has finitely many connected components.*

Proof. Since by Theorem 4.2.4 we know how to define integrality at finitely many primes over number fields, $O_{\mathbb{Q},\mathcal{W}}$ has a Diophantine definition over $\mathbb{Q}$. Therefore we can apply Corollary 11.2.4 to reach the desired conclusion. □

We can specialize Corollary 11.2.4 to obtain the following proposition, which will account for Mazur's second conjecture.

Proposition 11.2.6. *Let R be a subring of a number field K such that, for any affine algebraic set V defined over K, the topological closure of $V(R)$ has finitely many connected components. Then no infinite discrete (in archimedean* *topology) subset of R has a Diophantine definition over R. In particular, no infinite subset of $\mathbb{Z}^n$, where n is a positive integer, has a Diophantine definition over R.*

Using the definability of integrality at finitely many primes we can obtain another easy consequence of Corollary 11.2.4 and Proposition 11.2.6.

Corollary 11.2.7. *Let $\mathcal{S}$ be defined as in Corollary 11.2.5. Then there exists an affine algebraic set U such that the real closure of $U(O_{\mathbb{Q},\mathcal{S}})$ will have infinitely many components.*

Proof. By Theorem 4.2.4, $\mathbb{Z}$ has a Diophantine definition over $O_{\mathbb{Q},\mathcal{S}}$. Therefore we can apply Proposition 11.2.6 to reach the desired conclusion. □

Thus if we allow finitely many primes in the denominator in the closure, we will have algebraic sets over the resulting ring with infinitely many connected components. Similarly, if Conjecture 11.1.1 is true and we remove a finite number of primes from the denominator, all the varieties over the resulting rings will have finitely many components only, in the closure. The natural question is then how many primes we can remove from the denominator before we see algebraic sets with infinitely many components in the topological closure over the resulting rings. We will answer this question partially in this chapter. A more comprehensive answer will be provided in the next chapter, on Poonen's results.

11.3 First counterexamples

In this section we will describe some counterexamples to the ring version of Mazur's conjecture using norm equations. First we state a lemma whose proof follows from Part 5 of Proposition 6.2.1.

Lemma 11.3.1. *Let K be a number field. Let $\mathcal{W}_K \subset \mathcal{P}(K)$ be such that for some finite extension M of K all the primes of $\mathcal{W}_K$ remain prime in the extension M/K. Let $\mathcal{W}_M$ be the set of all the M-primes above the primes of $\mathcal{W}_K$. Then all the solutions $x \in O_{M,\mathcal{W}_M}$ to the equation*

$$\mathbf{N}_{M/K}(x) = 1 \tag{11.3.1}$$

are integral units.

Equipped with the lemma above, we can now produce an equation with an infinite set of integer solutions over a "large" subring of $\mathbb{Q}$.

Lemma 11.3.2. *Let M be any finite extension of $\mathbb{Q}$ of degree $n > 2$. Let $\mathcal{W}_{\mathbb{Q}} \subset \mathcal{P}(\mathbb{Q})$ be a set of $\mathbb{Q}$-primes not splitting in the extension $M/\mathbb{Q}$. Let $\{\omega_1, \ldots, \omega_n\} \subset O_M$ be an integral basis of M over $\mathbb{Q}$. Let $\{\omega_{i,j}, j = 1, \ldots, n\}$, $\omega_{i,1} = \omega_i$ be all the conjugates of ω_i over $\mathbb{Q}$. Then all the solutions $(a_1, \ldots, a_n) \in O_{\mathbb{Q},\mathcal{W}_{\mathbb{Q}}}$ to the equation*

$$\prod_{j=1}^{n}\sum_{i=1}^{n} a_i\omega_{i,j} = 1 \tag{11.3.2}$$

are actually in $\mathbb{Z}$. Furthermore the set of these solutions is infinite.

Proof. Let $\mathcal{W}_M$ contain all the M-primes lying above primes of $\mathcal{W}_{\mathbb{Q}}$. Then $x = \sum_{i=1}^{n} a_i\omega_i \in O_{M,\mathcal{W}_M}$. Further, the set $\{x_j = \sum_{i=0}^{n} a_i\omega_{i,j}, j = 1, \ldots, n\}$ contains all the conjugates of $x = x_1$ over $\mathbb{Q}$. Thus equation (11.3.2) is equivalent to equation (11.3.1), with $K = \mathbb{Q}$. Therefore, if $x = \sum_{i=1}^{n} a_i\omega_i$ is a solution to (11.3.2) then x is an integral unit of M. Since $\{\omega_1, \ldots, \omega_n\}$ is an integral basis, we must conclude that $a_i \in \mathbb{Z}$.

Conversely, if $x = \sum_{i=0}^{n} a_i\omega_i$ is a square of any integral unit of M then $(a_1, \ldots, a_n)$ are solutions to this equation. Since we have assumed the degree of the extension to be greater than 2, we can conclude that by the Dirichlet unit theorem (see Theorem 11.19, Chapter 1 of [37]), the unit group of M is of rank at least 1 and the solution set of (11.3.2) is infinite in $\mathbb{Z}^n$. □

We can now state our first counterexample for a subring of $\mathbb{Q}$.

Proposition 11.3.3. *For any $\varepsilon > 0$ there exists a set of rational primes $\mathcal{W}_{\mathbb{Q}}$ such that the Dirichlet density of $\mathcal{W}_{\mathbb{Q}}$ is greater than $1 - \varepsilon$ and there exists a variety V defined over $\mathbb{Q}$ such that the topological closure of $V(O_{\mathbb{Q},\mathcal{W}_{\mathbb{Q}}})$ in $\mathbb{R}$ has infinitely many connected components.*

Proof. It is enough to take M to be a cyclic extension of prime degree greater than ε^{-1}. Then by Lemma B.5.2 the set of primes splitting in the extension $M/\mathbb{Q}$ has density less than ε and we can apply Lemma 11.3.2 and Proposition 11.2.6. □

We will now prove analogous results for totally real number fields and their totally complex extensions of degree 2 using what we know about vertical Diophantine definability for these extensions.

Theorem 11.3.4. *Let K be a totally real field or a totally complex extension of degree 2 of a totally real field. Then for any $\varepsilon > 0$ there exists a set of primes*

$\mathcal{W}_K \subset \mathcal{P}(K)$ such that the Dirichlet density of $\mathcal{W}_K$ is greater than $1 - \varepsilon$ and there exists an affine algebraic set V defined over K such that $\overline{V(O_{K,\mathcal{W}_K})}$ has infinitely many connected components.

Proof. Since we have dealt with the case $K = \mathbb{Q}$ already, we can assume that K is a non-trivial extension of $\mathbb{Q}$. We consider the case of totally real fields first. First let E, K_G be as described in Corollary 7.6.3 with the additional assumption that

$$\frac{1}{p} > \varepsilon^{-1}$$

and $\mathcal{W}_K$ contains *all* the primes of K not splitting in the extension EK/K. Observe that, under this assumption, by Lemma B.5.2 the density of the set of all K primes not splitting in the extension EK/K is greater than $1 - \varepsilon$. Adding finitely many primes to $\mathcal{W}_K$ to form $\bar{\mathcal{W}}_K$, as in Corollary 7.6.3, will not change the density. Let $\mathcal{W}_{\mathbb{Q}}$ be the set of all the rational primes below the primes of $\mathcal{W}_K$ such that for every $q \in \mathcal{W}_{\mathbb{Q}}$ we have that $\mathcal{W}_K$ contains all the factors of q in K, and note that, owing to our assumption on p, by Lemmas B.4.7 and B.4.8 primes of $\mathcal{W}_{\mathbb{Q}}$ do not split in the extension $E/\mathbb{Q}$. Note also that $O_{K,\mathcal{W}_K} \cap \mathbb{Q} = O_{\mathbb{Q},\mathcal{W}_{\mathbb{Q}}}$.

Now let $\bar{\mathcal{W}}_{\mathbb{Q}}$ be the set of all the rational primes below the primes of $\bar{\mathcal{W}}_K$ such that, for every $q \in \bar{\mathcal{W}}_{\mathbb{Q}}$, we have that $\bar{\mathcal{W}}_K$ contains all the factors of q in K. Again we observe that $\mathbb{Q} \cap O_{K,\bar{\mathcal{W}}_K} = O_{\mathbb{Q},\bar{\mathcal{W}}_{\mathbb{Q}}}$. By construction, $\bar{\mathcal{W}}_{\mathbb{Q}}$ can differ from $\mathcal{W}_{\mathbb{Q}}$ by finitely many primes only. Therefore we claim that the following statements are true by Corollary 7.6.3, the definability of integrality at finitely many primes, and the transitivity of Dioph-generation:

$$O_{\mathbb{Q},\bar{\mathcal{W}}_{\mathbb{Q}}} \leq_{Dioph} O_{K,\bar{\mathcal{W}}_K},$$
$$O_{\mathbb{Q},\mathcal{W}_{\mathbb{Q}}} \leq_{Dioph} O_{\mathbb{Q},\bar{\mathcal{W}}_{\mathbb{Q}}},$$
$$O_{\mathbb{Q},\mathcal{W}_{\mathbb{Q}}} \leq_{Dioph} O_{K,\bar{\mathcal{W}}_K}.$$

Also, by Corollary 7.5.4 there exists an infinite set of rational integers Diophantine over $O_{\mathbb{Q},\mathcal{W}_{\mathbb{Q}}}$ and thus over $O_{K,\bar{\mathcal{W}}_K}$. □

The case where K is a totally complex extension of degree 2 of a totally real field is handled in a similar manner using Theorem 7.8.7.

We now turn our attention to extensions with one pair of non-real conjugate embeddings. There we do not have results analogous to Corollary 7.6.3 and Theorem 7.8.7, but we do know that rational integers have a Diophantine definition over the rings of integers of these fields (see Section 7.2). We will use an approach utilized in the result cited above to prove the following theorem.

Theorem 11.3.5. *Let K be a non-real number field with exactly one pair of non-real conjugate embeddings. Then there exists a set $\mathcal{W}_K \subset \mathcal{P}(K)$ such that the Dirichlet density of $\mathcal{W}_K$ is $1/2$ and for some affine variety V defined over K we have that $\overline{V(O_{K,\mathcal{W}_K})}$ has infinitely many connected components.*

Proof. Let $a \in O_K$ be as described in Notation 6.3.3 for the case where K has exactly one pair of conjugate non-real embeddings. Let $M = K(\sqrt{a^2-1})$ be a totally complex extension of degree 2 of K. Note that the density of the set of K-primes not splitting in the extension M/K is exactly $1/2$ by Lemma B.5.2. So let $\mathcal{W}_K \subset \mathcal{P}(K)$ be the set of primes not splitting in the extension M/K. Let $\mathcal{W}_M$ be the set of M-primes above the primes of W_K and observe that by Proposition 11.3.1 all the solutions to $\mathbf{N}_{M/K}(z) = 1$ in $O_{M,\mathcal{W}_M}$ are algebraic integers. However, in this case we can say a little bit more. By Lemma 6.3.5, we know that solutions to this norm equation form a multiplicative group of rank 1 and, modulo roots of unity, are powers of $\mu = a - \sqrt{a^2-1}$. Note that either μ or μ^{-1} is of absolute value greater than 1 by Lemma 6.3.4.

Assume without loss of generality that $|\mu| > 1$ and let $\mu^{rk} = x_k - \sqrt{a^2-1}y_k$ for some sufficiently large r such that $|\mu^{rk}| > 2^k$. Then $|x_k| = |(\mu^{rk} + \mu^{-rk})/2| > (2^k - 1)/2$. Therefore, for any $l \in \mathbb{N}$ and any neighborhood U of x_l, there exist only finitely many $m \in \mathbb{N}$ such that $x_m \in U$. In other words, the set

$$\big\{x \in O_{K,\mathcal{W}_K} \mid \exists x_0, y_0, y \in O_{K,\mathcal{W}_K},$$
$$x - \sqrt{a^2-1}\,y = (x_0 - \sqrt{a^2-1}\,y_0)^r,\ x_0^2 - (a^2-1)y_0^2 = 1\big\}$$

is discrete and the assertion of the theorem follows from Corollary 11.2.7. □

11.4 Consequences for Diophantine models

As we have seen in the first section of this chapter, the truth of the first Mazur conjecture implies that there is no Diophantine definition of $\mathbb{Z}$ over $\mathbb{Q}$. However, we know that Diophantine definitions are only one type of element in a large class of objects called Diophantine models. From Corollary 3.4.6 we also know that it would be enough to show that $\mathbb{Q}$ has a Diophantine model of $\mathbb{Z}$ to assert the Diophantine undecidability of $\mathbb{Q}$. Therefore, we would like to know whether the above-mentioned Mazur's conjecture precludes the existence of a Diophantine model of $\mathbb{Q}$ over $\mathbb{Z}$. Cornelissen and Zahidi showed that this was indeed the case in [6] using a more restrictive definition of a Diophantine model. Under their assumptions the image of a Diophantine set was a Diophantine set in the target ring. We could call this kind of Diophantine model *tight*. We reproduce below a slightly generalized version of their proof.

Proposition 11.4.1. *Let K be a number field and let $\mathcal{W}_K$ be a recursive (computable) set of primes of K and assume that for any affine algebraic set V defined over K we have that $\overline{V(O_{K,\mathcal{W}_K})}$ has finitely many connected components. Then $O_{K,\mathcal{W}_K}$ does not have a tight Diophantine model of $\mathbb{Z}$.* (We remind the reader that a definition of a recursive set of primes can be found in Section A.8.)

Proof. Let $P(x_1, \ldots, x_k, t_1, \ldots, t_m)$ be a Diophantine definition of a set $D = \phi(\mathbb{Z})$ over $(O_{K,\mathcal{W}_K})^k$. Let

$$V = \left\{(x_1, \ldots, x_k, t_1, \ldots, t_m) \in O_{K,\mathcal{W}_K}^{k+m} : P(x_1, \ldots, x_k, t_1, \ldots, t_m) = 0\right\}$$

and consider the map

$$f : V(O_{K,\mathcal{W}_K}) \rightarrow (O_{K,\mathcal{W}_K})^k$$

implemented by projection on the first k coordinates. Note that $f(V) = D$. By assumption and by Proposition 11.2.3 we have that $\bar{D}$, the closure of D in $\mathbb{R}$ if K is a real field or in $\mathbb{C}$ otherwise, will have finitely many connected components. Since $\phi(\mathbb{Z}) = D$ has infinitely many points, for at least one connected component C of $\bar{D}$ it is the case that $C \cap \phi(\mathbb{Z})$ must have more than one point, and the projection of C onto one of the coordinates, if K is real, or onto the imaginary or real part of one of the coordinates, if K is not real, will contain a non-trivial interval whose end points are rational numbers. (This is so because the only connected subsets of the real line containing more than one point are intervals.) Let a be the left-hand endpoint of this interval and let l be its length. Let $d_n = s \circ \phi(n)$, where s is either the projection onto the coordinate described above or the real part or the imaginary part of the projection, as necessary, and let

$$\tilde{Z} = \left\{n \in \mathbb{Z} \mid a + \frac{l}{2j+1} \leq d_n \leq a + \frac{l}{2j},\ j \in \mathbb{Z}_{>0}\right\}.$$

Since ϕ is computable and by Proposition A.8.8 we can compute effectively decimal expansions for real and (if necessary) imaginary parts of all the elements of K, we conclude that $\tilde{Z}$ is recursively enumerable and therefore $\tilde{Z}$ is a Diophantine subset of $\mathbb{Z}$ by the result of Matiyasevich, Robinson,

Davis, and Putnam. Further, $s(\phi(\mathbb{Z})) \cap [a, a+l]$ is dense in $[a, a+l]$. Indeed, $[a, a+l] \subseteq s(C) \subset s(\overline{\phi(\mathbb{Z})}) = s(\bar{D})$. Since D is dense in $\bar{D}$ and projection maps dense subsets into dense subsets, our claim is true. Thus, any interval $[a + l/(2j+1), a + l/2j]$ will have infinitely many points from $s(D)$ and therefore elements d_n, with $n \in \tilde{Z}$ by the definition of $\tilde{Z}$. Let $\tilde{D} = \{\phi(n) | n \in \tilde{Z}\}$. Then $\tilde{D}$ has a Diophantine definition over $(O_K)^k$ as the ϕ-image of a Diophantine subset of $\mathbb{Z}$. Let $\tilde{P}(x_1, \ldots, x_k, t_1, \ldots, t_m)$ be a Diophantine definition of $\tilde{D} = \phi(\tilde{Z})$, and let $\tilde{V}$ be the algebraic set defined by $\tilde{P}(x_1, \ldots, x_k, t_1, \ldots, t_m) = 0$. Then $s \circ f$, the projection from $\tilde{V}$ onto the first k coordinates, combined with the projection onto a real or imaginary part of a coordinate chosen as above, will produce a projection of $\tilde{V}$ onto a set whose closure has infinitely many components. Thus $\tilde{V}$ would have to have infinitely many components, in contradiction of our assumptions for this proposition. □

12

Results of Poonen

Poonen's theorem is arguably the most important development in the subject since Matiyasevich completed the proof of the original HTP in the late 1960s. One could say that for the first time the solution of HTP for $\mathbb{Q}$ has become visible over the distant horizon, though we still have to traverse an "infinite" distance, as will be explained below.

The result came out of the attempts to falsify the ring version of Mazur's conjecture (Conjecture 11.1.1) for a ring of rational $\mathcal{S}$-integers where $\mathcal{S}$ has natural density equal to 1. In the process of constructing a counterexample to the conjecture, Poonen constructed a (tight) Diophantine model of $\mathbb{Z}$ over such a ring. This result has moved us "infinitely" far away from where we started ($\mathbb{Z}$ and rings of rational $\mathcal{S}$-integers with finite $\mathcal{S}$), but since the set of allowed denominators in Poonen's theorem still misses being an infinite set of primes (though of natural density zero), we still have "infinitely" far to go.

In this chapter we will go over Poonen's proof, which appeared originally in [74], in some detail. We will start with the overall plan and then will try to sort out the rather challenging technical details.

12.1 A statement of the main theorem and an overview of the proof

A good place to start is a precise statement of the theorem, which is presented below.

Theorem 12.1.1. *There exist recursive sets of rational primes $\mathcal{T}_1$ and $\mathcal{T}_2$, both of natural density zero and with an empty intersection, such that for any set $\mathcal{S}$ of rational primes containing $\mathcal{T}_1$ and disjoint from $\mathcal{T}_2$ the following hold:*

1. *There exists an affine curve E over $O_{\mathbb{Q},\mathcal{S}}$ such that the topological closure of $E(O_{\mathbb{Q},\mathcal{S}})$ in $E(\mathbb{R})$ is an infinite discrete set.*
2. *$\mathbb{Z}$ has a (tight) Diophantine model over $O_{\mathbb{Q},\mathcal{S}}$.*
3. *Hilbert's Tenth Problem is undecidable over $O_{\mathbb{Q},\mathcal{S}}$.* (The computability (recursiveness) of $\mathcal{T}_2$ implies that $\mathcal{S} = \mathcal{P}(\mathbb{Q}) \setminus \mathcal{T}_2$ is computable, and therefore by Proposition A.8.6 we have that $O_{\mathbb{Q},\mathcal{S}}$ is computable. Thus over such a ring it makes sense to talk about the undecidability of HTP.)

In proving the theorem we will rely on the existence of an elliptic curve E defined over $\mathbb{Q}$ such that the following conditions are satisfied.

1. $E(\mathbb{Q})$ is of rank 1.
2. $E(\mathbb{R}) \cong \mathbb{R}/\mathbb{Z}$ as topological groups.
3. E does not have complex multiplication.

(An example of such a curve is given in Proposition B.6.2.)

Denote an infinite-order point of $E(\mathbb{Q})$ by P. For a non-zero integer n, let $(x_n(P), y_n(P))$ be the affine coordinates of $[n]P$ given by a Weierstrass equation of E (from now on fixed) of the form $y^2 = x^3 + a_2x^2 + a_4x + a_6$ (see Section 1, Chapter III of [113]) where $[n]$ denotes the nth multiple of P under the addition on E. We note for future reference that, given this form of Weierstrass equation, for a non-zero point $Q \in E$ with affine coordinates (x, y) we have that $(x, -y)$ are the affine coordinates of $-Q$. (See Section 2, Chapter III of [113] for more details on group law.)

The proof of the theorem will consist of the following steps:

1. showing that there exists a computable sequence of rational primes $l_1 < \cdots < l_n < \cdots$ such that $[l_j]P = (x_{l_j}, y_{l_j})$ and for all $j \in \mathbb{N}$ we have that $|y_{l_j} - j| < \frac{1}{10j}$;
2. proving the existence of infinite sets $\mathcal{T}_1$ and $\mathcal{T}_2$, as described in the statement of the theorem, such that for any set $\mathcal{S}$ of rational primes containing $\mathcal{T}_1$ and disjoint from $\mathcal{T}_2$ we have that $E(O_{\mathbb{Q},\mathcal{S}}) = \{[l_j]P\} \cup$ (finite set);
3. noting that $\{y_{l_j}\}$ is an infinite discrete set and thus is a counterexample to Mazur's conjecture for the ring; $O_{\mathbb{Q},\mathcal{S}}$;
4. showing that $\{y_{l_j}\}$ is a (tight) Diophantine model of $\mathbb{Z}$ over $\mathbb{Q}$.

To carry out the steps outlined above and, especially, to show that the sets $\mathcal{T}_1$ and $\mathcal{T}_2$ have the required densities, we will have to use a fair amount of material concerning elliptic curves. We have tried to separate out general properties (which can be found in the number theory appendix, Section B.6) from properties more or less unique to the situation at hand (which are discussed in this chapter). We will use the notation from Notation B.6.1 with $\mathbb{Q}$ replacing the arbitrary number field K. We will also use additional notation described below.

An attentive reader will note that the statement of the theorem refers to the natural density of the prime sets. The density that we have used so far is the Dirichlet density, however. For a definition of the natural density and its relation to the Dirichlet density, we refer the reader to Definition B.5.9 and Proposition B.5.10.

Finally, before taking the plunge, we remind the reader that everything we need to know about elliptic curves (and much more, of course) can be found in [113] and [114].

Notation 12.1.2.

- For each prime number l, let $a_l(P)$ be the smallest positive integer such that $\mathcal{S}_{l^{a_l(P)}}(P) \setminus \mathcal{S}_1(P) \neq \emptyset$. Such an $a_l(P)$ exists for every l and is equal to 1 for all but finitely many l's by Siegel's theorem, since $\mathcal{S}_0(\mathbb{Q}) \cup \mathcal{S}_1(P)$ is a finite set. ($\mathcal{S}_i(P)$ and $\mathcal{S}_0(\mathbb{Q})$ are defined in Notation B.6.1.)
- Let $\mathcal{L}(P) = \{l \in \mathcal{P}(K) : a_l 2(P) > 1\}$.
- Let $p_l(P) = \max \mathcal{S}_{l^{a_l(P)}}(P) \setminus \mathcal{S}_1(P)$.
- For rational prime numbers l, p, let $p_{lp} = \max(\mathcal{S}_{lp} \setminus (\mathcal{S}_l \cup \mathcal{S}_p))$ if the set difference is not empty. By Proposition 12.2.2 below, for sufficiently large $\max(l, p)$ the set difference will be non-empty.
- Let

$$\mu_l(P) = \sup_{X \in \mathbb{Z}, X \geq 2} \frac{\#\{p \in \mathcal{S}_l(P) : p \leq X\}}{\#\{p \in \mathcal{P}(\mathbb{Q}) : p \leq X\}}.$$

- For $X \in \mathbb{R}$ let $\pi_{\mathbb{Q}}(X) = \#\{p \in \mathcal{P}(\mathbb{Q}) : p \leq X\}$.

12.2 Properties of elliptic curves I: Factors of denominators of points

By the denominators of points we mean the denominators of the affine coordinates of non-zero multiples of P under our fixed Weierstrass equation, where the numerator and the denominator of the coordinates are relatively prime. We should note here that since the affine Weierstrass equation is monic in x and y, the set of primes which occur in the denominator of x is the same as the set of primes occurring in the denominator of y. The goal is to understand which primes to ban from denominators to eliminate the "extraneous" points from the solution set in the constructed ring.

The primes are divided into sets $\mathcal{S}_i$ as mentioned above. One can think of $\mathcal{S}_0$ as a set of generally inconvenient primes, such as primes where E does not have a good reduction, and $\mathcal{S}_i(P)$ for $i > 0$ can be thought of as the set of "relevant" primes appearing in the denominator of $[\pm i]P$.

The first proposition below deals with the primes that two different point denominators can have in common.

Proposition 12.2.1. *For any $m, n \in \mathbb{Z} \setminus \{0\}$, we have that $\mathcal{S}_m(P) \cap \mathcal{S}_n(P) = \mathcal{S}_{(m,n)}(P)$, where (m, n) is the* GCD *of m and n.*

Proof. Let $\mathfrak{q} \in \mathcal{S}_m(P) \cap \mathcal{S}_n(P)$. Then

$$\{m, n\} \subseteq \mathcal{G}_l = \{\text{non-zero } l \in \mathbb{Z} : \mathfrak{q} \mid \mathfrak{d}_l(P)\} \cup \{0\},$$

a subgroup of $\mathbb{Z}$ by Corollary B.6.5. Thus on the one hand $(m, n) \in \mathcal{G}_l$ or in other words $\mathfrak{q} \in \mathcal{S}_{(m,n)}$. Hence $(\mathcal{S}_m \cap \mathcal{S}_n) \subseteq \mathcal{S}_{(m,n)}$. On the other hand, by Corollary B.6.7 we have that $\mathcal{S}_{(m,n)} \subseteq \mathcal{S}_m \cap \mathcal{S}_n$. Thus, the proposition holds. □

The next proposition shows when we can expect new primes to appear in a point denominator.

Proposition 12.2.2. *If $l, m \in \mathcal{P}(\mathbb{Q})$ then for sufficiently large* $\max(l, m)$ *we have that*

$$\mathcal{S}_{lm} \setminus (\mathcal{S}_l \cup \mathcal{S}_m) \neq \emptyset.$$

(The subscript lm refers to the product of l and m.)

Proof. First of all we note that without loss of generality we can assume that $l \geq 2, m \geq 2$. Next suppose that $\mathcal{S}_{lm} \setminus (\mathcal{S}_l \cup \mathcal{S}_m) = \emptyset$. We want to evaluate $\text{ord}_{\mathfrak{q}} d_{lm}$ for $\mathfrak{q} \in \mathcal{S}_l \cup \mathcal{S}_m = \mathcal{S}_{lm}$. ("$d_{lm}$" is defined in Notation B.6.1.) Without loss of generality assume that $\mathfrak{q} \in \mathcal{S}_l$. Then by Proposition B.6.6 we have that $\text{ord}_{\mathfrak{q}} d_{lm} = \text{ord}_{\mathfrak{q}} d_l$ if $\mathfrak{q} \nmid m$ and $\text{ord}_{\mathfrak{q}} d_{lm} = \text{ord}_{\mathfrak{q}} d_l + 2$ if $\mathfrak{q} \mid m$. Thus, $d_{lm} \mid d_l d_m l^2 m^2$ and by Lemma B.6.9 it follows that

$$h_0(d_{lm}) \leq h_0(d_l) + h_0(d_m) + 2l + 2m. \tag{12.2.1}$$

(h_0 is defined in Notation B.6.1 also.)

Together, Lemma B.6.8 and (12.2.1) imply that for some positive constant c we have that

$$\begin{aligned} h_0(d_{lm}) &= (c - o(1))l^2 m^2, \\ h_0(d_l) &= (c - o(1))l^2, \\ h_0(d_m) &= (c - o(1))m^2. \end{aligned} \tag{12.2.2}$$

Given $\varepsilon > 0$, let $C(\varepsilon)$ be such that for any $n > C(\varepsilon)$ it is the case that $h_0(d_n) = (c - o(1))n^2$ with $|o(1)| < \varepsilon$. Also let $C \in \mathbb{R}$ be such that $h_0(d_n) < Cn^2$ for all $n \in \mathbb{N} \setminus \{0\}$. Choose ε small enough that

$$c > 3\varepsilon.$$

Next consider two cases, $m > C(\varepsilon)$ and $m < C(\varepsilon)$. In the first case choose $l > C(\varepsilon)$ and large enough that

$$c > 3\varepsilon + \frac{4}{l}.$$

Then our assumptions and (12.2.2) imply that

$$(c - o(1))l^2m^2 \leq (c - o(1))l^2 + (c - o(1))m^2 + 2l + 2m$$

and

$$(c - \varepsilon)l^2m^2 \leq (c + \varepsilon)l^2 + (c + \varepsilon)m^2 + 2l + 2m,$$

which after division by l^2m^2 becomes

$$(c - \varepsilon) \leq (c + \varepsilon)m^{-2} + (c + \varepsilon)l^{-2} + \frac{2}{lm^2} + \frac{2}{l^2m} < \frac{c + \varepsilon}{2} + \frac{2}{l},$$

contradicting the choice of l and m.

Suppose now that $m < C(\varepsilon)$. Next choose $l > C(\varepsilon)$ and large enough that

$$l > \frac{CC(\varepsilon)^2 + 2C(\varepsilon) + 2}{(3c - 5\varepsilon)}.$$

Finally, (12.2.1), (12.2.2), and our assumptions imply that

$$4(c - \varepsilon)l^2 \leq (c + \varepsilon)l^2 + CC(\varepsilon)^2 + 2C(\varepsilon) + 2l,$$
$$(3c - 5\varepsilon)l \leq CC(\varepsilon)^2 + 2C(\varepsilon) + 2,$$

and we have a contradiction with our choice of l. □

Remark 12.2.3. The denominators of the affine coordinates of the points of E form what is called an "elliptic divisibility sequence," and the properties listed above are characteristic of such sequences. For more information on these sequences we refer the reader to [31].

12.3 Properties of elliptic curves II: The density of the set of "largest" primes

In this section we will discuss probably the most difficult part of the proof of Theorem 12.1.1 – the part which will be crucial in showing that $\mathcal{T}_2$ can have natural density zero. As we have observed from the section above, except possibly for finitely many primes l it is the case that $\mathcal{S}_l$ contains a prime which is not in any $\mathcal{S}_q$ with q prime and $q \neq l$. Further, from Proposition 12.2.1 we know that $\mathcal{S}_l \subset \mathcal{S}_{kl}$ for any $k \in \mathbb{Z} \setminus \{0\}$. Thus, if we banned a prime unique to $\mathcal{S}_l$ from the denominators of the points in our ring, we would eliminate all the multiples of $[l]P$ from the solution set over the ring. Therefore it is important to know the density of the set of the largest primes "unique to $\mathcal{S}_l$." Our first task is to establish the relationship between a prime $p_l \in \mathcal{S}_l \setminus \mathcal{S}_1$ and a prime l. Our hope is that p_l is going to be "on average" large compared with l. (This will help to make the density of the p_l small.) We start with the following observation.

Lemma 12.3.1. *Let $l \in \mathcal{P}(\mathbb{Q})$ and suppose that $p \in \mathcal{S}_{l^{a_l(P)}}(P) \setminus \mathcal{S}_1(P)$. Then $l \mid \#E(\mathbb{F}_p)$.*

Proof. If $p \in \mathcal{S}_{l^{a_l(P)}}(P) \setminus \mathcal{S}_1(P)$ then p does not divide the discriminant of our Weierstrass equation and E has a good reduction at p (and thus $\tilde{E}$ is non-singular) with all the coefficients of our chosen affine Weierstrass equation integral at p. Further, $x_1(P)$ and $y_1(P)$ are integral at p, while $\text{ord}_p\, x_1([l^{a_l(P)}]P) < 0$, $\text{ord}_p\, y_1([l^{a_l(P)}]P) < 0$. Therefore, under reduction mod p, the image of P is not $\tilde{O}$, the image of O, while $[l^{a_l(P)}]\tilde{P} = \tilde{O}$. By the definition of $a_l(P)$, we must conclude that $E(\mathbb{F}_p)$ has an element of order $l^{a_l(P)}$ and therefore $l \mid \#E(\mathbb{F}_p)$. □

The lemma provides us with some information about the relationship of l and p_l, but to estimate the density of the set $\{p_l\}$ we need more. From a theorem of Hasse we know that $\#E(\mathbb{F}_p) \leq Cp$, where C is a constant independent of p (see Theorem 1.1, Chapter V of [113]). Thus we could hope that "on average" $\#E(\mathbb{F}_p)$ has many small factors and therefore that l is "much" smaller than $\#E(\mathbb{F}_p)$. The next two propositions will show just that. They are based on two results: the natural density version of the Chebotarev density theorem (Theorem 1 of [88]) and a result on the relationship between the action of the Galois group and the automorphism groups of torsion elements of the curve (Proposition B.6.14). The first of these propositions relates whether $E(\mathbb{F}_p)$ has a point of order l to the action of the Frobenius automorphism of p.

Proposition 12.3.2. *Let $p \neq l$ be rational prime numbers.* Let $\mathbb{F}_p$ *be a field of p elements. Let M be a Galois extension of $\mathbb{Q}$ containing all the elements of $E(\bar{\mathbb{Q}})[l]$, where $\bar{\mathbb{Q}}$ is the algebraic closure of $\mathbb{Q}$. Let $G(M/\mathbb{Q})$ be the Galois group of M over $\mathbb{Q}$. Assume that $p \notin \mathcal{S}_0(M)$ (see Notation B.6.1). Then $E(\mathbb{F}_p)$ has an element of order l if and only if, for some $\sigma \in$* Gal$(M/\mathbb{Q})$ *and some*

$Q \in E(\bar{\mathbb{Q}})[l] \setminus \{O\}$, we have that $\sigma(Q) = Q$ and σ is the Frobenius automorphism for some factor of p in M.

Proof. Let $\mathfrak{p}_M$ be a factor of p in M and let $\mathbb{F}_{\mathfrak{p}_M}$ be the residue field of $\mathfrak{p}_M$. Let $\tilde{\sigma}$ be the Frobenius automorphism of $\mathbb{F}_p$, i.e. $\tilde{\sigma}(x) = x^p$ for all $x \in \mathbb{F}_p$. Then, on the one hand, for any point $\tilde{Q} \in E(\mathbb{F}_{\mathfrak{p}_M})$ of order l we have that $\tilde{\sigma}(\tilde{Q}) = \tilde{Q}$ if and only if $\tilde{Q} \in E(\mathbb{F}_p)$. On the other hand, let $Q \in E(M) \setminus \{O\}$ be of order l with

$$\sigma(Q) \neq Q, \tag{12.3.1}$$

where σ is the Frobenius automorphism of $\mathfrak{p}_M$ over $\mathbb{Q}$. If we reduce (12.3.1) modulo $\mathfrak{p}_M$, we will obtain $\tilde{\sigma}(\tilde{Q}) \neq \tilde{Q}$ by Corollary B.6.12. Further, since by Corollary B.6.12 reduction modulo $\mathfrak{p}_M$ is an isomorphism of $E(M)[l]$ onto $E(\mathbb{F}_{\mathfrak{p}_M})[l]$, every $\tilde{P} \in E(\mathbb{F}_{\mathfrak{p}_M})$ of order l is the image of an order-l point in $E(M)[l]$. Thus $\tilde{\sigma}$ fixes an element of order l if and only if σ fixes an element of order l, and the assertion of the proposition follows. □

Remark 12.3.3. If $l = p$ and the Frobenius automorphism of a factor of p fixes some point of order p then, under reduction modulo the factor of p, the image of this point will remain fixed. However, we are no longer assured of the converse. That is, even if a point is fixed after the reduction we do not know whether it was fixed before the reduction.

In the next proposition we will examine directly the set of p's for which $\#E(\mathbb{F}_p)$ has "few" factors.

Proposition 12.3.4. *For a prime $p \notin \mathcal{S}_0(\mathbb{Q})$ and a positive constant C, define*

$$\mathcal{A}(p, C) = \{l \in \mathcal{P}(\mathbb{Q}) \setminus \mathcal{S}(\mathbb{Q}),\ l \,|\, \#E(F_p),\ l < C\},$$
$$f(p, C) = |\mathcal{A}(p, C)|.$$

Then, for any $t \geq 1$, the upper natural density of

$$\mathcal{B}(C, t) = \{p \in \mathcal{P}(\mathbb{Q}) : f(p, C) \leq t\}$$

tends to 0 *as $C \to \infty$.* (See Proposition B.6.14 for the definition of $\mathcal{S}(\mathbb{Q})$.)

Proof. Fix t, C. Let M be any Galois extension of $\mathbb{Q}$ such that M contains $E(\bar{\mathbb{Q}})[l]$ for all $l \leq C$. Next consider the natural homomorphism

$$\Lambda_{M,C} : \mathrm{Gal}(M/\mathbb{Q}) \to \prod_{l \in \mathcal{P}(\mathbb{Q}) \setminus \mathcal{S}(\mathbb{Q}), l \leq C} \mathrm{Aut}(E(\bar{\mathbb{Q}})[l])$$

and note that by Proposition B.6.14 this map is onto. Let $\sigma \in \text{Gal}(M/\mathbb{Q})$ and let $\Lambda_{M,C}(\sigma) = (\sigma_1, \ldots, \sigma_n)$, where $\{l_1, \ldots, l_n\}$ is the set of prime numbers less than C and not in $\mathcal{S}(\mathbb{Q})$. Let $\Sigma(C, t)$ denote the set

$$\left\{\sigma \in \text{Gal}(M/\mathbb{Q}),\ \Lambda_{M,C}(\sigma) \text{ has at most } t \text{ components with a fixed point } Q \neq O\right\}.$$

We want to determine

$$R(C, t) = \frac{|\Sigma(C, t)|}{|\text{Gal}(M/\mathbb{Q})|}. \tag{12.3.2}$$

Denote $(\sigma_1, \ldots, \sigma_n) : \sigma_i \in \text{Aut}(E(\bar{\mathbb{Q}}))[l_i]$ by $\bar{\sigma}$. Since $\Lambda_{M,C}$ is a surjective homomorphism, the ratio in (12.3.2) is equal to

$$\frac{\#\{\bar{\sigma} :\ \text{at most } t \text{ components have a fixed point } Q \neq O\}}{\#\{\bar{\sigma}\}}. \tag{12.3.3}$$

The ratio in (12.3.3) in turn gives us

$$\sum_{j=1}^{t} \frac{\#\{\bar{\sigma} :\ \text{exactly } j \text{ components have a fixed point } Q \neq O\}}{\#\{\bar{\sigma}\}}. \tag{12.3.4}$$

Given $i, j \in \{1, \ldots, n\}$ and a j-element subset $I_j \subseteq \{1, \ldots, n\}$, let

$$F_{i,I_j} = \{\bar{\sigma} : \forall i \in I_j, \exists Q \in E(\bar{\mathbb{Q}})[l_i] \setminus \{O\}, \sigma_i(Q) = Q\}$$

and let

$$G_{i,I_j} = \{\bar{\sigma} : \forall i \notin I_j, \forall Q \in E(\bar{\mathbb{Q}})[l_i] \setminus \{O\}, \sigma_i(Q) \neq Q\}.$$

Then the sum in (12.3.4) is equal to

$$\sum_{j=1}^{t} \sum_{I_j} \frac{|F_{i,I_j} \cap G_{i,I_j}|}{\#\{\bar{\sigma}\}}, \tag{12.3.5}$$

where I_j ranges over all the j-element subsets of $\{1, \ldots, n\}$. Continuing further, we observe that the sum from (12.3.5) is equal to

$$\sum_{j=1}^{t} \sum_{I_j} \left(\prod_{i \in I_j} \alpha_i\right) \left(\prod_{i \notin I_j} (1 - \alpha_i)\right) = \sum_{j=1}^{t} \sum_{I_j} \left(\prod_{i \in I_j} \frac{\alpha_i}{1 - \alpha_i}\right) \left(\prod_{i=1}^{n} (1 - \alpha_i)\right), \tag{12.3.6}$$

where for each $i = 1, \ldots, n$ we have that

$$\begin{aligned} \alpha_i &= \frac{\#\{\tau \in \text{Aut}(E(\bar{\mathbb{Q}}))[l_i] : \exists Q \in E(\bar{\mathbb{Q}})[l_i] \setminus \{O\}, \tau(Q) = Q\}}{|\text{Aut}(E(\bar{\mathbb{Q}}))[l_i]|} \\ &= \frac{1}{l_i} + \frac{1}{l_i^2} + o\left(\frac{1}{l_i^2}\right), \end{aligned}$$

by Proposition B.6.13. Now, by Proposition B.10.2, if we fix t and let $C \to \infty$ (or, alternatively, let $n \to \infty$) then $R(C, t) \to 0$.

Next we observe that $\Sigma(C, t)$ is closed under conjugation. In other words, if $\sigma \in \Sigma(C, t)$ then all the conjugates of σ in $\mathrm{Gal}(M/\mathbb{Q})$ are also in $\Sigma(C, t)$ since, for any point $P \in E(M)$ and any $\tau, \sigma \in \mathrm{Gal}(M/\mathbb{Q})$, we have that $\tau^{-1}\sigma\tau(\tau^{-1}(P)) = \tau^{-1}(P) \Leftrightarrow \sigma(P) = P$. Thus, since the order of the point is preserved by any action of the Galois group, for any natural number r we have that $\Lambda_{M,C}(\sigma)$ has exactly r components with non-O fixed points if and only if the same is true of $\Lambda(M, C)(\tau\sigma\tau^{-1})$. Therefore by the natural density version of the Chebotarev density theorem (see Theorem 1 of [88]), the density of primes of $\mathbb{Q}$ whose factors have a Frobenius automorphism in $\Sigma(C, t)$ is $R(C, t)$.

Next suppose that $p \in \mathcal{B}(C, t)$ and no factor of p is in $\mathcal{S}_0(M)$. (We remind the reader that $\mathcal{S}_0(M)$, defined in Notation B.6.1, contains all the factors of primes in $\mathcal{S}_0(\mathbb{Q})$ together with primes ramifying in the extension $M/\mathbb{Q}$.) Then we consider two cases: $p \nmid \#E(\mathbb{F}_p)$ and $p \mid \#E(\mathbb{F}_p)$. In the first case, exactly $t_0 \leq t$ primes $l \notin \mathcal{S}(\mathbb{Q}), l \neq p, l \leq C$, divide $E(\mathbb{F}_p)$. Therefore by the Sylow theorem, for exactly $t_0 \leq t$ primes $l \notin \mathcal{S}(\mathbb{Q}), l \neq p$, we have that $E(\mathbb{F}_p)$ has an element of order l and therefore, by Proposition 12.3.2, for at exactly $t_0 \leq t$ primes $l \notin \mathcal{S}(\mathbb{Q}), l \neq p, l \leq C$, a Frobenius automorphism σ of a factor of p fixes a point, of order l, not equal to O. Therefore p has a factor in M with a Frobenius automorphism σ such that $\Lambda_{M,C}(\sigma)$ has exactly $t_0 \leq t$ components σ_i with fixed points different from O. Thus $p \in \Sigma(C, t)$.

Similarly, if $p \mid \#E(\mathbb{F}_p)$ then we have exactly $t_0 < t$ primes $l \notin \mathcal{S}(\mathbb{Q}), l \neq p, l \leq C$, that divide $E(\mathbb{F}_p)$. Since Proposition 12.3.2 does not cover the case $l = p$ (see Remark 12.3.3), it is possible that $p \in \Sigma(C, t-1)$. However, since $\Sigma(C, t-1) \subseteq \Sigma(C, t)$, we reach the same conclusion in both cases. Thus in any case p belongs to the set whose natural density is $R(C, t)$.

Finally, let t be given and choose C large enough that $R(C, t) < \varepsilon$. Note that we can do this before we choose the field M by using (12.3.6). We just need to arrange that n, which is the number of primes not in $\mathcal{S}(\mathbb{Q})$ and less than C, is large enough. Then we can choose M as above and conclude that the set $\mathcal{B}(C, t)$ minus primes having factors in $\mathcal{S}_0(M)$ has upper density less than ε. Since $\mathcal{S}_0(M)$ is finite, we conclude that $\mathcal{B}(C, t)$ has upper density less than ε. □

Proposition 12.3.5. *The set $\mathcal{U}(P) = \{p_l(P), l \in \mathcal{P}(\mathbb{Q})\}$ has natural density zero.*

Proof. We will show that for every $\varepsilon > 0$ the set $\mathcal{U}(P)$ has upper natural density less than ε. So let ε be given. Choose an integer t such that $2^{2-t} < \varepsilon/2$. Choose

$C > 0$ such that the set $\mathcal{B}(C, t)$ (defined in Proposition 12.3.4) has upper natural density less than $\varepsilon/2$. Such a C exists, by Proposition 12.3.4. It remains to show that the set

$$\mathcal{U} \cap \overline{\mathcal{B}(C,t)} = \{p_l : l \in \mathcal{P}(\mathbb{Q}) \wedge f(p_l, C) > t\}$$

has upper natural density less than or equal to $\varepsilon/2$. Suppose that $p_l \in \overline{\mathcal{B}(C,t)}$. Then by Lemma 12.3.1 we have that $l \,|\, \#E(\mathbb{F}_{p_l})$ and $\#E(\mathbb{F}_{p_l})$ is divisible by at least t other primes. Hence $2^t l \leq \#E(\mathbb{F}_{p_l})$. Further, by a theorem of Hasse mentioned above,

$$\#E(\mathbb{F}_{p_l}) \leq p_l + 1 + 2\sqrt{p_l} \leq 4p_l.$$

Thus $l \leq 2^{2-t} p_l \leq \varepsilon p_l/2$. Consequently, for each $p_l \in \overline{\mathcal{B}(C,t)}$ there exists a distinct $l \in \mathcal{P}(\mathbb{Q})$ such that $l \leq \varepsilon p_l/2$. Therefore, by the prime number theorem (Theorem 4, Section 5, Chapter XV of [46]),

$$\#\left\{p_l \in \overline{\mathcal{B}(C,t)} : p_l \leq X\right\} \leq \pi_{\mathbb{Q}}\left(\tfrac{1}{2}\varepsilon X\right) = \frac{\varepsilon X}{2\log \frac{1}{2}\varepsilon X} + o\left(\frac{X}{\log X}\right)$$

as $X \to \infty$. Therefore the upper natural density of $\mathcal{U} \cap \overline{\mathcal{B}(C,t)}$ is less than or equal to $\varepsilon/2$. Consequently, the natural density of $\mathcal{U}$ is 0. □

12.4 Properties of elliptic curves III: Finite sets looking big

In this section we will prove a technical proposition which will allow us to make sure that $\mathcal{T}_1$ is of natural density zero. The proposition considers certain properties of the "denominator" sets $\mathcal{S}_i$. Since these sets are finite, their natural density is, of course, zero. We are, however, interested in how large the ratio computing the density gets for most primes. It turns out, not very surprisingly, that "on average" this ratio is always arbitrarily close to zero.

Proposition 12.4.1. *For any $\varepsilon > 0$, the natural density of the set $\{l : \mu_l(P, \epsilon) > \varepsilon\}$ is zero.*

Proof. If $l \in \mathcal{P}(\mathbb{Q})$ and $\mu_l(P) > \varepsilon$ then there exists $X_l(P, \varepsilon) \geq 2$ in $\mathbb{Z}$ such that

$$\frac{\#\{p \in \mathcal{S}_l(P) \setminus \mathcal{S}_1(P) : p \leq X_l(P,\varepsilon)\}}{\pi_{\mathbb{Q}}(X_l(P,\varepsilon))} > \varepsilon.$$

For $M \in \mathbb{Z}$, $M \geq 2$, let

$$U_M(P, \varepsilon) = \{l \in \mathcal{P}(\mathbb{Q}) : \mu_l(P) > \varepsilon \wedge X_l(P, \varepsilon) \in [M, 2M)\}.$$

Then, if $l \in U_M(P, \varepsilon)$ we have that

$$\begin{aligned}\#\{p \in \mathcal{S}_l(P) \setminus \mathcal{S}_1(P) : p \leq 2M\} &\geq \#\{p \in \mathcal{S}_l(P) \setminus \mathcal{S}_1(P) : p \leq X_l(P, \varepsilon)\} \\ &> \varepsilon \pi_{\mathbb{Q}}(X_l(P, \varepsilon)) \geq \varepsilon \pi_{\mathbb{Q}}(M).\end{aligned}$$

Since by Proposition 12.2.1, for $l \neq p \in \mathcal{P}(\mathbb{Q})$, we have that $\mathcal{S}_l \cap \mathcal{S}_p = \mathcal{S}_1$, it is also the case that

$$\begin{aligned}\pi_{\mathbb{Q}}(2M) &\geq \sum_{l \in U_M(P,\varepsilon)} \#\{p \in \mathcal{S}_l(P) \setminus \mathcal{S}_1(P) : p \leq 2M\} \\ &\geq \varepsilon \pi_{\mathbb{Q}}(M) \#\{l \in U_M(P, \varepsilon)\}.\end{aligned}$$

Thus, by the prime number theorem, $\#\{l \in U_M(\varepsilon, P)\} = O(1)$ as $M \to \infty$. Next, suppose that N is an integer such that $2^{k-1} \leq N < 2^k$. Then on the one hand

$$\begin{aligned}\#\{l \in \mathcal{P}(\mathbb{Q}) &: \mu_l(P) > \varepsilon \wedge X_l(P, \varepsilon) \leq N\} \\ &\leq \sum_{i=1}^{k-1} \#\{l \in \mathcal{P}(\mathbb{Q}) : \mu_l > \varepsilon \wedge 2^i \leq X_l(P, \varepsilon) < 2^{i+1}\} \\ &= \sum_{i=1}^{k-1} \#\{l \in U_{2^i}(\varepsilon, P)\} = O(k) = O(\log N). \qquad (12.4.1)\end{aligned}$$

On the other hand, if $\mu_l(P) > \varepsilon$ then by the definition of $X_l(P, \varepsilon)$ we have that

$$\pi_{\mathbb{Q}}(X_l(P, \varepsilon)) < \frac{\#\{p \in \mathcal{S}_l \setminus \mathcal{S}_1\}}{\varepsilon} \leq \frac{\log_2 d_l}{\varepsilon} = O(l^2),$$

by Lemma B.6.8, as $l \to \infty$. By the prime number theorem, $\pi_{\mathbb{Q}}(X_l(P, \varepsilon)) = O(X_l / \log X_l)$ as $l \to \infty$. Further, by Lemma B.10.1,

$$X_l < Cl^2 \log l \qquad (12.4.2)$$

for some positive constant C as $l \to \infty$. Combining (12.4.1) and (12.4.2), we obtain the following inequality for $Y \in \mathbb{R}$:

$$\begin{aligned}\#\{l \in \mathcal{P}(\mathbb{Q}) &: (l \leq Y) \wedge (\mu_l(P) > \varepsilon)\} \\ &\leq \#\{l \in \mathcal{P}(\mathbb{Q}) : (\mu_l(P) > \varepsilon) \wedge (X_l(P, \varepsilon) \leq CY^2 \log Y)\} \\ &\leq \bar{C} \log(Y^2 \log Y) = o(\pi_{\mathbb{Q}}(Y)),\end{aligned}$$

where $\bar{C}$ is a positive constant and $Y \to \infty$. □

Remark 12.4.2. Note that the supremum $\mu_l(P)$ is attained for some $X \leq \max \mathcal{S}_l(P)$, and therefore $\mu_l(P)$ is computable (recursive) for each l, given P.

12.5 Properties of elliptic curves IV: Consequences of a result of Vinogradov

Below we state a result of Vinogradov which allows the y-coordinates of certain multiples of a chosen point of infinite order to get arbitrarily close to integers.

Proposition 12.5.1. *Let $\alpha \in \mathbb{R} \setminus \mathbb{Q}$. Let $J \subseteq [0, 1]$ be an interval. Then the natural density of the set of primes $\{l \in \mathcal{P}(\mathbb{Q}) : (l\alpha \bmod 1) \in J\}$ is equal to the length of J.* (See Chapter XI of [118].)

From this result we derive the following corollary for our elliptic curve.

Corollary 12.5.2. *Let E be an elliptic curve defined over $\mathbb{Q}$ such that $E(\mathbb{R}) \cong \mathbb{R}/\mathbb{Z}$ as topological groups. Let P be any point of infinite order. Then for any interval $J \subset \mathbb{R}$ whose interior is non-empty, the set $\{l \in \mathcal{P}(\mathbb{Q}) | y_1([l]P) \in J\}$ has positive density.*

Proof. Under our assumptions we have an isomorphism $E(\mathbb{R}) \to \mathbb{R}/\mathbb{Z}$ as topological groups. Under this isomorphism a point of infinite order must be mapped into an irrational number. Since every real number occurs as a y-coordinate of some point in $E(\mathbb{R})$, the set of all points of $E(\mathbb{R})$ projecting y-coordinates onto J is non-empty and open. Finally, this open subset of $E(\mathbb{R})$ will correspond under the above-mentioned isomorphism to a non-empty open subset of $\mathbb{R}/\mathbb{Z}$. □

12.6 Construction of the sets $\mathcal{T}_1(P)$ and $\mathcal{T}_2(P)$ and their properties

In this section we construct (in an effective manner) the sets $\mathcal{T}_1$ and $\mathcal{T}_2$ and make sure that they live up to expectations.

Lemma 12.6.1. Construction of the sequence $\{l_i(P)\}$ *We define a sequence of rational prime numbers $\{l_i\} = \{l_i(P)\}$ in the following inductive manner.*

or, in other words, the order of P mod p is l_i. Thus, since the set $\{l_i\}$ is computable, it is enough to compute the order of P mod p and ascertain whether this order belongs to $\{l_i\}$.

As before we need to consider separately the computability of the three sets comprising $\mathcal{T}_2(P)$.

- $\mathcal{T}_{2a}(P)$ is computable. If $p \in \mathcal{T}_{2a}(P)$ then $p = p_l$ for some $l \notin \{l_i\}$. Thus as above it is enough to compute the order of P mod p and establish that $l \notin \{l_i\}$.
- $\mathcal{T}_{2b}(P)$ is computable. If $p \in \mathcal{T}_{2b}(P)$ then $p = p_{l_i l_j}$, where i, j are positive integers and $1 \leq j \leq i < \log_2 p$, by requirement 4 of Lemma 12.6.1. Thus it is enough to determine the largest element of each set $\mathcal{S}_{l_i l_j} \setminus (\mathcal{S}_{l_i} \cup \mathcal{S}_{l_j})$ for $1 \leq j \leq i < \log_2 p$.
- $\mathcal{T}_{2c}(P)$ is computable. The proof of this assertion is similar to the proof above but uses requirement 5 of Lemma 12.6.1. □

Proposition 12.6.5. *Both $\mathcal{T}_1(P)$ and $\mathcal{T}_2(P)$ are of natural density* 0.

Proof. We start with $\mathcal{T}_1(P)$. The upper natural density of this set is equal to

$$\limsup_{X\to\infty} \frac{\#\left\{p \in \mathcal{S}_0(\mathbb{Q}) \cup \left(\bigcup_{i=1}^{\infty} \mathcal{S}_{l_i}(P)\right) : p \leq X\right\}}{\#\{p \in \mathcal{P}(\mathbb{Q}) : p \leq X\}}$$
$$\leq \limsup_{X\to\infty} \left(\frac{\#\{p \in \mathcal{S}_0(\mathbb{Q}) : p \leq X\}}{\#\{p \in \mathcal{P}(\mathbb{Q}) : p \leq X\}} + \sum_{i=1}^{\infty} \frac{\#\{p \in \mathcal{S}_{l_i}(P) : p \leq X\}}{\#\{p \in \mathcal{P}(\mathbb{Q}) : p \leq X\}} \right)$$
$$\leq \sum_{i=r}^{\infty} \mu_{l_i} \leq \sum_{i=r}^{\infty} 2^{-i} = 2^{-r+1},$$

where r is any positive integer. Thus the density must be zero.

Next we consider the upper natural density of $\mathcal{T}_{2b}(P)$. Note that, by requirement 4 of Lemma 12.6.1, for each $q \in \mathcal{T}_{2b}(P)$ we can find a distinct pair of natural numbers (i, j) with $j \leq i \leq \log_2 q$. Thus for any positive real number X,

$$\#\{q \in \mathcal{T}_{2b} : q \leq X\} \leq \#\{(i, j),\ \ i, j \in \mathbb{N},\ \ 1 \leq i, j \leq \log X\}.$$

Consequently, the upper natural density of $\mathcal{T}_{2b}(P)$ is equal to

$$\limsup_{X\to\infty} \frac{\#\{q \in \mathcal{T}_{2b} : q \leq X\}}{\#\{p \in \mathcal{P}(\mathbb{Q}) : p \leq X\}}$$
$$\leq \limsup_{X\to\infty} \frac{\#\{(i, j),\ \ i, j \in \mathbb{N},\ \ 1 \leq j \leq i \leq \log X\}}{\#\{p \in \mathcal{P}(\mathbb{Q}) : p \leq X\}}$$
$$= \limsup_{X\to\infty} \frac{O(\log^2 X)}{O(X/\log X)} = 0.$$

By requirement 5 of Lemma 12.6.1, the upper natural density of $\mathcal{T}_{2c}(P)$ is similarly equal to

$$\begin{aligned}
&\limsup_{X\to\infty} \frac{\#\{p_{ll_i} \leq X,\ \ i \in \mathbb{N},\ \ i \geq 1,\ l \in \mathcal{L}(P)\}}{\#\{p \in \mathcal{P}(\mathbb{Q}) : p \leq X\}} \\
&\leq \limsup_{X\to\infty} \frac{|\mathcal{L}(P)|\#\{i \in \mathbb{N},\ 1 \leq i \leq \log X\}}{\#\{p \in \mathcal{P}(\mathbb{Q}) : p \leq X\}}. \\
&= \limsup_{X\to\infty} \frac{O(\log X)}{O(X/\log X)} = 0.
\end{aligned}$$

Finally, $\mathcal{T}_{2a}(P) \subset \mathcal{U}(P)$, where $\mathcal{U}(P)$ was defined in Proposition 12.3.5. Since by this proposition the natural density of $\mathcal{U}(P)$ is zero, the natural density of $\mathcal{T}_{2a}(P)$ is also zero. □

The next two propositions will show that if we exclude all primes from $\mathcal{T}_2$ and allow all primes from $\mathcal{T}_1$ in the denominators, the only points of E which will "survive" will be points of the form $[\pm l_i]P$ together with a finite set of points.

Proposition 12.6.6. *Let $\mathcal{W} \subset \mathcal{P}(\mathbb{Q})$ be such that $\mathcal{T}_1(P) \subseteq \mathcal{W}$ and $\mathcal{T}_2(P) \cap \mathcal{W} = \emptyset$. Then there exists a finite set A of natural numbers such that for any $m \in \mathbb{Z}$ we have $[m]P \in E(O_{\mathbb{Q},\mathcal{W}})$ only if $\exists i \in \mathbb{Z}_{>0} : m = \pm l_i$ or $m \in A$.*

Proof. Let

$$m = \pm \prod_{l \in \mathcal{P}(\mathbb{Q})} l^{b(l)},$$

where $b(l) = 0$ for all but finitely many primes. Suppose that $[m]P \in E(O_{\mathbb{Q},\mathcal{W}})$ and for some $i \in \mathbb{Z}_{>0}$ it is the case that $b(l_i) \neq 0$. We claim that in this case $b(l_i) = 1$, and for all $j \neq i$ we have that $b(l_j) = 0$. Indeed, suppose that $b(l_i) > 1$. Then $p_{l_i^2} \mid d_m$ and by Corollary B.6.7 and the construction of $\{l_i\}$, $\mathcal{T}_1(P)$, and $\mathcal{T}_2(P)$, we have to conclude that

$$p_{l_i l_i} \in \left(\mathcal{S}_{p_{l_i^2}} \setminus \mathcal{S}_1\right) \subseteq (\mathcal{S}_m \setminus \mathcal{S}_1) \subset \mathcal{W} \cap \mathcal{T}_2(P) = \emptyset.$$

Thus, if $b(l_i) \neq 0$ then $b(l_i) = 1$. Next we note that a similar argument excludes the case of $b(l_i) = b(l_j) = 1$ simultaneously, for some $i, j \in \mathbb{Z}_{>0}$.

Suppose now that for some $l \in \mathcal{P}(\mathbb{Q}) \setminus \{l_i, i \in \mathbb{Z}_{>0}\}$ we have that $b(l) \neq 0$. We claim that in this case $b(l) \leq a_l(P) - 1$. Indeed, if $b(l) \geq a_l(P)$ then

$$p_l \in \left(\mathcal{S}_{l^{a_l(P)}} \setminus \mathcal{S}_1\right) \subseteq (\mathcal{S}_m \setminus \mathcal{S}_1) \subset \mathcal{W} \cap \mathcal{T}_2(P) = \emptyset.$$

Since Notation 12.1.2 tells us that $a_l(P) > 1$ for $l \in \mathcal{L}$ only, from the discussion above we can now conclude that $m = l_i^{b(l_i)} \prod_{l \in \mathcal{L}} l^{b(l)}$, where $i \in \mathbb{Z}_{>0}$, $b(l_i) = 0$ or1, and $b(l) \leq a(l) - 1$ for $l \in \mathcal{L}$. Now it suffices to show that if $b(l) > 0$ for some $l \in \mathcal{L}$ then $b(l_i) = 0$. Suppose not; then, just as above, we have

$$p_{l_i l} \in \left(\mathcal{S}_{p_{l_i l}} \setminus \mathcal{S}_1\right) \subseteq (\mathcal{S}_m \setminus \mathcal{S}_1) \subset \mathcal{W} \cap \mathcal{T}_2(P) = \emptyset.$$

Now the assertion of the proposition follows from the fact that $\mathcal{L}$ is a finite set. □

It is also clear from the definition of $\mathcal{W}$ and $\mathcal{T}_1(P)$ that the following statement is true.

Proposition 12.6.7. *For all $i \in \mathbb{Z}_{>0}$ we have that $[\pm l_i]P \in E(O_{K,\mathcal{W}})$, where $\mathcal{W}$ is defined as in Proposition 12.6.6.*

We are almost ready to proceed with the final part of the proof of Theorem 12.1.1. We need just one more proposition and corollary to deal with the torsion elements of E.

Proposition 12.6.8. *Let $E/\mathbb{Q}$ be a curve as in Proposition B.6.2. Let $r \geq 1$ be the size of the torsion group of E. Let $Q \in E(\mathbb{Q})$ be a generator of $E(\mathbb{Q})$ modulo the torsion group. Let $P = [r]Q$. Let $\mathcal{W}$* be *defined as in Proposition 12.6.6. Then the set $\{y_{\pm l_i}(P) | i \in \mathbb{N}, i \geq 1\}$ is Diophantine over $E(O_{\mathbb{Q},\mathcal{W}})$.*

Proof. First of all we note that $\{(x, y) \in [r]E(\mathbb{Q})\}$ is a Diophantine subset of $\mathbb{Q}$. Second, let $T \in [r]E(\mathbb{Q})$. Then $T = [r]T'$, where $T' = [k]Q +_E V$ and where $k \in \mathbb{Z}$, "$+_E$" is addition on E, V is an element of the torsion group, and Q is a generator of $E(\mathbb{Q})$ modulo the torsion group. Then $T = [k]([r]Q) = [k]P$. At the same time, it is clear that every multiple of P is in $[r]E(\mathbb{Q})$. Next we observe that, since $O_{\mathbb{Q},\mathcal{W}}$ is Dioph-regular, $\mathbb{Q} \leq_{Dioph} O_{\mathbb{Q},\mathcal{W}}$ and therefore $[r]E(\mathbb{Q}) \cap O_{\mathbb{Q},\mathcal{W}}$ is Diophantine over $O_{\mathbb{Q},\mathcal{W}}$, by the "going up and then down" method. Finally, by Propositions 12.6.6 and 12.6.7 it follows that $[r]E(\mathbb{Q}) \cap O_{\mathbb{Q},\mathcal{W}} = \{\pm[l_i]P, i \in \mathbb{Z}_{>0}\} \cup$ (finite set). Since we can eliminate the points from the finite set by explicitly listing several "not equalities," we conclude that the assertion of the proposition is true. □

Corollary 12.6.9. *Let P, E be as in Proposition 12.6.8. Then the set $\{y_{l_i}(P) | i \in \mathbb{Z}_{>0}\}$ is Diophantine over $E(O_{\mathbb{Q},\mathcal{W}})$.*

Proof. To show that the corollary is true, it is enough to note that we just need to select positive elements of the sets $\{y_{\pm l_i}(P) | i \in \mathbb{N}, i \geq 1\}$. As before this can be done in a Diophantine manner, by Corollary 5.1.2. □

12.7 Proof of Poonen's theorem

Let $\mathcal{W}$ again be defined as in Proposition 12.6.6. We will show that the conditions of Theorem 12.1.1 are satisfied over $O_{\mathbb{Q},\mathcal{W}}$.

1. Since $|y_{l_i}(P) - i| < 1/10i$, the set $D = \{y_{l_i}(P), i \in \mathbb{Z}_{>0}\}$ is clearly discrete and therefore provides a counterexample for the first Mazur conjecture for the ring $O_{\mathbb{Q},\mathcal{W}}$.
2. By Proposition 3.4.7, to show that $O_{\mathbb{Q},\mathcal{W}}$ has a Diophantine model of $\mathbb{Z}$ it is enough to show that, under the proposed computable mapping of $\mathbb{Z}$ into the ring, the graphs of addition and multiplication are Diophantine. To accomplish this, we will first show that the sets

$$D+ = \left\{ \left(y_{l_i}, y_{l_j}, y_{l_k} \right) \in D^3 : k = i + j, \ \ k, i, j \in \mathbb{Z}_{>0} \right\}$$

and

$$D_2 = \left\{ \left(y_{l_i}, y_{l_k} \right) \in D^2 : k = i^2, i \in \mathbb{Z}_{>0} \right\}$$

are Diophantine over $O_{\mathbb{Q},\mathcal{W}}$. To see that $D+$ is a Diophantine subset of $O_{\mathbb{Q},\mathcal{W}}$ observe that for $i, j, k \in \mathbb{Z}_{>0}$ it is true that

$$k = i + j \quad \Leftrightarrow \quad \left| y_{l_i} + y_{l_j} - y_{l_k} \right| < \frac{1}{3}.$$

Indeed, suppose that $k = i + j$; then

$$\begin{aligned} \left| y_{l_i} + y_{l_j} - y_{l_k} \right| &= \left| y_{l_i} - i + y_{l_j} - j - y_{l_k} + k \right| \\ &\leq |y_{l_i} - i| + |y_{l_j} - j| + |k - y_{l_k}| \\ &< \frac{1}{10i} + \frac{1}{10j} + \frac{1}{10k} < \frac{1}{3}. \end{aligned}$$

Conversely, suppose that

$$\left| y_{l_i} + y_{l_j} - y_{l_k} \right| < \frac{1}{3}. \tag{12.7.1}$$

Then

$$\left| y_{l_i} - i + i + y_{l_j} + j - j - k + k - y_{l_k} \right| < \frac{1}{3},$$

$$|i + j - k| < \frac{1}{3} + |y_{l_i} - i| + |y_{l_j} - j| + |k - y_{l_k}| < \frac{2}{3}.$$

Since $i, j, k \in \mathbb{Z}$, we must conclude that $i + j - k = 0$.

Next, we want to show that for $i \in \mathbb{Z}_{>0}$ it is the case that

$$k = i^2 \quad \Leftrightarrow \quad \left|y_{l_i}^2 - y_{l_k}\right| < \frac{2}{5}.$$

So, suppose that $k = i^2$. Then

$$\left|y_{l_i}^2 - i^2 + k - y_{l_k}\right| < \frac{1}{10i}\left|2i + \frac{1}{10i}\right| + \frac{1}{10k} \leq \frac{2}{10} + \frac{1}{100} + \frac{1}{10} < \frac{2}{5}.$$

Conversely, suppose that

$$\left|y_{l_i}^2 - y_{l_k}\right| < \frac{2}{5}. \tag{12.7.2}$$

Then, by an argument similar to the one used above,

$$\begin{aligned}
|i^2 - k| &\leq \frac{2}{5} + \left|y_{l_i}^2 - i^2\right| + \left|y_{l_k} - k\right| \\
&< \frac{2}{5} + \left|y_{l_i} + i\right|\frac{1}{10i} + \frac{1}{10k} \\
&\leq \frac{2}{5} + \left(2i + \frac{1}{10i}\right)\frac{1}{10i} + \frac{1}{10k} \\
&\leq \frac{2}{5} + \frac{1}{5} + \frac{1}{5} = \frac{4}{5} < 1.
\end{aligned}$$

Thus, we conclude that $i^2 - k = 0$. Finally, we note that, by Corollary 5.1.2, the inequalities (12.7.1) and (12.7.2) are Diophantine over $\mathbb{Q}$ and consequently over $O_{\mathbb{Q},\mathcal{W}}$. Thus both $D+$ and D_2 are Diophantine. As in the case of function fields, if squares and sums form Diophantine sets then so do products, since $xy = \frac{1}{2}((x + y)^2 - x^2 - y^2)$.

What we have constructed so far is really a Diophantine model of $(\mathbb{Z}_{>0}, +, \cdot)$ over $O_{\mathbb{Q},\mathcal{W}}$. To obtain a Diophantine model of $(\mathbb{Z}, +, \cdot)$ we can adopt one of the following two strategies.

1. Extend the model of $\mathbb{Z}_{>0}$ to a model of $\mathbb{Z}$. Here we need to select images for 0 and negative integers. Note that $y_{-l_i} = -y_{l_i}$, given the form of our Weierstrass equation. Therefore it is natural to send $-i$ to $-y_{l_i}$ and 0 to 0.
2. Construct a model of $(\mathbb{Z}, +, \cdot)$ over $(\mathbb{Z}_{>0}, +, \cdot)$ by, for example, using pairs of positive integers to represent arbitrary integers: for $a > 0$, let $(1, a)$ represent a, let $(2, a)$ represent $-a$, and let $(3, 1)$ represent 0.

We leave the details of these constructions to the reader.

Remark 12.7.1. We finish our discussion by mentioning two papers which attempt different approaches to the Diophantine problem of $\mathbb{Q}$. The first paper, [70], is by Pheidas and attempts to use points on an elliptic curve to construct a Diophantine model of $\mathbb{Z}$, but in a manner different from Poonen's. Given a rank-1 elliptic curve over $\mathbb{Q}$ and a generator P, Pheidas proposes to map n to $[n]P$. Here the difficulty lies in showing that such a map will make the image of the multiplication graph Diophantine. In fact it is not at all clear that it does.

Taking a different route, Cornelissen and Zahidi in [7] revisit first-order definability results by Julia Robinson and attempt to update them.

13
Beyond global fields

The questions and problems raised by the solution of Hilbert's Tenth Problem extend to many other objects besides the global fields which are the subject of this book. In particular, the questions we have raised are just as relevant for infinite algebraic extensions of global fields, for fields of positive characteristic and of transcendence degree greater than 1, and for function fields of characteristic 0. A detailed and substantial discussion of existential definability and decidability over these objects is beyond the scope of this book, but in this chapter we will briefly survey extensions of Hilbert's Tenth Problem to some objects mentioned above, so that an interested reader can be directed to the original sources. Before proceeding, we would like to note that many detailed surveys of the subjects discussed in this chapter can be found in [20].

13.1 Function fields of positive characteristic and of higher transcendence degree or over infinite fields of constants

The most general result concerning rational function fields of positive characteristic was obtained by H. K. Kim and F. W. Roush in [42]. They proved the following proposition.

Theorem 13.1.1. *Let K be a rational function field over a field C of constants of characteristic $p > 0$. Assume that C does not contain the algebraic closure of a finite field. Then HTP is undecidable over K.*

The proof of the theorem followed essentially the same line as the proof of the Diophantine undecidability of function fields over finite fields of constants described in Chapter 10. Thus its two main ingredients were the Diophantine definability of pth power equations and the Diophantine definability of integrality at a single prime. The proof of the existential definability of pth power equations in the paper [69] by Pheidas applies to any rational function field, but Kim and Roush had to come up with a new existential definition of integrality at a single prime. For that proof they used the strong Hasse norm principle and a function field version of Hilbert's class field. While the original idea of using the strong Hasse norm principle to define integrality at finitely many primes probably belongs to Rumely, who used it in [85] in the context of first-order definability questions, Kim and Roush were the first to use the strong Hasse norm principle explicitly for Diophantine definability.

The result of Kim and Roush was partially lifted to non-rational function fields by Eisenträger and the present author in [22], [102], [104], and [109]. It should be mentioned here that ideas in Prunescu from [77] played an important role in the proofs in [104]. Perhaps the most general results concerning non-rational function fields are in [109] and can be stated in the following manner.

Theorem 13.1.2. *Let M be any function field of characteristic $p > 0$ such that the algebraic closure C of a finite field in M has an extension of degree p. Let L be any field finitely generated over C and linearly disjoint from M over C. Let $K = ML$. Then the Diophantine problem of K is undecidable.*

This theorem has an important corollary.

Corollary 13.1.3. *Let M be a field finitely generated over a finite field. Then HTP is undecidable over M.*

Let us finish this section with two related questions, which so far have eluded researchers in the area.

Question 13.1.4. *Is it possible to give a Diophantine definition of order at a single prime over a function field of positive characteristic over a field of constants which is algebraically closed?*

Question 13.1.5. *Is HTP undecidable over a function field of positive characteristic over a field of constants which is algebraically closed?*

If we are to pursue the second question along the road we have traveled before, with respect to function fields of positive characteristic, then we will have to answer Question 13.1.4 first. However, it is quite conceivable that Question 13.1.4 has a negative answer, while HTP is still undecidable over function fields over algebraically closed fields of constants. So perhaps a different approach is warranted here.

13.2 Algebraic extensions of global fields of infinite degree

Since HTP clearly becomes decidable over the field of all algebraic numbers (the algebraic closure of $\mathbb{Q}$), when we consider infinite algebraic extensions we might expect the situation to change, as indeed it does. We do have decidability in some sufficiently large extensions. First, however, we describe the few Diophantine undecidability results that we have for infinite algebraic extensions of $\mathbb{Q}$.

Perhaps the first person to consider in print the problem of showing that HTP is undecidable in some rings whose quotient fields are infinite extensions of $\mathbb{Q}$ was Denef in [18]. He pointed out that if one could find an elliptic curve defined over $\mathbb{Q}$ such that it had the same positive rank over $\mathbb{Q}$ as over some infinite totally real extension of $\mathbb{Q}$, then one could use such a curve to give a definition of $\mathbb{Z}$ over the ring of algebraic integers of this totally real infinite extension. We do have examples of such elliptic curves, though the complete picture concerning the phenomenon according to which elliptic curves keep the same rank under infinite extensions is far from clear. For more information on the subject see, for example, [59] and [54].

Using a refinement of the methods developed for finite extensions, the present author constructed Diophantine definitions of $\mathbb{Z}$ in "small" and "large" rings of algebraic numbers whose fraction fields were totally real infinite extensions of $\mathbb{Q}$. (See [102] and [110] for more details.)

When one considers HTP as a problem of determining the decidability of some existential theory, it is clear that proving the existential undecidability of some ring implies that the full theory of this ring is also undecidable. Of course the reverse happens in the case of decidability. There, clearly one gets a stronger result by showing that the full theory of a ring is decidable. Thus in many cases, some of them listed below, the decidability of HTP over a ring is a consequence of the fact that the ring's full first-order theory is decidable.

Perhaps the most famous decidability result concerning HTP is due to Rumely. In [86], he showed that HTP was decidable in the ring of all algebraic integers. The proof relied on what became known as Rumely's local–global principle, stating that a variety has a smooth integral point in the algebraic closure of $\mathbb{Q}$ if and only if it has a smooth point in every localization of the ring of all algebraic integers. This result was a generalization of a more restricted version of the local–global principle obtained by Cantor and Roquette in [2]. A similar result was obtained by Moret-Bailly in [61] but using completely different methods.

Rumely's results were later strengthened in different ways. In particular, van den Dries showed in [115] that the first-order theory of the ring of all algebraic integers is decidable. Van den Dries and Macintyre extended this decidability result to many localizations of the ring of all algebraic integers and the ring of all integral functions in [116]. A similar result was obtained by Prestel and Schmid in [75].

In [61] Moret-Bailly, and in [35] Green, Pop, and Roquette, showed that Rumely's results apply to smaller (though still large) fields. Additional versions of both the local–global principle and the decidability results were later obtained by Moret-Bailly (see [62] and [63]), by Jarden and Razon (see [38], [39]), by Darnière (see [11] and [10]), by Prestel and Schmid (see [76]), and by Ershov (see, for example, [27], [28], [29], [30]).

Using different methods, Fried, Haran, and Völklein showed in [32] that the field of totally real numbers has a decidable first-order theory. This result is made more remarkable by the fact that Julia Robinson proved that the first-order theory of the ring of the totally real integers is not decidable (see [83]).

As we said above, this list of results is far from being exhaustive and is intended just to serve as a guide for further reading. We also would like to mention a very nice survey article on the subject by Darnière (see [9]), which we recommend to the interested reader.

13.3 Function fields of characteristic 0

The issues of existential definability prove to be much harder to understand over function fields of characteristic 0 than over function fields of positive characteristic. One problem which has been solved over large classes of function fields of positive characteristic but which remains elusive for many function fields is the problem of the Diophantine definability of order at finitely many primes. Before we state the main results concerning Diophantine undecidability over function fields of characteristic 0, we need to state a definition.

Definition 13.3.1. Let K be a field of characteristic 0. Then it is called formally real if -1 is not a sum of squares. If it is also the case that every odd-degree polynomial has a root in K, then K is called real closed.

The first result concerning function fields was due to Denef, who proved in [16] that HTP is undecidable over rational function fields over formally real fields of constants. Note that this result covers rational function fields of any finite transcendence degree over any real subfield of $\mathbb{R}$ since a rational function field over a formally real field of constants is itself formally real. This result was important not only for what it said about the Diophantine problem of a large class of rational function fields but also because its proof introduced an elliptic curve,

$$(T^3 + aT + b)Y^2 = X^3 + aX + b, \qquad a, b \in \mathbb{Q},$$

(later named the Manin–Denef curve) as a tool for constructing Diophantine models of integers. The proof used the fact that, assuming that the elliptic curve $y^2 = x^3 + ax + b$ has no complex multiplication, the Manin–Denef curve was of rank 1 over $K(T)$ for any field K of characteristic 0. The other part necessary, under Denef's method, for the construction of a Diophantine model was the existential definition of order at any one prime. (This was where the "formally real" condition was used.)

The results of Denef were extended by Kim and Roush, who showed in [44] the Diophantine unsolvability of any rational function field whose constant field could be embedded in a $\mathfrak{p}$-adic field. As Denef, Kim, and Roush constructed a Diophantine model of integers over the fields in question using the Manin–Denef elliptic curve, they devised a different method for defining existentially integrality at a single prime using quadratic forms. Kim and Roush also proved, in [41], that HTP was undecidable over $\mathbb{C}(t_1, t_2)$. This proof again constructed a Diophantine model of $\mathbb{Z}$ with the help of an elliptic curve of rank 1. A very nice exposition of their method can be found in [71].

The next step was taken by Karim Zahidi who showed in [120] that integers have a Diophantine definition in hyperelliptic function fields over real closed fields of constants. The proof of this result was obtained via a sophisticated refinement of the argument used by Denef. For a while, further progress over function fields was stalled by the lack of knowledge concerning elliptic curves of rank 1 over function fields which are not rational. This obstacle has been surmounted by Moret-Bailly, who has shown in [60] that every function field of characteristic 0 has an elliptic curve of rank 1. This result allowed Moret-Bailly to show that HTP is undecidable over any function field over a formally real field of constants or over a constant field which is a subfield of a $\mathfrak{p}$-adic field. (The last result was obtained independently by Kirsten Eisenträger in [25].) Finally, Moret-Bailly's result on elliptic curves has also served as a foundation for the proof by Eisenträger in [23] extending the Kim and Roush results on $\mathbb{C}(t_1, t_2)$ to any function field over $\mathbb{C}$ in two or more variables.

The case of an arbitrary function field of transcendence degree 1 over $\mathbb{Q}$ remains open, the difficulty being centered on the existential definition of the order. In particular, we still do not know the Diophantine status of $\mathbb{C}(t)$

and its algebraic extensions. (As a matter of fact, the status of the first-order theory is unknown for these fields.)

When it comes to rings of functions of characteristic 0 the situation is better, since over these rings there is a natural way to define order using primes which are not inverted. In particular, Moret-Bailly has shown in [60] that HTP is undecidable in any semilocal holomorphy ring of functions of characteristic 0. (Zahidi proved the analogous result in [121] for rational function fields.) The present author proved that $\mathbb{Z}$ has a Diophantine definition in any ring of S-integers of a function field of characteristic 0 in [92] and also in bigger rings in [96] and [105].

Appendix A: Recursion (computability) theory

This appendix contains some basic information on recursive functions, recursive sets, recursively enumerable sets, and relativized versions of these concepts. We also discuss briefly the recursiveness of some algebraic objects. We use [84] and [33] as our main references for the material below.

Terminology. Before proceeding we would like to note the following, concerning the terminology in this appendix. During the past ten years the terms "recursion theory," "recursive functions," "recursive sets," and "recursively enumerable sets" have fallen out of fashion and have been partially supplanted by "computability theory," "computable functions," "computable sets," and "computably enumerable sets" respectively. We will use both types of term in what follows, in Section A.1 the latter and in the remaining short Sections A.2–A.8 the former.

A.1 Computable (recursive) functions

We will use the definition of computable or recursive functions from [33]. First, however, we need to define the "least" operator μ.

Definition A.1.1. Let $R(x_1, \ldots, x_m, y)$ be a relation on $\mathbb{N}$ such that, for each m-tuple

$$\bar{x} = (x_1, \ldots, x_m) \in \mathbb{N}^m,$$

there exists $y \in \mathbb{N}$ for which $R(x_1, \ldots, x_m, y)$ is true. Then, for any $\bar{x} \in \mathbb{N}^m$, define $(\mu y)R(x_1, \ldots, x_m, y)$ to be the smallest natural number y such that $R(x_1, \ldots, x_m, y)$ is true.

Definition A.1.2. Let $\mathcal{F}_n$ be the set of all functions from $\mathbb{N}^n$ to $\mathbb{N}$. Let $\mathcal{F} = \bigcup_{n=1}^{\infty} \mathcal{F}_n$. Then the set of computable or recursive functions is the smallest subset of $\mathcal{F}$ that contains the following functions, which we will call the basic functions.

1. $f(x) = 0$.
2. the successor function $f(x) = x + 1$.
3. the projection function $\pi_i(x_1, \ldots, x_n) = x_i$, where $n \in \mathbb{N}, i \in \{1, \ldots, n\}$,

   which is closed under the following operations.

   1. *Composition.* For computable functions $g \in \mathcal{F}_m$, $h_1, \ldots, h_m \in \mathcal{F}_n$, the function $h \in \mathcal{F}_n$ defined by
   $$h(x_1, \ldots, x_n) = g(h_1(x_1, \ldots, x_n), \ldots, h_m(x_1, \ldots, x_n))$$
   is also computable.
   2. *Inductive definition.* For computable functions $f_0 \in \mathcal{F}_n$, $g \in \mathcal{F}_{n+2}$, the function $f \in \mathcal{F}_{n+1}$ defined by
   $$f(x_1, \ldots, x_n, 0) = f_0(x_1, \ldots, x_n),$$
   $$f(x_1, \ldots, x_n, y + 1) = g(x_1, \ldots, x_n, y, f(x_1, \ldots, x_n, y))$$
   is also computable.
   3. *Minimization.* Let $R(x_1, \ldots, x_m, y)$ be a relation with a computable characteristic function and such that for all $(x_1, \ldots, x_m) \in \mathbb{N}^m$ there exists $y \in \mathbb{N}$ with $R(x_1, \ldots, x_m, y)$. Then the function $f \in \mathcal{F}_m$ defined by $f(x_1, \ldots, x_m) = (\mu y)R(x_1, \ldots, x_m, y)$ is also computable.

There are other equivalent definitions of computable or recursive functions using Turing machines and rudimentary programming languages. (See for example [13].) In practice, however, the definition given above as well as alternative definitions are cumbersome to use. For that reason a proof that a given function or set is computable or recursive is often rendered informally by appealing to Church's thesis, which states that the set of recursive functions is precisely the set of functions whose values can be computed algorithmically. For a discussion of Church's thesis see Section 1.7 of [84], among other texts. In this book we have attempted to provide wherever feasible formal proofs of computability using our definition of computable functions. We have produced informal versions of the argument along the lines of Church's thesis where formalization of the proof is straightforward but would be labor intensive.

Definition A.1.3. Let $A \subset \mathbb{N}^m$ be a set whose characteristic function is computable. Then A will be called computable.

Lemma A.1.4. *The following functions from the natural numbers to the natural numbers or to subsets of the natural numbers are computable:*

1. *addition* $f(x, y) = x + y$ *and multiplication* $f(x, y) = xy$*;*
2. *the minimum function* $\min(x, y)$*;*
3. *the maximum function* $\max(x, y)$*;*
4. *the exponential function* $H(x_1, x_2) = x_1^{x_2}$*;*
5. *the sign function,* $\text{sgn}(0) = 0$ *and* $\text{sgn}(x + 1) = 1$*;*
6. *the absolute-value function* $|x - y|$*;*
7. *the predecessor function* $\Pr(x) = \max(0, x - 1)$*;*
8. *the truncated difference* $\text{TD}(x, y) = \max(0, x - y)$*;*
9. *summation*

$$F(x_1, \ldots, x_m, a, y) = \sum_{i=a}^{y} f(x_1, \ldots, x_m, i)$$

and product

$$H(x_1, \ldots, x_m, a, y) = \prod_{i=a}^{y} f(x_1, \ldots, x_m, i),$$

assuming that $f(x_1, \ldots, x_m, i)$ *is computable;*

10. *bounded minimization, defined as follows. Let* $R(x_1, \ldots, x_m, y)$ *be a relation with a computable characteristic function. Then the function* $f(x_1, \ldots, x_m, a, z) = \mu_{a<y\leq z} R(x_1, \ldots, x_m, y)$ *is defined to be equal to the smallest* y *such that* $a < y \leq z$ *and* $R(x_1, \ldots, x_m, y) = 1$ *if such a* y *exists and to* z *otherwise.*

See Sections 8.3 and 8.4 of [33] for a proof.

Remark A.1.5. We will also use bounded minimization with $a \leq y \leq z$, $a < y < z$, and $a \leq y < z$. It is not hard to show that the slightly different bounds for the variable do not change the computable status of bounded minimization.

Using closure under composition, one can easily show that the following functions are computable.

Corollary A.1.6.

1. *The sum and product of two computable functions are computable.*
2. *All polynomial functions are computable.*
3. *If* f, g *are computable functions then*

$$F(x_1, \ldots, x_n) = f(x_1, \ldots, x_n)^{g(x_1,\ldots,x_n)}$$

is computable.

With a slight effort we also obtain the following corollary of Lemma A.1.4.

Corollary A.1.7.

1. *Let $f(x_1, \ldots, x_n)$ be a computable function. Then the relation*
$$R(x_1, \ldots, x_n, y) = 1 \Leftrightarrow f(x_1, \ldots, x_n) = y$$
is computable.
2. *Let $\{R_i(z_1, \ldots, z_l), i = 1, \ldots, k\}$ be a finite collection of computable relations. Then the relation $R(z_1, \ldots, z_l) = \bigwedge_{i=1}^{k} R_i(z_1, \ldots, z_l)$ is also computable.*

Proof.

1. Define the characteristic function $\chi(x_1, \ldots, x_n, y)$ for R in the following manner. Let $\chi(x_1, \ldots, x_n, y) = 1 - \mathrm{sgn}|f(x_1, \ldots, x_n) - y|$. This function is computable by Lemma A.1.4 and by Corollary A.1.6.
2. This part follows from the fact that a product of computable functions is computable from Corollary A.1.6. □

The next lemma tells us that computable sets are closed under simple set operations.

Lemma A.1.8. *The unions, intersections, cartesian products, and complements of computable sets are computable.*

Proof. Let A, B be computable sets with characteristic functions χ_A and χ_B respectively. Then $\chi_{A\cap B}(n) = \chi_A(n)\chi_B(n)$ is the characteristic function of the intersection of A and B, while $\chi_{A\cup B}(n) = \chi_A(n) + (1 - \chi_A(n))\chi_B(n)$ is the characteristic function of the union. Both functions are computable by Corollary A.1.6. Further, if $(x, y) \in A \times B$ then we can define $\chi_{A\times B}(x, y) = \chi_A(x)\chi_B(y)$. Thus $\chi_{A\times B}(x, y)$ is computable by Corollary A.1.6 also. Finally, if $\chi_A(n)$ is a characteristic function of A then $1 - \chi_A(n)$ is the characteristic function of the complement. □

Next we observe that functions defined by computable clauses are computable.

Lemma A.1.9. *Let $A_1, \ldots, A_n \subset \mathbb{N}^m$ be computable sets constituting a partition of $\mathbb{N}^m$. Let $f_1, \ldots, f_n \in \mathcal{F}^m$ be computable functions. Then the function*
$$h(x_1, \ldots, x_m) = f_i(x_1, \ldots, x_m), \text{ if } (x_1, \ldots, x_m) \in A_i,$$
is computable.

Proof. Let χ_{A_i} be the characteristic function of A_i and define

$$h(x_1, \dots, x_m) = \sum_{i=1}^{n} \chi_{A_i}(x_1, \dots, x_m) f_i(x_1, \dots, x_m).$$

Next we note that the function $h(x_1, \dots, x_m)$ is computable by Corollary A.1.6. □

Lemma A.1.10. *Finite sets are computable.*

Proof. First we show that the set consisting of one element is computable. Let $A = \{a\}$. Then $\chi_A(x) = 1 - \text{sgn}(|x - a|)$, which is clearly computable by Definition A.1.2 and Lemma A.1.4. Next we proceed by induction. Let $\chi_{\{a_1,\dots,a_{n-1}\}}$ be a computable characteristic function of a set of $n - 1$ elements and let $\chi_{\{a_n\}}$ be the characteristic function of $\{a_n\}$. Then $\max(\chi_{\{a_1,\dots,a_{n-1}\}}(x), \chi_{\{a_n\}}(x)) = \chi_{\{a_1,\dots,a_n\}}(x)$ is computable, also by Lemma A.1.4. □

It is a bit more difficult to show that the following proposition holds.

Proposition A.1.11. *The following sets and functions are computable:*

1. *the set of pairs of positive integers (m, n) such that $m \mid n$;*
2. *the set of prime numbers;*
3. *the function $\phi(n)$ computing the nth prime number in the ascending list of all prime numbers. We set $\phi(0) = 0$;*
4. *the function $\rho(n, m)$ computing the exponent of the nth prime number in the prime factorization of a given positive integer m. We set $\rho(n, 0) = \rho(0, m) = 0$;*
5. *the function $\mu(n)$ computing the largest prime dividing a given integer $n \geq 2$. We set $\mu(0) = 0$ and $\mu(1) = 1$.*
6. *the function $\nu(n)$ computing the number of distinct prime factors of a positive integer. We set $\nu(0) = \nu(1) = 0$.*
7. *the function $g(n)$ defined as follows: $g(n) = 0$ if n is not a prime number; $g(n)$ is the sequence number of n in the ascending sequence of all prime numbers if n is a prime number;*
8. *the function computing (m, n), the GCD of m and n for $m \geq 1$, $n \geq 1$. We set $(m, 0) = 0$, $(0, n) = 0$;*
9. *the set $\{(m, n) | m < n\}$;*
10. *the function computing the result of integer division, $n \div m = q$, where $m \neq 0$, $n = mq + r$, $0 \leq r < m$. We set $n \div 0 = 0$.*

Proof.

1. Assuming that $m \geq 1, n \geq 1$, we can let

$$\chi(m, n) = \operatorname{sgn}(n + 1 - \mu_{1 \leq y < n+1}(my = n)).$$

Then $\chi(m, n) = 1$ if and only if $m \mid n$. By Lemmas A.1.4, A.1.8, A.1.9, A.1.10, and Remark A.1.7, $\chi(m, n)$ is computable.

2. Let $\chi_{\text{primes}}(n) = 1 - \operatorname{sgn}(n - \mu_{2 \leq y < n}(y \mid n))$ for $n \geq 2$. Set $\chi_{\text{primes}}(0) = \chi_{\text{primes}}(1) = 0$. Then $\chi(n)$ is computable, by Lemmas A.1.4, A.1.8, A.1.9, A.1.10, and Part 1 of this lemma.
3. We construct an inductive definition for ϕ: let $\phi(1) = 2$ and let

$$\phi(n) = \big(\mu m\big)\big(\max(m, \phi(n-1) + 1) = m \wedge m \text{ is a prime number}\big).$$

Note that $\phi(n)$ is computable by Definition A.1.2, Lemma A.1.4, Corollary A.1.7, and Part 2 of this lemma.

4. Define $\rho(n, m) = \mu_{1 \leq z \leq m}(\phi(n)^{|m-z|} \mid m)$, where $\phi(n)$, as above, is the nth prime in the ascending sequence of primes. Observe that $\rho(n, m)$ is computable by Lemmas A.1.4, A.1.8, A.1.9, A.1.10, and Parts 1 and 3 of this lemma.
5. For $n > 1$ define

$$\mu(n) = n - \mu_{0 \leq z < n+1}(n - z \text{ is a prime number} \wedge (n - z) \mid n).$$

As above, $\mu(n)$ is computable by Lemmas A.1.4, A.1.8, A.1.9, A.1.10, Corollary A.1.7, and Parts 1 and 3 of this lemma.

6. For $n > 1$ define $\nu(n) = \sum_{i=2}^{n} \chi_{\text{primes}}(i)\chi(i, n)$ where, for $i \geq 1$, we have that $\chi(i, n) = 1$ if $i \mid n$ and $\chi(i, n) = 0$ otherwise. Now $\nu(n)$ is computable by Lemmas A.1.4, A.1.8, A.1.9, A.1.10, and Parts 1 and 2 of this lemma.
7. Let $g(n) = \chi_{\text{primes}}(n)\mu_{1 \leq k \leq n}(n = \phi(k))$. Then $g(n)$ is computable by Lemma A.1.4 and Part 2 of this lemma.
8. For $m \neq 0, n \neq 0$ define

$$(m, n) = (m + n) - \mu_{1 \leq k < m+n}\big(((m + n - k) \mid m) \wedge ((m + n - k) \mid n)\big).$$

Then (m, n) is computable by Lemmas A.1.4, A.1.8, A.1.9, A.1.10, Corollary A.1.7, and Part 1 of this lemma.

9. Since $m < n$ is equivalent to $m = \min(m, \Pr(n))$, the computability of this relation follows from Lemma A.1.4.
10. For $m \neq 0$ let $n \div m = \mu(q)\big((mq < n \vee mq = n) \wedge (n - mq < m)\big)$. Then $n \div m$ is computable by Lemmas A.1.4, A.1.8, A.1.9, A.1.10, Corollary A.1.7, and Part 9 of this lemma. □

Lemma A.1.12. *Let $G(X_1, \ldots, X_m)$ be a computable function. Then, for any $A \in \mathbb{N}$,*

$$G(X_1, \ldots, X_{i-1}, A, X_{i+1}, \ldots, X_m)$$

is computable.

Proof. This lemma is easily proved by induction starting with basic functions. □

Next we expand the definition of computable functions to functions whose range is the set of all the finite sequences of non-negative integers. To be more formal, let $\mathcal{N} = \bigcup_{i=1}^{\infty} \mathbb{N}^i$ and consider the following definition.

Definition A.1.13. Let $g : \mathbb{N} \to \mathcal{N}$ be such that the following conditions are satisfied.

- Let $h : \mathbb{N} \to \mathbb{N}$ be defined by $h(n) = (\mu m)(g(n) \in \mathbb{N}^m)$. Then h is computable.
- For $n \in \mathbb{N}$, for $1 \leq i \leq h(n)$ let $f(i, n)$ be the ith coordinate of $g(n)$. For $i > h(n)$ define $f(i, n) = f(i_0, n)$, where $1 \leq i_0 \leq h(n)$ and $i \equiv i_0 \mod h(n)$. Then $f(i, n)$ is computable.

Then call g computable.

Given this expanded definition of computable functions, one can now state the following lemma, whose proof we leave to the reader.

Lemma A.1.14.

1. *For $n > 1$ let $\bar{F}(n, m) = (a_1, \ldots, a_m)$, where $a_i \geq 0$, $n = p_1^{a_1} \cdots p_l^{a_l}$, and $p_1 = 2 \leq \cdots \leq p_l$ is the listing of the first l prime numbers ordered so that $l \geq m$ and is as small as possible subject to this condition. Set $\bar{F}(0, m) = 0$, $\bar{F}(1, m) = 1$.*
2. *Let $G_m : \mathbb{N}^m \to \mathbb{N}$ be defined by $G_m(a_1, \ldots, a_m) = \prod_{i=1}^{m} p_i^{a_i}$, where $p_1 = 2, \ldots, p_m$ are the first m prime numbers in order.*

Then $\bar{F}$ and G_m, for all m, are computable. Further,

$$\bar{F}(G_m(a_1, \ldots, a_m), m) = (a_1, \ldots, a_m)$$

and

$$G_m \circ \bar{F}(n, m) = n,$$

for all n whose largest prime factor is less than or equal to p_m.

A.2 Recursively enumerable sets

We now define recursively or computably enumerable sets, abbreviated as r.e. or c.e. sets.

Definition A.2.1. Let $A \subseteq \mathbb{N}^k$, $k \in \mathbb{Z}_{>0}$. Then we will call A recursively or computably enumerable if A is empty or the range of some recursive function.

Next we observe that it is not hard to show the following.

Lemma A.2.2. *A recursive set is an r.e. set. Further, in the case of an infinite set it can be enumerated in strictly ascending order.*

Proof. If A is an empty recursive set then it is an r.e. set by the definition of r.e. sets. Now let A be a finite recursive set $\{a_1 < \cdots < a_n\}$. For $i \in \{1, \ldots, n\}$, let χ_i be the characteristic function of the set $\{m \in \mathbb{N} : m \equiv i \bmod n\}$. Then χ_i is recursive by Proposition A.1.11. Next let $\xi_A(m) = \sum a_i \chi_i(m+1)$. Then $\xi_A(m)$ is recursive by Corollary A.1.6 and lists the members of A.

Suppose now that A is an infinite recursive set with a recursive characteristic function $\chi_A(n)$. We construct inductively a recursive function $\xi_A(n)$ to list the members of A in ascending order. Let $\xi_A(0) = (\mu k)(\chi_A(k) = 1)$. Let

$$\xi_A(n) = (\mu k)(k > \xi_A(n-1) \wedge \chi_A(k) = 1).$$

Thus $\xi_A(n)$ is recursive by Corollary A.1.7 and lists the members of A in ascending order. □

The converse of Lemma A.2.2 is not true and its negation is the logical foundation for proving the unsolvability of Hilbert's Tenth Problem.

Proposition A.2.3. *There exist r.e. sets which are not recursive.* (See Section 1.9 of [84] for the proof.)

Further, we have the following property of r.e. sets.

Lemma A.2.4. *Let $A \subset \mathbb{N}^k$ be an r.e. set. Let $a_1, \ldots, a_l \in \mathbb{N}$. Let $i_1, \ldots, i_l \in \{1, \ldots, k\}$. Then $B = \{(x_1, \ldots, x_k) \in A : x_{i_1} = a_1 \wedge \cdots \wedge x_{i_l} = a_l\}$ is also an r.e. set.*

Proof. We have to consider two cases: B is finite, possibly empty, and B is infinite. In the former case B is recursive and therefore r.e. by Lemma A.2.2. In the latter case let $\xi_A : \mathbb{N} \to A$ be a recursive function listing the members of

A. Then define $\xi_B : \mathbb{N} \to B$ in the following manner:

$$\xi_B(0) = \xi_A(\mu s)\big(\pi_{i_1}(\xi_A(s)) = a_1, \ldots, \pi_{i_l}(\xi_A(s)) = a_l\big),$$
$$\xi_B(n) = \xi_A(\mu s)\big(s \geq n \wedge \pi_{i_1}(\xi_A(s)) = a_1 \wedge \cdots \wedge \pi_{i_l}(\xi_A(s)) = a_l\big),$$

where π_i, the projection onto the ith coordinate, is one of the basic functions from Definition A.1.2. As in many cases above, ξ_B is recursive by Corollary A.1.7. □

A.3 Turing and partial degrees

We will now relativize the notions of recursiveness and enumerability.

Definition A.3.1. Let $A \subset \mathbb{N}^m$, $B \subset \mathbb{N}^l$ for some positive integers m and l. Let χ_B be the characteristic function of B. Suppose that χ_A, the characteristic function of A, can be constructed from the basic functions and χ_B by a finite number of applications of composition, induction, and minimization. Then we will say that A is Turing reducible to B and write $A \leq_T B$.

Definition A.3.2. Let $A \subset \mathbb{N}^m$, $B \subset \mathbb{N}^l$ for some positive integers m and l. Suppose also that, for any function $\bar{f} : \mathbb{N} \to B$ enumerating B, there exists a function $\bar{g} : \mathbb{N} \to A$ enumerating A, constructed from the basic functions and $\bar{f}$ by a finite number of applications of composition, induction, and minimization. Then we say that A is enumerably reducible to B and write $A \leq_e B$.

It is clear that both Turing reducibility and enumeration reducibility are transitive and reflexive and therefore can be used to form equivalence relations. The equivalence classes corresponding to Turing reducibility are called *Turing degrees* while the classes corresponding to enumeration reducibility are called *partial degrees*. We should also note here that neither type of reducibility implies the other. See Section 9.7 of [84] for more information on the matter.

Proposition A.3.3. *There exist infinitely many partial degrees.*

Proof. Suppose that on the contrary there are only finitely many partial degrees. Let $A_1, \ldots, A_n$ be sets representing each of these degrees and let $\xi_1, \ldots, \xi_n$ be functions listing $A_1, \ldots, A_n$ respectively. Consider all the functions that can be constructed from $\xi_1, \ldots, \xi_n$ using the rules described in Definition A.1.2. Then there are only countably many of these functions. Let $\{f_i, i \in \mathbb{N}\}$ be a listing of all such functions. Let $\xi(n) = f_n(n) + 1$. Then $\xi(n)$ is not on the list and thus cannot be constructed from $\xi_1, \ldots, \xi_n$ using the rules from Definition A.1.2. Thus $\xi(n)$ lists a set that is not relatively enumerable with respect to any of the A_i. Consequently, this set cannot belong to the same partial degree as any of the A_i. □

A.4 Degrees of sets of indices, primes, and products

Given the above definitions of Turing and enumeration reducibilities we now make the following proposition.

Proposition A.4.1. *Let $A \subset \mathbb{Z}_{>0}$ and let p_i be the ith prime number in the ascending list of all prime numbers. Finally, let*

$$\mathcal{P} = \{n \in \mathbb{N} | \exists i \in A, n = p_i\}.$$

Then A and $\mathcal{P}$ are Turing and enumeration equivalent.

Proof. Without loss of generality we can assume that neither set is empty. Next let $g(n)$ be the recursive function defined in Proposition A.1.11, Part 7. Then we can set $\chi_{\mathcal{P}}(m) = \chi_A(g(m))$, since $0 \notin A$. Further, $\chi_A(n) = \chi_{\mathcal{P}}(\phi(n))$, where $\phi(n) = p_n$ is also a recursive function defined in Proposition A.1.11. Thus $\mathcal{P} \equiv_T A$.

Now let $\xi_A(n)$ be the function enumerating A. Then $\phi(\xi_A(n))$ will produce an effective listing of $\mathcal{P}$. Conversely, let $\xi_{\mathcal{P}}(n)$ be a function listing $\mathcal{P}$. Then $g(\xi_{\mathcal{P}}(n))$ will list A. Consequently, $\mathcal{P} \equiv_e A$. □

We will now relate sets of primes to sets of their products.

Proposition A.4.2. *Let $\mathcal{P}$ be a set of prime numbers. Let U consist of all the finite products of elements of $\mathcal{P}$. Then $U \equiv_T \mathcal{P}$ and $U \equiv_e \mathcal{P}$.*

Proof. First of all, without loss of generality we can assume that neither set is empty. Next let the functions $\phi(m)$, $\rho(m)$, $\mu(n, m)$, $\nu(n)$, and $g(m)$ be defined as in Proposition A.1.11, and let the function $\mathrm{sgn}(x)$ be defined as in Proposition A.1.4. Now consider the following function:

$$\chi_U(m) = \prod_{i=1}^{g(\mu(m))} \big(\chi_{\mathcal{P}}(\phi(i))\,\mathrm{sgn}(\rho(i, m)) + (1 - \mathrm{sgn}(\rho(i, m)))\big). \quad \text{(A.4.1)}$$

Suppose that all the factors of m are elements of $\mathcal{P}$. Then for each term in the product one of the following statements will be true.

1. $\phi(i)$ occurs in the factorization of m with a non-zero exponent and therefore, by assumption, $\phi(i) \in \mathcal{P}$. Then $\rho(i,m) > 0$, $\operatorname{sgn}(\rho(i,m)) = 1$, $\chi_{\mathcal{P}}(\phi(i)) \operatorname{sgn}(\rho(i,m)) = 1$, and $1 - \operatorname{sgn}(\rho(i,m)) = 0$.
2. $\phi(i)$ does not occur in the factorization of m. Then $\rho(i,m) = 0$, $\operatorname{sgn}(\rho(i,m)) = 0$, $\chi_{\mathcal{P}}(\phi(i)) \operatorname{sgn}(\rho(i,m)) = 0$, and $1 - \operatorname{sgn}(\rho(i,m)) = 1$.

Thus in either case each term in the product is equal to unity and so $\chi_U(m) = 1$. Suppose now that, for some $j \in \mathbb{N}^+$, the prime p_j occurs with a non-zero exponent in the factorization of m, while $\chi_{\mathcal{P}}(p_j) = 0$. In this case $j \leq g(\mu(m))$ and thus the product contains a term with $i = j$. Note that in this case $\rho(j,m) > 0$, $\operatorname{sgn}(\rho(j,m)) = 1$, and $\chi_{\mathcal{P}}(p_j) \operatorname{sgn}(\rho(j,m)) = 0$, while at the same time $1 - \operatorname{sgn}(\rho(j,m)) = 0$. Thus the jth term in the product is equal to zero and therefore the product is equal to zero. Hence the function $\chi_U(m)$ defined in (A.4.1) is actually the characteristic function of U and we conclude that $U \leq_T A$.

Conversely, let $\chi_{\text{primes}}(m)$ be, as before, the characteristic function of the set of all prime numbers. Then, by Proposition A.1.11, $\chi_{\text{primes}}(m)$ is a recursive function. Hence, it is not difficult to see that $\chi_A(m) = \chi_U(m)\chi_{\text{primes}}(m)$.

Now let $\xi_{\mathcal{P}}(n)$ be any function listing $\mathcal{P}$. Then let

$$\xi_U(m) = \prod_{i=1}^{\rho(1,m)} \xi_{\mathcal{P}}(i)^{\rho(i+1,m)}.$$

As m runs through $\mathbb{N}$, we have that $(\rho(2,m), \ldots, \rho(\rho(1,m)+1, m))$ runs through all the finite sequences of natural numbers. Indeed, let $(a_1, \ldots, a_l)$ be a finite sequence. Then if we let $m = 2^l \prod_{i=2}^{l+1} p_i^{a_{i-1}}$, we see that

$$(\rho(2,m), \ldots, \rho(\rho(1,m)+1, m)) = (a_1, \ldots, a_l).$$

Thus, $\xi_U(m)$ will run through all the elements of U. Consequently, $U \leq_e \mathcal{P}$.

Conversely, let $\xi_U(n)$ be any function listing the elements of U and let $p_0 \in \mathcal{P}$. Then let $\xi_{\mathcal{P}}(m) = \xi_U(m)\chi_{\text{primes}}(m) + p_0(1 - \chi_{\text{primes}}(m))$ and we can conclude that $\mathcal{P} \leq_e U$. □

A.5 Recursive algebra

In this section we will extend the notions of recursive sets and functions to algebraic objects.

Definition A.5.1. Let R be a countable ring or a field. Let $J : R \to \mathbb{N}$ be an injective function such that $J(R)$ is recursive and the functions translating addition, multiplication, and subtraction are recursive (in the case of a field, division is also translated by a recursive function). Then J will be called a recursive presentation of R.

Let $A \subset R$. Then A will be called recursive (r.e) if $J(A)$ is recursive (r.e.) under some recursive presentation of R. Similarly for $A, B \subseteq R$ we will say that $A \cong_T B$ ($A \cong_e B$) if $J(A) \cong_T J(B)$ ($J(A) \cong_e J(B)$) under some recursive presentation J of R.

Let $f : R^n \to R$ be a function. Then f will be called recursive if under some recursive presentation of R, the translation of f is recursive.

Proposition A.5.2. *Let R, J be as in Definition A.5.1. Then translations of all polynomial functions, if R is a ring, and the translation of all rational functions, if R is a field, are recursive.*

Proof. The proof of this proposition is analogous to the proof of Lemma 3.2.2. □

A.6 Recursive presentation of $\mathbb{Q}$

In this section we will describe a recursive presentation of $\mathbb{Q}$ in some detail. We will start, though, with a presentation of $\mathbb{Z}$.

Proposition A.6.1. *Consider the map $J : \mathbb{Z} \to \mathbb{N}$ defined by $J(m) = 2^a 3^{|m|}$, where $a = 0$ if $m \geq 0$ and $a = 1$ otherwise. Then the following statements are true.*

1. *J is a recursive presentation of $\mathbb{Z}$.*
2. *For any $A \subseteq \mathbb{N}$ we have that $J(A) \equiv_T A$ and $J(A) \equiv_e A$.*
3. *The absolute-value function is a recursive function on $\mathbb{Z}$.*
4. *Let $A \subset \mathbb{N}$. Let $B \subset \mathbb{Z}$ be defined by $B = \{x \in \mathbb{Z} | |x| \in A\}$. Then $J(B) \equiv_T J(A)$ and $J(B) \equiv_e J(A)$.*

Proof.
1. First of all, the unique factorization theorem assures us that J is injective. To see that $J(\mathbb{Z})$ is a recursive subset of $\mathbb{N}$, note that

$$J(\mathbb{Z}) = \big\{m \in \mathbb{N} : m = 1 \vee [(4 \nmid m) \wedge (\mu(m) = 3)]\big\}, \qquad \text{(A.6.1)}$$

where $\mu(m)$ is the function defined in Proposition A.1.11, Part 5. In other words we can describe an element of $J(\mathbb{Z})$ in the following manner: either it is equal to

1 or it is a natural number not divisible by 4 and 3 is the largest prime dividing it. Using Section A.1 we easily deduce that (A.6.1) is a recursive set.

Next we need to show that addition, subtraction, and multiplication are recursive under J. Let

$$A_{o,o},\ A_{o,e},\ A_{e,o},\ A_{e,e},\ A_{o,e,1},\ A_{o,e,2},\ A_{e,o,1},\ A_{e,o,2} \subset \mathbb{N}^2$$

be defined as follows:

$$\begin{aligned}
A_{o,o} &= \{(m,n), 2 \nmid m, 2 \nmid n\},\\
A_{e,e} &= \{(m,n), 2 \mid m, 2 \mid n\},\\
A_{o,e} &= \{(m,n), 2 \nmid m, 2 \mid n\},\\
A_{e,o} &= \{(m,n), 2 \mid m, 2 \nmid n\},\\
A_{o,e,1} &= \{(m,n) \in A_{o,e},\ 2m \geq n\},\\
A_{o,e,2} &= \{(m,n) \in A_{o,e},\ 2m < n\},\\
A_{e,o,1} &= \{(m,n) \in A_{o,e},\ m > 2n\},\\
A_{e,o,2} &= \{(m,n) \in A_{e,o},\ m \leq 2n\}.
\end{aligned}$$

Using Section A.1 again, we conclude that these sets are recursive. Further, it is easy to see that

$$\{A_{o,o},\ A_{e,e},\ A_{o,e,1},\ A_{o,e,2},\ A_{e,o,1},\ A_{e,o,2}\}$$

is a partition of $\mathbb{N}^2$. Therefore Lemma A.1.9 tells us that we can define the following recursive (computable) function:

$$P_+(m,n) = \begin{cases} 2 \cdot 3^{(\rho(2,m)+\rho(2,n))} & \text{if } (m,n) \in A_{e,e},\\ 3^{(\rho(2,m)+\rho(2,n))} & \text{if } (m,n) \in A_{o,o},\\ 3^{|\rho(2,m)-\rho(2,n)|} & \text{if } (m,n) \in A_{o,e,1} \cup A_{e,o,2},\\ 2 \cdot 3^{|\rho(2,m)-\rho(2,n)|} & \text{if } (m,n) \in A_{e,o,1} \cup A_{o,e,2}, \end{cases}$$

where $\rho(2,n)$ is the exponent of 3 in the prime factorization of n. It is clear that $J(x+y) = P_+(J(x), J(y))$. Next let

$$\text{Minus}(m) = \begin{cases} 2m & \text{if } 2 \nmid m,\\ m/2 & \text{if } 2 \mid m. \end{cases}$$

Then Minus(m) is a recursive function by Section A.1. Also, by construction, Minus$(J(x)) = J(-x)$. Now define $P_\times$ by

$$P_\times(m,n) = \begin{cases} 3^{(\rho(2,m)\rho(2,n))} & \text{if } (m,n) \in A_{e,e} \cup A_{o,o},\\ 2 \cdot 3^{(\rho(2,m)\rho(2,n))} & \text{if } (m,n) \in A_{o,e} \cup A_{e,o} \end{cases}$$

By the same considerations as above, it is clear that $P_\times$ is a recursive function and, by construction,

$$P_\times(J(x), J(y)) = J(xy)$$

for $x, y \in \mathbb{Z}$. Thus, the first assertion of the proposition holds.

2. Next we note that $J(\mathbb{N}) = \{m \in \mathbb{N} : (m = 1) \vee ((2 \nmid m) \wedge (\mu(m) = 3))\}$ and, therefore, $J(\mathbb{N})$ is recursive.

Now let $A \subseteq \mathbb{N}$ and let χ_A be a characteristic function of $A \subseteq \mathbb{N}$. Also let $\chi_{J(\mathbb{N})}$ be the characteristic function of $J(\mathbb{N})$ and observe that it is recursive by the argument above. Then $\chi_{J(A)}(n) = \chi_{J(\mathbb{N})}(n)\chi_A(\rho(2, n))$. Conversely, if $\chi_{J(A)}$ is the characteristic function of $J(A)$ then $\chi(A)(n) = \chi_{J(A)}(3^n)$. Further, let $\xi_A(n)$ be a function listing A; then $\xi_{J(A)} = 3^{\xi_A(n)}$ will list $J(A)$. Conversely, let $\xi_{J(A)}(n)$ be a function listing $J(A)$; then $\rho(2, \xi_{J(A)}(n))$ will list A.

3. Let $J(x) = n$. We set $\text{abs}(n) = 3^{\rho(2,n)}$. Note that $\text{abs}(n)$ is a recursive function by Section A.1, and it is clear that $\text{abs}(J(x)) = J(|x|)$.

4. Here we note that

$$\chi_{J(B)}(n) = \chi_{J(A)}(\text{abs}(n))$$

and

$$\chi_{J(A)}(n) = \chi_{J(B)}(n)\chi_{J(\mathbb{N})}(n),$$

where $\chi_{J(A)}$, $\chi_{J(B)}$, and $\chi_{J(\mathbb{N})}$ are the characteristic functions of $J(A)$, $J(B)$, and $J(\mathbb{N})$ respectively. Thus we have Turing equivalence.

Now let $\xi_{J(A)}(n)$ be a function listing $J(A)$. Then define $\xi_{J(B)}(2m) = \xi_{J(A)}(m)$ and $\xi_{J(B)}(2m + 1) = 2\xi_{J(A)}(m)$. Using Section A.1 we can ascertain that $\xi_{J(B)}(n)$ can be constructed from $\xi_{J(A)}(n)$ and basic functions, by the rules stipulated in Definition A.1.2. Finally, given a function $\xi_{J(B)}(n)$ enumerating $J(B)$ we can obtain $\xi_{J(A)}(n)$, a function enumerating $J(A)$, by setting $\xi_{J(A)}(n) = \text{abs}(\xi_{J(B)}(n))$. □

We now expand our recursive presentation to $\mathbb{Q}$.

Definition A.6.2. Define $J_{\mathbb{Q}} : \mathbb{Q} \to \mathbb{N}$ as follows. For $m \in \mathbb{Z} \setminus \{0\}$, $n \in \mathbb{N} \setminus \{0\}$, $(m, n) = 1$, set $J_{\mathbb{Q}}(m/n) = J(m) \cdot 5^{n-1}$. Also set $J_{\mathbb{Q}}(0) = J(0) = 1$.

Proposition A.6.3. *$J_{\mathbb{Q}}$ is a recursive presentation of $\mathbb{Q}$ extending the map $J : \mathbb{Z} \to \mathbb{N}$ defined in Proposition A.6.1.*

Proof. First of all we note that $J_{\mathbb{Q}}$ is injective by the fundamental theorem of arithmetic. The fact that $J_{\mathbb{Q}}$ extends J from A.6.1 is clear since $J_{\mathbb{Q}}(m/1) = J(m)$. Next observe that

$$J_{\mathbb{Q}}(\mathbb{Q}) = \{m \in \mathbb{N} : (m = 1) \vee [(3 \mid m) \wedge (4 \nmid m) \wedge (\mu(m) \leq 5)]\}$$

and this set is recursive by Section A.1 as before.

Next we need to show that the field operations are also recursive. We will start with functions that can produce the reduced numerator and denominator of an element of $\mathbb{Z}$. Define recursive functions

$$\text{Num}(n) = 2^{\rho(1,n)} 3^{\rho(2,n)}$$

and

$$\text{Denom}(n) = 3^{\rho(3,n)+1}.$$

Note that the image of both functions is clearly in $J(\mathbb{Z})$, and if $n \in J_{\mathbb{Q}}(\mathbb{Q})$ then

$$J_{\mathbb{Q}}\big(J^{-1}(\text{Num}(n)) / J^{-1}(\text{Denom}(n))\big) = n.$$

Further, observe that for $m \in J_{\mathbb{Q}}(\mathbb{Q})$ we have that $m \leq 2 \cdot 5^{\rho(2,N)+\rho(2,D)}$, where $N = \text{Num}(m)$ and $D = \text{Denom}(m)$. Now given n_1, n_2 let

$$\begin{aligned} N_+ &= \text{Num}_+(n_1, n_2) \\ &= P_+\big(P_\times(\text{Num}(n_1), \text{Denom}(n_2)),\ P_\times(\text{Num}(n_2), \text{Denom}(n_1))\big). \end{aligned}$$

Let

$$D_+ = \text{Denom}_+(n_1, n_2) = P_\times(\text{Denom}(n_1), \text{Denom}(n_2)).$$

Here $P_+, P_\times$ are translations of addition and multiplication under J. Next let

$$\begin{aligned} P_{+,\mathbb{Q}}(n_1, n_2) = \mu_{1 \leq m \leq 2 \cdot 5^{\rho(2,N_+)+\rho(2,D_+)}}\big(m \in J_{\mathbb{Q}}(\mathbb{Q}) \wedge P_\times(N_+, \text{Denom}(m)) \\ = P_\times(D_+, \text{Num}(m))\big). \end{aligned}$$

Next let

$$\text{Minus}_{\mathbb{Q}}(n) = 2^{\rho(1,\text{Minus}(\text{Num}(n)))} 3^{\rho(2,n)} 5^{\rho(3,n)}.$$

Let

$$N_\times = \text{Num}_\times(n_1, n_2) = P_\times(\text{Num}(n_1), \text{Num}(n_2))$$

and let

$$D_\times = \text{Denom}_\times(n_1, n_2) = P_\times(\text{Denom}(n_1), \text{Denom}(n_2)).$$

Further, let

$$P_{\times,\mathbb{Q}}(n_1, n_2) = \mu_{1 \le m \le 2\cdot 5^{\rho(2,N_\times)+\rho(2,D_\times)}}\big(m \in J_{\mathbb{Q}}(\mathbb{Q}) \wedge P_\times(N_\times, \text{Denom}(m)) \\ = P_\times(D_\times, \text{Num}(m))\big)$$

Our last task is to define the reciprocals. Define

$$\text{Oneover}(n) = 2^{\rho(1,n)}3^{\rho(3,n)+1}5^{\rho(2,n)-1}$$

and note that for $n \in J_{\mathbb{Q}}(\mathbb{Q})$ we always have that $\rho(2,n) \ge 1$ since $3|n$. □

We can now continue the discussion we started with Propositions A.4.1 and A.4.2.

Proposition A.6.4. *Let $I \subseteq \mathbb{N}$. Let $\mathcal{A}_I = \{p_i, i \in I\}$. Let U_I be the set of natural numbers with prime factors in $\mathcal{A}_I$. Then $J_{\mathbb{Q}}(O_{\mathbb{Q},U_I}) \cong_e I$ and $J_{\mathbb{Q}}(O_{\mathbb{Q},U_I}) \cong_T I$.*

Proof. Since $U_I \cong_T \mathcal{A}_I \cong_T I$ and $U_I \cong_e \mathcal{A}_I \cong_e I$ by Propositions A.4.1 and A.4.2, it is enough to show that U_I is Turing and enumeration equivalent to $O_{\mathbb{Q},U_I}$. Let χ_{U_I} be the characteristic function of U_I. Then $\chi_{J_{\mathbb{Q}}(O_{\mathbb{Q},U_I})}(n) = \chi_{U_I}(\rho(3,n)+1)$ is the characteristic function of $J_{\mathbb{Q}}(O_{\mathbb{Q},U_I})$. Conversely, let $\chi_{J_{\mathbb{Q}}(O_{\mathbb{Q},U_I})}(n)$ be the characteristic function of $J_{\mathbb{Q}}(O_{\mathbb{Q},U_I})$. Then $\chi_{U_I}(n) = \chi_{J_{\mathbb{Q}}(O_{\mathbb{Q},U_I})}(3 \cdot 5^{n-1})$. □

One of the most important properties of the recursive presentation above is that it preserves all the Turing and enumeration degrees of subsets of $\mathbb{N}$. This fact is not accidental. As a matter of fact, this will be true of any recursive presentation of $\mathbb{Z}$ and $\mathbb{Q}$. Since this fact will play an important role in our discussion of the undecidability of HTP, we prove below that a general version of this phenomenon holds.

Proposition A.6.5. *Let R (F) be a finitely generated ring (or field). For $i = 1, 2$, let $J_i : R \to \mathbb{N}$ $(J_i : F \to \mathbb{N})$ be two recursive presentations of R (F). Then there exists a recursive function $g : \mathbb{N} \to \mathbb{N}$ such that $g \circ J_1 = J_2$.*

Proof. We will sketch a proof for the case of a finitely generated ring of characteristic 0. The positive-characteristic case and the case of a finitely generated field are similar. Let $x_1 = 1, \ldots, x_n$ be a set of generators of R. Let $m \in J_1(R)$ be given. Then by systematically iterating ring operations on $x_1, \ldots, x_n$ we will construct a polynomial $f(x_1, \ldots, x_n) \in \mathbb{Z}[x_1, \ldots, x_n]$ such that $J_1(f)(J_1(x_1), \ldots, J_1(x_n)) = m$. Then we can set

$$g(m) = J_2(f)(J_2(x_1), \ldots, J_2(x_n)).$$

□

If we apply Proposition A.6.5 to $R = \mathbb{Z}$ and $F = \mathbb{Q}$, we obtain the following.

Corollary A.6.6. *Let $J : \mathbb{Z} \to \mathbb{N}$ be a recursive presentation of $\mathbb{Z}$. Then for any subset $A \subset \mathbb{N}$, $A \equiv_T J(A)$ and $A \equiv_e J(A)$. Further, under any recursive presentation J of $\mathbb{Q}$, we have that $J(\mathbb{Z})$ is recursive.*

A.7 Recursive presentation of other fields

Having dealt in great detail with a recursive presentation of $\mathbb{Q}$ we will present a more abbreviated account of the construction of recursive presentations of other fields.

Proposition A.7.1. *Let F be a countable field and let $j : F \to \mathbb{N}$ be a recursive presentation of F. Let t be transcendental over F. Then there exists a recursive presentation $J : F[t] \to \mathbb{N}$ such that $J|_F = \psi \circ j$, where $\psi : \mathbb{N} \to \mathbb{N}$ is recursive. Further, under J there exist the following recursive functions.*

- $\deg : \mathbb{N} \to \mathbb{N}$ *such that for any $P(t) \in F[t]$ we have that* $\deg \circ J(P(t)) = \deg(P(t))$.
- $C : \mathbb{N} \times \mathbb{N} \to \mathbb{N}$ *such that for any $a_0, \ldots, a_n \in F$ we have that* $C(J(\sum_{i=0}^n a_i t^i, m)) = J(a_m)$, *where $a_m = 0$ for $m >$* $\deg(\sum_{i=0}^n a_i t^i)$.
- $GCD : \mathbb{N} \times \mathbb{N} \to \mathbb{N}$, *where for any $P_1(t), P_2(t) \in F[t]$ we have that* $GCD(J(P_1(t)), J(P_2(t))) = J((P_1(t), P_2(t)))$.

Proof. Let $P(t) = \sum_{i=0}^n a_i t^i \in F[t]$. Then define $J(t) = \prod_{i=0}^n p_{i+1}^{j(a_i)}$, where $p_1 = 2 < p_2 < \cdots < p_{n+1}$ are the first $n + 1$ rational primes. By Proposition A.1.11, since $j(F)$ is recursive, $J(F[t])$ is a recursive set and the functions deg and C are recursive. Defining ring operations and the greatest common divisor of two polynomials is a straightforward but rather tedious process requiring the use of Proposition A.1.11 again and again. We will leave the details as an exercise for the reader. □

Corollary A.7.2. *Let F be a countable field and let $j : F \to \mathbb{N}$ be a recursive presentation of F. Let t be transcendental over F. Then there exists a* *recursive presentation* $\bar{J} : F(t) \to \mathbb{N}$ *such that* $J|_F = \psi \circ j$, *where* $\psi : \mathbb{N} \to \mathbb{N}$ *is recursive. Further, under* $\bar{J}$ *there exist the following recursive functions,* $\text{Num} : \mathbb{N} \to \mathbb{N}$ *and* $\text{Denom} : \mathbb{N} \to \mathbb{N}$, *such that for any* $f \in F(t)$ *we have that*

- $\bar{J}^{-1}(\text{Num}(\bar{J}(f))),\ \bar{J}^{-1}(\text{Denom}(\bar{J}(f))) \in F[t]$;
- $\bar{J}^{-1}(\text{Denom}(\bar{J}(f)))$ *is monic;*
- $\left(\bar{J}^{-1}(\text{Num}(\bar{J}(f))),\ \bar{J}^{-1}(\text{Denom}(\bar{J}(f)))\right) = 1$ *in* $F[t]$;
- $f = \bar{J}^{-1}(\text{Num}(\bar{J}(f)))/\bar{J}^{-1}(\text{Denom}(\bar{J}(f)))$.

Proof. Let $f \in F(t)$; then f has a unique representation as $P(t)/Q(t)$, where $P(t), Q(t) \in F[t]$ are relatively prime in the polynomial ring and $Q(t) \neq 0$ is monic. Thus we can define $\bar{J}(f) = 2^{J(P(t))}3^{J(Q(t))}$, where J is the presentation of $F[t]$ defined in Proposition A.7.1. Again, we leave the straightforward but lengthy proof that the resulting presentation satisfies all the conditions stated in the corollary to the interested reader. □

We will next prove a general proposition about constructing a recursive presentation of a fraction field, given a recursive presentation of a ring. First we need a new definition.

Definition A.7.3. Let R be a countable ring. We will call R strongly recursive if R is recursive and under some recursive presentation of R we have that the set $D = \{(x, y) : \exists z \in R,\ x = yz\}$ is recursive. Such a presentation of R will also be called strongly recursive. (Observe that D is a Diophantine subset of R^2.)

By virtue of the division algorithm, it is clear that $\mathbb{Z}$ and any polynomial ring over a recursive field are strongly recursive. This property is in general enough for the construction of a recursive presentation of the fraction field while preserving the recursive status of the ring.

Proposition A.7.4. *Let R be a recursive ring under a presentation j. Let F be its fraction field. Then there exists a recursive presentation J of F such that $J(R)$ is also recursive if and only if R is strongly recursive. Further, if R is strongly recursive then under some recursive presentation J of its fraction field F there exist recursive functions $\Lambda_1, \Lambda_2 : \mathbb{N} \to \mathbb{N}$ such that, for all $x \in F$,*

$$x = \frac{J^{-1}(\Lambda_1(J(x)))}{J^{-1}(\Lambda_2(J(x)))}, \qquad \Lambda_1(J(x)), \Lambda_2(J(x)) \in J(R), \qquad \Lambda_2(J(x)) \neq 0.$$

Proof. We will first show that if R is strongly recursive then the required presentation of F exists. The construction we present here is in some sense the usual construction of the fraction field of the ring using the equivalence classes of fractions. To simplify the discussion, instead of constructing a map $J : F \to \mathbb{N}$, we will construct a map $J : F \to \mathbb{N}^2$ satisfying all the requirements. We remind the reader that we can always move back and forth from $\mathbb{N}^k$ to $\mathbb{N}$ via recursive functions with recursive inverses, using Lemma A.1.14.

So let $\phi : \mathbb{N} \to j(R)^2$ be a recursive listing of $j(R)^2$. (It exists by Lemma A.2.2 and Lemma A.1.8.) We will denote $\phi(n)$ by (a_n, b_n). Below we list steps that will result in the construction of J. We will define a recursive function $H : \mathbb{N} \to j(R^2)$, which essentially will be the presentation of F.

Assume that $H(1), \ldots, H(n)$ have been defined already. (In the case $n = 1$ the list is empty.) If $b_{n+1} = j(0)$ then set $H(n+1) = (j(0), j(0))$. Otherwise check whether for some $i = 1, \ldots, n$ we have that $P_\times(a_i, b_{n+1}) = P_\times(a_{n+1}, b_i)$, where $P_\times$ is, as usual, the translation of multiplication over R under j. If such an i is found then set $H(n+1) = H(i)$. If such an i is not found then check whether $(a_n, b_n) \in j(D)$. If this is the case, then find c such that $P_\times(c, b_n) = a_n$ and set $H(n+1) = (c, j(1))$. Finally, if $(a_{n+1}, b_{n+1}) \notin j(D)$ set $H(n+1) = (a_{n+1}, b_{n+1})$.

We claim that H is recursive (by the construction above) and that there exists a map $J : F \to H(\mathbb{N})$ such that J is a recursive presentation of F. Let $z \in F$. Then $z = x/y$, $y \neq 0$, $x, y \in R$. Let $(a_n, b_n) = (j(x), j(y))$. Then define $J(z) = H(n)$. The construction above implies that J is well defined. (We leave the details of the proof of the above claim to the reader.) Next observe that $J(R)$ is recursive. Indeed, given $(m, k) \in \mathbb{N}^2$ we can determine effectively whether $(m, k) \in J(R)$. First of all we determine whether $(m, k) \in j(R)^2$ and if so determine $n \in \mathbb{N}$ such that $(m, k) = (a_n, b_n)$. If $(m, k) \notin j(R)^2$ then $(m, k) \notin J(F)$. Next, assuming that $(m, k) = (a_n, b_n)$, we compute $H(n)$ and check whether $H(n) = (m, k)$. If the answer is "yes" and $k \neq j(0)$ then $(m, k) \in J(F)$. If either $H(n) \neq (m, k)$ or $k = j(0)$ then $(m, k) \notin J(F)$. Further, note that $J(R)$ is also recursive. Indeed, given a pair $(m, k) \in J(F)$ it is enough to check whether $k = j(1)$ to determine whether $(m, k) \in J(R)$. Also, by construction, for each $z \in F$ we have $J(z) = (\Lambda_1(J(z)), \Lambda_2(J(z)))$.

Next we need to show that all the operations over F are recursive. We will show that this is true for addition, and analogous proofs could be produced for the remaining field operations. Given (m_1, k_1), (m_2, k_2), do the following to compute the sum. Let

$$l = P_+(P_\times(m_1, k_2),\ P_\times(m_2, k_1)), u = P_\times(k_1, k_2)),$$

where P_+, $P_\times$ are the translation of addition and multiplication over R under j. Note that $(l, u) \in j(R)^2$. So we can find n such that $(a_n, b_n) = (l, u)$. Then set the sum of (m_1, k_1) and (m_2, k_2) equal to $H(n)$.

Suppose now that R is an arbitrary recursive ring and there exists a recursive presentation $J : F \to \mathbb{N}$ such that $J(R)$ is recursive. Then let $(m, k) \in J(R)^2$. We can determine whether $(m, k) \in J(D)$ by checking whether $P_/(m, k) \in J(R)$. □

Remark A.7.5. In the proof above we presented algorithms informally. We remind the reader that informal descriptions of algorithms and their relation to the formal descriptions were discussed immediately following Definition A.1.2.

Remark A.7.6. While the proposition above plays a role in tying together generation and Diophantine undecidability, in general it does not provide enough information about the recursive presentation of the fraction field. For this reason we have chosen to construct the recursive presentations of rational numbers and rational functions directly.

Our next task is to discuss a presentation of finite field extensions.

Proposition A.7.7. *Let F be a countable field and let $j : F \to \mathbb{N}$ be a recursive presentation of F. Let G be a finite extension of F of degree n. Let $\Omega = \{\omega_1, \ldots, \omega_n\}$ be a basis of G over F. Then there exists a recursive presentation $J : G \to \mathbb{N}$ such that J restricted to F is equal to $\psi \circ j$, where $\psi : \mathbb{N} \to \mathbb{N}$ is recursive, and there exist recursive coordinate functions $C_1, \ldots, C_n : \mathbb{N} \to \mathbb{N}$ such that for any element $x \in G$ we have that $x = \sum_{i=1}^{n} J^{-1} \circ C_i \circ J(x)\omega_i$ with $C_i \circ J(x) \in J(F)$, for all $i = 1, \ldots, n$.*

Proof. This proposition can be proved utilizing a construction that we used to extend weak presentations in Proposition 3.2.4. □

As an immediate corollary of Proposition A.6.3, Corollary A.7.2, and Proposition A.7.7 we get the following result.

Corollary A.7.8. *Global fields are recursive.*

Remark A.7.9. From now on we will assume that number fields are given under a recursive presentation J, as described in Proposition A.7.7, with respect to some integral basis Ω of the field over $\mathbb{Q}$. (Such a basis always exists. See Definition B.1.29 and Proposition B.1.30.)

Proposition A.7.7 has another useful corollary.

Corollary A.7.10. *Let F, G, J be as in Proposition A.7.7. Given $x \in F$, let $A_0(x), \ldots, A_n(x)$ be the coefficients of the characteristic polynomial of x over*

G. Then each $A_i(x)$ is a recursive function under J, depending on the chosen basis of F over G only.

Proof. Indeed, let $x \in G$, $x = \sum_{i=1}^{n} a_i\omega_i$, where $\Omega = \{\omega_1, \ldots, \omega_n\}$. Then for each i we have that $A_i(x)$ is a fixed polynomial in $a_1, \ldots, a_n$. Since $J(a_i) = C_i(J(x))$, where the C_i are the recursive functions defined in Proposition A.7.7 and all the polynomial functions are translated by recursive functions, by Proposition A.5.2 the assertion holds. □

We will now deal with a case of countable transcendence degree.

Proposition A.7.11. *Let F be a countable field and let $j : F \to \mathbb{N}$ be a recursive presentation of F. Let $\{t_i, i \in \mathbb{N}\}$ be a set of variables algebraically independent over F. Then $F(t_1, \ldots, t_k, \ldots)$ also has a recursive presentation.*

Proof. We will describe the encoding and leave it to the reader to prove that the resulting presentation is recursive. Let $f = P/Q$, where

$$P = \sum_{(i_1,\ldots,i_k)} A_{i_1,\ldots,i_k} t_1^{i_1} \cdots t_k^{i_k},$$

$$Q = \sum_{(l_1,\ldots,l_s)} A_{l_1,\ldots,l_s} t_1^{l_1} \cdots t_s^{l_s},$$

where P, Q are relatively prime in the polynomial ring $F[t_1, \ldots]$. Assume that if we sort the terms of Q into a lexicographical order using powers of the t_i then the first term has coefficient 1. Clearly, every element of the field has a unique representation of this form. For a monomial $A_{i_1,\ldots,i_k} t_1^{i_1} \cdots t_k^{i_k}$, let

$$J_1\big(A_{i_1,\ldots,i_k} t_1^{i_1} \cdots t_k^{i_k}\big) = 2^{\,j(A_{i_1,\ldots,i_k})} 3^{i_1} \cdots p_{k+1}^{i_k}.$$

For a polynomial $P = \sum_{r=1}^{u} M_r$, where M_r is a monomial, let $J_2(P) = \prod_{r=1}^{u} p_r^{J_1(M_r)}$. Finally, let $J(f) = 2^{J_2(P)} 3^{J_2(Q)}$. □

Proposition A.7.12. *Let K be a recursive field. Then there exists an algebraic closure of K which is recursive.* (See Theorem 7 of [78].)

Proposition A.7.13. *Let G be a countable field. Then G is a subfield of a recursive field.*

Proof. We have to consider two possibilities: G has characteristic 0 or G has a positive characteristic. First assume that G has characteristic 0. Then G contains

$\mathbb{Q}$ and consequently G is a subfield of the algebraic closure of $\mathbb{Q}(t_1, \ldots)$, which is computable by Propositions A.7.11 and A.7.12.

If G has characteristic $p > 0$ then the argument above applies with $\mathbb{Q}$ replaced by a finite field of p elements. □

A.8 Representing sets of primes and rings of $\mathcal{S}$-integers in number fields

In this section we will address the issue of representing recursive prime sets in number fields. We have already discussed these sets when the field in question is $\mathbb{Q}$. We will now do it for other number fields.

Let K be a number field. Let $\Omega = \{\omega_1, \ldots, \omega_n\}$ be an integral basis of K over $\mathbb{Q}$. (See Definition B.1.29 for a definition of integral basis.) Let

$$J : K \to \mathbb{N}$$

be a presentation described in Proposition A.7.7 with respect to basis Ω, $\mathbb{Q}$ being the field below under presentation from Proposition A.6.3. Fix a recursive enumeration of $J(K)$, i.e. a recursive bijection $\Phi : \mathbb{N} \to J(K)$. (For example, using an ordering of integers $0, -1, 1, -2, 2, \ldots$ one can order the elements of K using the lexicographical ordering with respect to Ω-coordinates of the field elements.) For each prime $\mathfrak{p}$ fix the smallest (under the chosen ordering) element α of K satisfying the following conditions:

- $\text{ord}_{\mathfrak{p}}\, \alpha = 1$;
- $\text{ord}_{\mathfrak{q}}\, \alpha = 0$ for any conjugate $\mathfrak{q}$ of $\mathfrak{p}$ over $\mathbb{Q}$;
- $\alpha = \sum_{i=1}^{n} a_i \omega_i,\ a_i \in \mathbb{Z},\ |a_i| < p^2$, where p is the rational prime below $\mathfrak{p}$.

Such an α exists for every $\mathfrak{p}$ by Proposition B.2.4. We will represent each prime $\mathfrak{p}$ by a pair $(\alpha(\mathfrak{p}), p)$, where $\alpha(\mathfrak{p})$ satisfies the conditions above and p is the rational prime below $\mathfrak{p}$ in $\mathbb{Q}$. Now call a set of K-primes $\mathcal{W}_K$ recursive if the corresponding set of pairs is recursive and call it r.e. if the corresponding set of pairs is r.e. (These attributes are simultaneously possible.) This identification has several desirable properties, described below.

Proposition A.8.1. *Let K, J be as above. For a rational prime q let*

$$S(q) = 2^{J(\alpha(\mathfrak{q}_1))} 3^{J(\alpha(\mathfrak{q}_2))} \cdots p_k^{J(\alpha(\mathfrak{q}_k))},$$

where $\mathfrak{q}_1, \ldots, \mathfrak{q}_k$ are all the distinct factors of q in K and $p_1 = 2, \ldots, p_k$ are the first k rational primes. For n not equal to a prime, let $S(n) = 1$. Then $S : \mathbb{N} \to \mathbb{N}$ is a recursive function.

Proof. We will present an informal algorithm to compute $S(n)$. Let $\omega_1, \ldots, \omega_n$ be an integral basis of K over $\mathbb{Q}$. For each expression of the form $\sum_{i=1}^{n} a_i\omega_i$, with $a_i \in \mathbb{Z}$ and $|a_i| < p^2$, do the following.

1. Compute the $\mathbb{Q}$-norm of this sum. (This a recursive operation by Corollary A.7.10.)
2. If the norm is divisible by q, set the sum aside.

Now let $B = \{\beta_1, \ldots, \beta_k\}$ be the set of all the elements set aside. Repeat the following procedure until B is empty.

1. Choose elements of B whose norm has the lowest order at q. Among the elements with norms of the lowest order, choose the element with the lowest sequence number under the chosen ordering of K. Denote this element by γ.
2. For every element $\beta_i \neq \gamma$ currently in the set, compute the norm of $\gamma + \beta_i$.
3. If the norm is divisible by q then remove β_i from the set.
4. Remove γ from B and put it into a set which we will denote by F.

We claim that at the end of this process, F will contain exactly

$$\alpha(\mathfrak{q}_1), \alpha(\mathfrak{q}_2), \ldots, \alpha(\mathfrak{q}_k),$$

where $\mathfrak{q}_1, \ldots, \mathfrak{q}_k$ are all the factors of q in K. Indeed, by Proposition B.2.4 we know that B contains $\alpha(\mathfrak{q}_i)$ for each factor of q. Further, any element of B which is divisible by at least two distinct factors of q or by the square of a factor of q will have a norm of order at q greater than that of at least one $\alpha(\mathfrak{q}_i)$ at any stage of the process. Further, the norm of $\gamma + \beta_i$ is divisible by q if and only if both γ and β_i are divisible by the same factor of q. Thus elements of B divisible by at least two distinct factors of q or by the square of a factor of q will be removed from B at some point. Elements of B having exactly one factor of q in their divisors but not having the lowest sequence number will also be removed from B when the element with a lower sequence number and the same factor of q in the divisor is chosen. □

As an immediate corollary of the proposition above we have the following statement.

Corollary A.8.2. *Let $\mathcal{W}_{\mathbb{Q}}$ be a set of rational primes, and let $\mathcal{W}_K$ be the set of all K-primes above $\mathcal{W}_{\mathbb{Q}}$. Then $\mathcal{W}_K \equiv_T \mathcal{W}_{\mathbb{Q}}$ and $\mathcal{W}_K \equiv_e \mathcal{W}_{\mathbb{Q}}$.*

Remark A.8.3. In what follows we will continue to use the presentation J of the number field K that we have discussed above.

Proposition A.8.4. *Let K be a number field of degree n over $\mathbb{Q}$. Let $\omega_1, \ldots, \omega_n$ be an integral basis of K over $\mathbb{Q}$. Let $\mathcal{W}_K$ be a recursive (r.e.) set of K-primes. Then the following sets are also recursive (r.e.):*

1. *the set of K-integers whose divisors are a product of powers of elements of $\mathcal{W}_K$ (we will denote this set by $O(\mathcal{W}_K)$);*
2. $C_{\mathbb{Q}}(O(\mathcal{W}_K)) = \{(a_1, \ldots, a_n) \in \mathbb{Z}^n | \sum_{i=1}^n a_i\omega_i \in O(\mathcal{W}_K)\}$.

Proof. As above, we present an informal algorithm computing the listing or characteristic function of the sets involved.

1. We will consider the case of recursive $\mathcal{W}_K$ first. Given $\alpha \in O_K$ do the following.
   (a) Compute the norm of α. Determine what primes divide the norm. Let $q_1, \ldots, q_m$ be the list of these primes.
   (b) For each q_i compute $\alpha(\mathfrak{q}_{i,j})$ for each factor $\mathfrak{q}_{i,j}$ of q_i in K.
   (c) Use the characteristic function of the set $\{(q_i, \alpha(\mathfrak{q}_{i,j}))\}$ to determine for each j whether $\mathfrak{q}_{i,j} \in \mathcal{W}_K$.
   (d) If, for some i, no factor of q_i is in $\mathcal{W}_K$ then $\alpha \notin O(\mathcal{W}_K)$.
   (e) Otherwise for each i, j compute the $\mathbb{Q}$-norm of $\alpha + \alpha(\mathfrak{q}_{i,j})$. If for some i, j we have that $\mathfrak{q}_{i,j} \notin \mathcal{W}_K$ but the $\mathbb{Q}$-norm of $\alpha + \alpha(\mathfrak{q}_{i,j})$ is divisible by q_i then $\alpha \notin O(\mathcal{W}_K)$.
   (f) Conclude that $\alpha \in O(\mathcal{W}_K)$.

   Now assume that $\mathcal{W}_K$ is r.e. and let $f(n)$ be a recursive function listing pairs $(p, \alpha(\mathfrak{p}))$, where $\mathfrak{p} \in \mathcal{W}_K$. Given an element $\alpha \in O_K$ use the procedure described above to compute all pairs $(q, \alpha(\mathfrak{q}))$ such that $\mathfrak{q}$ occurs in the divisor of α. Next check whether all the pairs have been listed; if so then α should be included in the listing of $O(\mathcal{W}_K)$. Otherwise, it should be stored in a "waiting" list.
2. Given $(a_1, \ldots, a_n)$, check to see whether $\sum_{i=1}^n a_i\omega_i \in O(\mathcal{W}_K)$ when $\mathcal{W}_K$ is recursive and check to see whether $\sum_{i=1}^n a_i\omega_i$ has been listed already if $O(\mathcal{W}_K)$ is r.e. □

Remark A.8.5. Using the same approach as in the proof above we can show that $O(\mathcal{W}_K)$ and $C_{\mathbb{Q}}(O(\mathcal{W}_K))$ are actually Turing and enumerably equivalent to $\mathcal{W}_K$.

We are now ready for the following assertions.

Proposition A.8.6. *Let K be a number field of degree n over $\mathbb{Q}$. Let $\mathcal{W}_K$ be a set of primes of K. Then $O_{K,\mathcal{W}_K}$ is recursive (r.e.) if and only if $\mathcal{W}_K$ is recursive (r.e.).*

Proof. First assume that $\mathcal{W}_K$ is recursive (r.e.). Given $\alpha \in K$, compute the coefficients $A_r(\alpha)$, $r = 0, \ldots, n-1$, of the characteristic polynomial of α over $\mathbb{Q}$ and determine the rational primes $q_1, \ldots, q_l$ dividing the denominators of these coefficients. (This can be done effectively by A.7.10.) For each prime q_i compute $\alpha(\mathfrak{q}_{i,j})$ for each K-factor $\mathfrak{q}_{i,j}$ of q_i. (This part is recursive by A.8.1.) Further, for $i = 1, \ldots, l$ compute $a_i = -\min_{0 \le r < n}(\text{ord}_{q_i} A_r(\alpha))$. For each pair (i, j), let $\beta_{i,j}(\alpha) = \prod_{k \ne j} \alpha(\mathfrak{q}_{i,k})$. Next, for all i, j we compute the coefficients $A_r(\beta_{i,j}^{na_i}\alpha)$ of the characteristic polynomial of $\beta_{i,j}^{na_i}\alpha$ over $\mathbb{Q}$. Note that if $\text{ord}_{\mathfrak{q}_{i,j}}\alpha < 0$ then $\text{ord}_{\mathfrak{q}_{i,j}}\beta_{i,j}^{na_i}\alpha < 0$ and for some r we will have $\text{ord}_{q_i} A_r(\beta_{i,j}^{na_i}\alpha) < 0$. Conversely, for $k \ne j$ we have that $\text{ord}_{\mathfrak{q}_{i,k}} \alpha > -na_i$ and therefore $\text{ord}_{\mathfrak{q}_{i,k}} \alpha\beta_{i,j}^{na_i} > 0$. Thus $\alpha\beta_{i,j}^{na_i}$ is integral at q_i if and only if α is integral at $\mathfrak{q}_{i,j}$. Hence, by following this effective procedure we will be able to establish which $\mathfrak{q}_{i,j}$ occur in the denominator of the divisor of α. Once this is done, if $\mathcal{W}_K$ is recursive then for each (i, j) we can check to see whether the pair $(q_i, \alpha(\mathfrak{q}_{i,j})) \in \mathcal{W}_K$. If $\mathcal{W}_K$ is r.e. then we can wait for the pair to be listed.

Now suppose that $O_{K,\mathcal{W}_K}$ is recursive (r.e.). Then given a prime $\mathfrak{p}$ of K, we can conduct a systematic search of the field until we find an integral element β whose divisor is a power of $\mathfrak{p}$ only. Such an element exists since the divisor $\mathfrak{p}^{h_K}$, where h_K is the class number of K, is principal. We can recognize such an element, since p, the prime below $\mathfrak{p}$, will be the only prime dividing its norm, all the coefficients of the characteristic polynomial of β over $\mathbb{Q}$ will be integral, and the $\mathbb{Q}$-norm of $\beta + \alpha(\mathfrak{q})$, where $\mathfrak{q}$ is any other factor of p in K, will not be divisible by p. Next we can check whether $1/\beta$ is in $O_{K,\mathcal{W}_K}$ (or has already been listed).

Finally, we note that as above one can adjust this proof to show that $O_{K,\mathcal{W}_K}$ and $\mathcal{W}_K$ are Turing and enumerably equivalent. □

Proposition A.8.7. *Let K be a number field. Let $\mathcal{W}$ be a set of rational primes with a factor of relative degree d in K and let $\mathcal{W}_K$ be the set of K primes of relative degree d over $\mathbb{Q}$. Then $\mathcal{W}$ and $\mathcal{W}_K$ are both recursive sets of primes.*

Proof. Given a rational prime p, do the following.

- Compute $\alpha(\mathfrak{p}_i)$ for each factor $\mathfrak{p}_i$ of p.
- Compute $N_{K/\mathbb{Q}}(\alpha(\mathfrak{p}_i))$ and determine $f_i = \text{ord}_p N_{K/\mathbb{Q}}(\alpha(\mathfrak{p}_i)) = f(\mathfrak{p}_i/p)$ for all i. If for some i we have $f_i = d$ then $p \in \mathcal{W}$, otherwise $p \notin \mathcal{W}$.

The procedure for determining whether a pair $(p, \alpha(\mathfrak{p}_i))$ is in the set representing $\mathcal{W}_K$ is analogous. □

Proposition A.8.8. *Let K be a number field. Then, if K is real there exists a recursive function $f : K \times \mathbb{N} \to \mathbb{N}$ producing a decimal expansion of the elements of the field. If K is not real, there are recursive functions $f_i : K \times \mathbb{N} \to \mathbb{N}, i = 1, 2$, producing decimal expansions of the real and imaginary parts of the elements of the break field.*

Proof. Let $\omega \in \Omega$ (an integral basis of K over $\mathbb{Q}$). Let $h(T)$ be its monic irreducible polynomial over $\mathbb{Q}$. If ω is real then let $a_1 < b_1 \in \mathbb{Q}$ be such that $\omega \in (a_1, b_1)$ and $[a, b]$ contains no other real root of h. If ω is not real, let $a_1 < b_1, a_2 < b_2 \in \mathbb{Q}$ be such that $\omega \in B = \{z \in \mathbb{C} \mid \Re z \in (a_1, b_1), \Im z \in (a_2, b_2)\}$ and the topological closure $\bar{B}$ of B contains no other root of h. Let r be a positive rational number greater than the distance from any other root of $h(x)$ to thc boundary of the region. Next let $g(x) = h(x)/(x - \omega)$. Then, for any c in the chosen region, $|g(c)| > r^{d-1}$ where $d = \deg f$. Finally note that

$$|\omega - c| = \left|\frac{h(c)}{g(c)}\right| < r^{1-d}|f(c)|.$$

Thus by a systematic search of the rational elements of the region, we can effectively construct the decimal expansion of ω if it is real, or decimal expansions of its real and imaginary parts otherwise.

Since for every element of K we can produce effectively (i.e. there is a recursive (algorithmic) procedure to compute) its rational coordinates with respect to Ω, this means that knowing the decimal expansions of the elements of the basis is enough to produce the decimal expansions for all the elements of the field. □

Appendix B: Number theory

This appendix contains the algebraic and number-theoretic facts which we use in this book.

B.1 Global fields, valuations, and rings of $\mathcal{W}$-integers

We start with the definition of a global field.

Definition B.1.1. A global field is a finite extension of $\mathbb{Q}$ (a number field) or a finite extension of a rational function field over a finite field of constants.

We remind the reader that throughout the book we assume that all the number fields are subfields of $\mathbb{C}$, and for each $p > 0$ we have fixed an algebraic closure for a rational function field over a constant field of p elements. Thus any two global fields of the same characterstic in the book are assumed to be subfields of the same algebraically closed field.

In the function field case we need a couple more definitions.

Definition B.1.2. Let C be a field, let t be transcendental over C, and let K be a finite extension of $C(t)$. Then the algebraic closure of C in K is called the constant field of K or the field of constants of K.

Remark B.1.3. In general a function field is a finite extension of a rational function field. What makes a function field "global" is the fact that the constant field is finite.

We will distinguish a special class of function field extensions.

Definition B.1.4. Let K be a function field over a constant field C. Let C_1 be an algebraic extension of C. Then the field $K_1 = C_1K$ will be called a constant field extension of K. (Here we again remind the reader that, as we stated in Chapter 1, whenever we use a compositum of fields they are assumed to be subfields of the same algebraically closed field. In this case the compositum is simply the smallest field containing all the fields in the compositum.)

Next we define a non-archimedean valuation of a global field.

Definition B.1.5. Let K be a global field. Then let $v : K \to \mathbb{Z} \cup \{\infty\}$ be a map satisfying the following properties.

1. $v : K^* \to \mathbb{Z}$ is a homomorphism from the multiplicative group of K into $\mathbb{Z}$ as a group under addition.
2. $v(0) = \infty$.
3. For all $x, y \in K$ we have that $v(x + y) \geq \min(v(x), v(y))$.
4. If K is a function field then for any constant c we have that $v(c) = 0$.
5. There exists $x \in K$ such that $v(x) \neq 0$.

Then v is called a (non-trivial) discrete or non-archimedean valuation of K or a prime of K. If $x \in K$ then $v(x)$ is also denoted by $\operatorname{ord}_v x$. For a general discussion of valuations and primes of global fields the reader is referred to the following: Chapters 1 and 2 of [37]; Chapters 1–3 of [47]; Chapters 2 and 3 of [33]; and Chapter 1 of [3].

Next we list some important properties of valuations. We will leave their proof to the reader.

Proposition B.1.6. *Let K be a global field and let v be a valuation of K. Let*

$$R(v) = \{x \in K, v(x) \geq 0\}.$$

Then $R(v)$ is a discrete valuation ring (i.e. a local principal ideal domain) whose fraction field is K and whose unique maximal ideal is

$$M(v) = \{x \in R(v) | v(x) > 0\}.$$

(The identification of v with $M(v)$ explains the dual terminology.) *$R(v)$ is called the valuation ring of v.*

Conversely, let $R \subset K$ be a ring such that for every $x \in K$ either $x \in R$ or $x^{-1} \in R$. Then R is a discrete valuation ring such that for some valuation v of K, we have that $R(v) = R$. (For a discussion of discrete valuation rings see, for example, Section 3, Chapter 1 of [37].)

From this proposition we immediately derive the following property of non-archimedean valuations.

Corollary B.1.7. *Let K be a global field and v a non-archimedean valuation. Let $a, b \in K$ be such that $v(a) < v(b)$. Then $v(a + b) = v(a)$.*

The definition below tells us when two valuations are essentially the same.

Definition B.1.8. Let K be a global field. Then two valuations v_1 and v_2 of K will be called equivalent if $M(v_1) = M(v_2)$. (Note that, since $R(v)$ is a local ring for all v, $M(v_1) = M(v_2) \Leftrightarrow R(v_1) = R(v_2)$.)

The next definition introduces us to an important object associated with valuations.

Definition B.1.9. Let $K, v, M(v), R(v)$ be as in Proposition B.1.6. Then the field $F(v) = R(v)/M(v)$ is called the residue field of v.

The two propositions below constitute a starting point for our investigation of extensions of valuations. The first tells us that every valuation above comes from below. The second proposition lists several properties connecting valuations above to valuations below.

Proposition B.1.10. *Let M/K be a finite extension of global fields. Let w be a non-archimedean valuation of M and let v be its restriction to K. Then v is a valuation of K.*

Proof. The only way v can fail to be a valuation, as defined above, is for $v(x)$ to be identically zero on K. This case is excluded by Statement C in Section 4.1 of [80]. □

Proposition B.1.11. *Let M/K be a finite extension of global fields. Let v be a valuation of K. Then there exists a valuation w of M extending a valuation of K equivalent to v. Let $\bar{v}$ be the restriction of w to M. Then the following statements are true.*

1. *$F(v)$ is isomorphic to a subfield of $F(w)$ and under this isomorphism $[F(w) : F(v)]$ is finite. This degree is called the relative degree of w over v and is denoted by $f(w/v)$.*
2. *$\bar{v}(K^*) \subseteq w(M^*)$ and the index of $\bar{v}(K^*)$ in $w(M^*)$ is finite. This index is called the ramification of w over v and is denoted by $e(w/v)$.*
3. *For any valuation v there are only finitely many valuations w in M such that the restriction of w to K is a valuation equivalent to v. Let $w_1, \ldots, w_k$ be all such valuations of M. Then $\sum_{i=1}^{k} e(w_i/v) f(w_i/v) = [M : K]$. Further, if the extension is Galois, for all $i, j = 1, \ldots, k$ we have that $e(w_i/v) =$*

$e(w_j/v)$ and $f(w_i/v) = f(w_j/v)$, implying that for all i it is the case that $f(w_i/v) \mid [M : K]$, $e(w_i/v) \mid [M : K]$, and $k \mid [M : K]$.

4. *The integral closure $R_M(v)$ of $R(v)$ in M is equal to $\bigcap_{i=1}^{k} R(w_i)$.*
5. *$R_M(v)M(v) = \prod_{i=1}(M(w_i) \cap R_M(v))^{e(w_i/v)}$.*

For the proof of this proposition see Lemma 6.5 and Corollary 6.7 in Chapter 1 of [37], Theorem 2 in Chapter 4 of [80], and Theorem 1 in Section 1, Chapter IV of [3].

It is not hard to show that relative degrees and ramification degrees also have the following properties, whose proof we leave to the reader.

Proposition B.1.12. *Let $K \subset E \subset F$ be a finite extension of global fields. Let w_F be a non-archimedean valuation of F and let w_E and w_K be its restrictions to E and K respectively. Then $e(w_F/w_K) = e(w_F/w_E)e(w_E/w_K)$ and $f(w_F/w_K) = f(w_F/w_E)f(w_E/w_K)$.*

B.1.13. Alternative terminology The discussions above can be rephrased in terms of ideals rather than valuations since we can "reconstruct" v from its ideal $M(v)$. For any $x \in K^* \cap M(v)$, set $\bar{v}(x) = \{\min n : x \in M(v)^n \setminus M(v)^{n+1}\}$. For all $x \in R(v) \cap K^* \setminus M(v)$, set $\bar{v}(x) = 0$. For any $x \in K \setminus R(v)$ define $\bar{v}(x) = -\bar{v}(x^{-1})$. Finally, let $\bar{v}(0) = \infty$. Then $\bar{v}$ is equivalent to v. We will usually denote $M(v)$ by the characters "$\mathfrak{p}$," "$\mathfrak{P}$," "$\mathfrak{q}$," or "$\mathfrak{Q}$." Instead of writing "$v(x)$" we will normally write "$\text{ord}_{\mathfrak{p}}\, x$" (or "$\text{ord}_{\mathfrak{P}}\, x$," "$\text{ord}_{\mathfrak{q}}\, x$," "$\text{ord}_{\mathfrak{Q}}\, x$" respectively).

Given an extension M/K of global fields, instead of saying that a valuation w of M is an extension of a valuation $\bar{v}$ of K equivalent to a given valuation v of K, we will say that a prime $\mathfrak{p}_M$ of M is a factor or lies above a given prime $\mathfrak{p}_K$ of K. We will also say that $\mathfrak{p}_K$ lies below $\mathfrak{p}_M$ in K.

If an M-prime $\mathfrak{p}_M$ lies above a K-prime $\mathfrak{p}_K$ and $e(\mathfrak{p}_M/\mathfrak{p}_K) > 1$ then we will say that $\mathfrak{p}_M$ is "ramified" over $\mathfrak{p}_K$ or K, and we will say that $\mathfrak{p}_K$ is "ramified" in the extension M/K. If $[M : K] > 1$ and the ramification degree is equal to the degree of the extension, we will say that $\mathfrak{p}_K$ and $\mathfrak{p}_M$ are "totally ramified" in the extension. If a K-prime $\mathfrak{p}_K$ has only one prime $\mathfrak{p}_M$ of M above it in M and $\mathfrak{p}_M$ is unramified over $\mathfrak{p}_K$, we will say that $\mathfrak{p}_K$ "does not split" in the extension or that $\mathfrak{p}_K$ "is inert" in the extension, otherwise we will say that $\mathfrak{p}_K$ "splits" in the extension. Finally, if for every M-prime $\mathfrak{p}_M$ lying above a K-prime $\mathfrak{p}_K$ the relative degree is 1 then we will say that $\mathfrak{p}_K$ "splits completely." If, furthermore, $\mathfrak{p}_K$ is unramified in the extension, we will say that $\mathfrak{p}_K$ "splits completely into distinct factors."

To facilitate further discussion we introduce several more terms.

B.1.14. More terminology If K is a global field, v is a non-archimedean valuation of K, and $x \in K$ then we will say that x has a zero at v if $v(x) > 0$. We will say that x has a pole at v if $v(x) < 0$. Finally we will say that x is a unit at v if $v(x) = 0$. If $v(x) = \{\min n : x \in M(v)^n \setminus M(v)^{n+1}\}$ as above, and $v(x) = n > 0$, we will say that x has a zero of order n. If $v(x) = -n < 0$, we will say that x has a pole of order n at v.

This terminology is standard for function fields only but for the sake of brevity we will use it for number fields also.

The next two propositions will identify all the discrete valuations of a global field, starting with the case of rational numbers and rational functions.

Proposition B.1.15. *Let $a \in \mathbb{Z}$, $a \neq 0$. Let p be a rational prime. Let $b \in \mathbb{Z}$ be such that $a = p^m b$ and $(p, b) = 1$. Then define $v_p(a) = \operatorname{ord}_p a = m$. For a non-zero rational number $x = a_1/a_2$, with $a_2 \neq 0$ and $(a_1, a_2) = 1$, define $v_p(x) = v_p(a_1) - v_p(a_2)$. Then v_p is a valuation of $\mathbb{Q}$. Conversely, if v is a discrete valuation of $\mathbb{Q}$ then v is equivalent to v_p for some p.* (See Theorem 1 in Section 1.3 of [80].)

Proposition B.1.16. *Let C be a field. Let t be transcendental over C. Let $h(t), g(t) \in C[t] \setminus \{0\}$, let $P(t) \in C[t]$ be a monic irreducible polynomial of non-zero degree, and assume that $h(t) = g(t)P(t)^m, (P(t), g(t)) = 1$. Then define $v_{P(t)}(h(t)) = m$. Further, define $v_\infty(h(t)) = -\deg(h(t))$. If $f(t) = h_1(t)/h_2(t)$, where $h_1(t), h_2(t) \in C[t] \setminus \{0\}$ and are relatively prime to each other, then set $v_{P(t)}(f(t)) = v_{P(t)}(h_1(t)) - v_{P(t)}(h_2(t))$ and similarly set $v_\infty(f(t)) = v_\infty(h_1(t)) - v_\infty(h_2(t))$. Then $v_{P(t)}$ and v_∞ are valuations of $C(t)$. Further, any valuation of $C(t)$, trivial on C, is equivalent to $v_{P(t)}$ for some monic irreducible polynomial $P(t)$ or v_∞.* (See Section 3, Chapter 1 of [3] for the proof of this proposition.)

Note that from Proposition B.1.10 we know that any non-archimedean valuation of a global field must be an extension of a valuation on a rational field. Thus, knowing all the non-archimedean valuations of rational fields provides us with a lot of information about all the valuations of a generic global field. Among other things one can now easily prove the following results.

Proposition B.1.17. *The residue field of any non-archimedean valuation of a global field is finite. In the case of a global function field it is a finite extension of the field of constants.*

Proposition B.1.18. *Let x be an element of a global field K. Then for all but finitely many non-archimedean primes $\mathfrak{p}$ of K we have that* $\operatorname{ord}_{\mathfrak{p}} x = 0$.

Proposition B.1.17gives rise to the following definition.

Definition B.1.19. *Let $\mathfrak{p}$ be a prime of a global function field K over a field of constants C. Let $R_{\mathfrak{p}}$ be the residue field of $\mathfrak{p}$. Then $[R_{\mathfrak{p}} : C]$ is called the degree of $\mathfrak{p}$. (This degree should not be confused with the relative degree defined earlier.)*

The valuations of global fields form the rings which are the subject of discussion in this book.

Definition B.1.20. Let K be a global field and let $\mathcal{W}$ be a set of its non-archimedean valuations. Then set

$$O_{K,\mathcal{W}} = \{x \in K \mid \operatorname{ord}_{\mathfrak{p}} x \geq 0, \forall \mathfrak{p} \notin \mathcal{W}\}.$$

If K is a number field then $\mathcal{W}$ can be empty and in this case $O_{K,\mathcal{W}} = O_K$ is the ring of algebraic integers of K. In the case of function fields, we will never consider an empty $\mathcal{W}$ because the resulting ring would contain constants only. If $\mathcal{W}$ contains all non-archimedean valuations of K then the ring $O_{K,\mathcal{W}} = K$. In general we will consider subsets $\mathcal{W}$ which could be finite or infinite. In the case of finite $\mathcal{W}$ the ring $O_{K,\mathcal{W}}$ is called a "ring of $\mathcal{W}$-integers." If $\mathcal{W}$ is infinite or finite and K is a function field then $O_{K,\mathcal{W}}$ is called a holomorphy ring, but unfortunately there is no corresponding accepted term for the case of infinite $\mathcal{W}$ for number fields. Therefore we will extend the use of the term "$\mathcal{W}$-integers" to rings with infinite $\mathcal{W}$.

The following propositions will describe important properties of the rings of $\mathcal{W}$-integers.

Proposition B.1.21. *Let K be a global field. Let $\mathcal{W}_K$ be a set of primes of K. Then $O_{K,\mathcal{W}_K}$ is integrally closed.* (See Section 2, Chapter 1 of [37] for a definition of an integrally closed ring.)

Proof. Suppose that $x \in K \setminus O_{K,\mathcal{W}_K}$ and is integral over $O_{K,\mathcal{W}_K}$. Then for some $\mathfrak{p} \notin \mathcal{W}_K$ we have that $\operatorname{ord}_{\mathfrak{p}} x < 0$. Since x is integral over $O_{K,\mathcal{W}_K}$, for some $a_0, \ldots, a_{n-1} \in O_{K,\mathcal{W}_K} \subset R(\mathfrak{p})$ with $\operatorname{ord}_{\mathfrak{p}} a_i \geq 0$ we have that $x^n + a_{n-1}x^{n-1} + \cdots + a_0 = 0$. Next observe that

$$\operatorname{ord}_{\mathfrak{p}} x^n = n \operatorname{ord}_{\mathfrak{p}} x < \min_{i=0,\ldots,n-1} (\operatorname{ord}_{\mathfrak{p}} a_i x^i) \leq \operatorname{ord}_{\mathfrak{p}}(a_{n-1}x^{n-1} + \cdots + a_0)$$

and therefore, by Corollary B.1.7, $\text{ord}_{\mathfrak{p}}(x^n + a_{n-1}x^{n-1} + \cdots + a_0) < 0$, which is impossible. Thus our assumption on the existence of x as described above is incorrect. □

Then next proposition shows us that the integral closure of a ring of $\mathcal{W}$-integers in a finite extension is a ring of the same type.

Proposition B.1.22. *Let M/K be an extension of global fields. Let $\mathcal{W}_K$ be a subset of primes of K. Let $\mathcal{W}_M$ be the set of all primes of M lying above K-primes in $\mathcal{W}_K$. Then the integral closure of $O_{K,\mathcal{W}_K}$ in M is $O_{M,\mathcal{W}_M}$.*

Proof. In view of Proposition B.1.11, it is enough to show that being integral over $O_{K,\mathcal{W}_K}$ is equivalent to being integral over the valuation rings of every prime $\mathfrak{p} \notin \mathcal{W}_K$. Indeed, let x be integral over $O_{K,\mathcal{W}_K}$. Then for some $a_0, \ldots, a_{n-1} \in O_{K,\mathcal{W}_K}$,

$$x^n + a_{n-1}x^{n-1} + \cdots + a_0 = 0. \tag{B.1.1}$$

But for any prime $\mathfrak{p} \notin \mathcal{W}_K$, the valuation ring $R(\mathfrak{p})$ contains $O_{K,\mathcal{W}_K}$. Thus x is integral over $R(\mathfrak{p})$.

Conversely, if x is integral over every $R(\mathfrak{p})$ with $\mathfrak{p} \in \mathcal{W}_K$ then any conjugate of x over K also has this property. Since the integral closure of a ring in a bigger ring (in this case the bigger ring is the Galois closure of M over K) is also a ring (see the corollary to Proposition 2.2 in Chapter I of [37]), the coefficients of the monic irreducible polynomial of x over K are also integral over $O_{K,\mathcal{W}_K}$ and are elements of K. Thus, x satisfies a polynomial equation of the form (B.1.1) with $a_0, \ldots, a_{n-1} \in O_{K,\mathcal{W}_K}$ and therefore is integral over $O_{K,\mathcal{W}_K}$. Consequently, the integral closure of $O_{K,\mathcal{W}_K}$ is $\bigcap_{\mathfrak{p}\notin\mathcal{W}_K} \bigcap_{\mathfrak{P}\in\mathcal{P}(M),\mathfrak{P}|\mathfrak{p}} R(\mathfrak{P}) = O_{M,\mathcal{W}_M}$. (Here $\mathcal{P}(M)$ denotes the set of all (non-archimedean) primes of M.) □

Our next goal is to describe a notion of the divisor for an element of a global field.

Definition B.1.23. Let K be a global field. Let $x \in K$. Let $\mathcal{P}(K)$ denote the set of all the non-archimedean primes of K. Then let $\mathfrak{D}_K(x) = \prod_{\mathfrak{p}\in\mathcal{P}(K)} \mathfrak{p}^{\text{ord}_{\mathfrak{p}} x}$ denote the K-divisor of x. (Note that by Proposition B.1.18, this product is effectively finite.) In general, a formal product $\mathfrak{D} = \prod_{i=1}^{k} \mathfrak{p}_i^{a_i}$, where $\mathfrak{p}_i$ is a prime of K and $a_i \in \mathbb{Z}$, will be called a divisor of K. If all the a_i are actually positive then the divisor will be called integral. It is not hard to see that all algebraic integers of K have integral divisors.

For the case where K is a function field, this is a traditional definition of divisors. (See for example Section 3.1 of [33].) In the case of a number field, the usual terminology would refer to a fractional ideal of the field and to the factorization of that fractional ideal into the product of prime ideals. (See Section 4, Chapter 1 of [37].) However, in our case there are advantages in using the uniform terminology for both cases as in Definition 1, Chapter 10 [80]).

When $\mathfrak{D}$ is an integral divisor and $\mathcal{W}$ is a set of primes of the field not containing the factors of $\mathfrak{D}$, we can identify $\mathfrak{D}$ with a unique ideal of $O_{K,\mathcal{W}}$, i.e. the ideal of $O_{K,\mathcal{W}}$ which is the intersection $\bigcap_{i=1}^{k} \mathfrak{p}_i^{a_i} O_{K,\mathcal{W}}$.

In the case of function fields we can also define the degree of integral divisors.

Definition B.1.24. Let $\mathfrak{D}$ be an integral divisor of a function field K. Then define the degree of $\mathfrak{D}$, denoted by $\deg_K \mathfrak{D}$, to be $\sum_{\mathfrak{p}\in\mathcal{P}(K),\mathrm{ord}_{\mathfrak{p}}\mathfrak{D}>0} \deg \mathfrak{p}$. (The degree of a function field prime was defined in B.1.19.)

Using the notion of a divisor and its degree, we can define the notion of height for elements of function fields.

Definition B.1.25. Let K be a global function field. Let $x \in K$. Let

$$\mathfrak{D} = \prod_{\mathfrak{p}\in\mathcal{P}(K),\mathrm{ord}_{\mathfrak{p}}x>0} \mathfrak{p}^{\mathrm{ord}_{\mathfrak{p}}x}.$$

Then define the height of x to be the degree of $\mathfrak{D}$. We will refer to $\mathfrak{D}$ as the zero divisor of x.

Remark B.1.26. The product formula (see Chapter 12 of [1]) implies that the height can also be measured using the pole divisor of the element, i.e. we can define the height of x to be $\deg_K(\prod_{\mathfrak{p}\in\mathcal{P}(K),\mathrm{ord}_{\mathfrak{p}}x<0} \mathfrak{p}^{-\mathrm{ord}_{\mathfrak{p}}x})$.

The rings of $\mathcal{W}$-integers are the only integrally closed subrings of global fields. This follows from the following proposition.

Proposition B.1.27. *Let K be a global field. Let R be an integrally closed subring of K such that the fraction field of R is K. Then for some $\mathcal{W} \subseteq \mathcal{P}(K)$, we have that $R = O_{K,\mathcal{W}}$.*

Proof. Let $\mathcal{W} = \{\mathfrak{p} \in \mathcal{P}(K) : \exists x \in R, \mathrm{ord}_{\mathfrak{p}} x < 0\}$. We will show that $R = O_{K,\mathcal{W}}$, treating the cases of number fields and function fields separately.

First let K be a number field and observe that $O_K \subseteq R$ because $\mathbb{Z} \subset R$ and O_K is the integral closure of $\mathbb{Z}$ in K. Next note that if $\mathfrak{p} \in \mathcal{P}(K)$ and there exists $x \in R$ with $\text{ord}_\mathfrak{p} x < 0$ then there exists $y \in R$ such that y has a pole at $\mathfrak{p}$ only and no zeros. Indeed, first of all, by the Strong Approximation Theorem (see Theorem B.2.1), there exists $z \in O_K$ such that for all $\mathfrak{q} \neq \mathfrak{p}$ with $\text{ord}_\mathfrak{q} x < 0$, we have that $\text{ord}_\mathfrak{q} z > -\text{ord}_\mathfrak{q} x$, while $\text{ord}_\mathfrak{p} z = 0$. Note that $xz \in R$, $\text{ord}_\mathfrak{p} xz < 0$, and xz has no other poles. Next consider $(xz)^h$, where h is the class number of K, and observe that $(xz)^h = \alpha/\beta$, where α, β are integers which are relatively prime to each other, β having a zero at $\mathfrak{p}$ only. (See Section 5, Chapter 1 of [37] for a definition of class number.) By the Strong Approximation Theorem again, there exist $a, b \in O_K$ such that $a\alpha + b\beta = 1$. Now consider

$$a(zx)^h + b = \frac{a\alpha + b\beta}{\beta} = \frac{1}{\beta} \in R.$$

If $\mathcal{W} = \emptyset$ then $R = O_K$ and we are done. If $\mathcal{W}$ is not empty then $R \subseteq O_{K,\mathcal{W}}$. Next, for each $\mathfrak{p}$ that can occur in the denominator of the divisor of an element of R, let β, constructed as above, be called $\beta(\mathfrak{p})$. Now let $w \in O_{K,\mathcal{W}}$. Note that for some positive natural number b, we have that $w^{bh} = \gamma/\delta$, $\delta = \prod \beta(\mathfrak{p})^{a(\mathfrak{p})}$, where $\gamma \in O_K$, $a(p) \in \mathbb{N}$, and $1/\beta(\mathfrak{p}) \in R$. Thus, $w^{bh} \in R$ and, since R is integrally closed, $w \in R$. Therefore $O_{K,\mathcal{W}} \subseteq R$ and consequently $R = O_{K,\mathcal{W}}$.

We now consider the case where K is a function field. First of all we observe that, given our assumptions on R, we cannot have an empty $\mathcal{W}$. Indeed, since the fraction field of R is K, it must contain a non-constant element t. A non-constant element must have at least one pole (and at least one zero) (see Corollary 3 in Section 4, Chapter 1 of [3]). Next let C be the constant field of K and consider $C[t] \subset C(t) \subseteq K$. Let $\mathfrak{q}_{C(t)}$ be the degree valuation of $C(t)$ as described in Proposition B.1.16. Then observe that $C[t] = O_{C(t),\{\mathfrak{q}_{C(t)}\}}$. Indeed, any polynomial has a pole at the degree valuation only. Next note that any rational function which is not a polynomial will have a pole at some other valuation, namely a valuation corresponding to an irreducible polynomial dividing the reduced denominator of the rational function. Thus, $C[t]$ is integrally closed by Proposition B.1.21 and its integral closure in R is a ring of $\mathcal{S}$-integers where $\mathcal{S}$ contains all the factors of $\mathfrak{q}_{C(t)}$ in K by Proposition B.1.22. In this context $O_{K,\mathcal{S}}$ is often called a ring of integral functions, by analogy with the rings of integers of number fields. From this point on, the proof proceeds exactly as in the number field case; now $O_{K,\mathcal{S}}$ plays the role that O_K plays in the number field case. We will leave the remaining details to the reader and note only that the proof will require the function field version of the Strong Approximation Theorem (Theorem B.2.1). □

Using Proposition B.1.11 and the Strong Approximation Theorem, one can also show that the following proposition is true. We leave the details to the reader.

Proposition B.1.28. *Let K be a number field and let $\mathcal{W}$ be a set of its primes. Let $\mathfrak{I}$ be an integral divisor of K (i.e. $\mathfrak{I}$ has a trivial denominator) and assume that no factor of $\mathfrak{I}$ is in $\mathcal{W}$. Then $O_{K,\mathcal{W}}/\mathfrak{I}$ is a finite ring. (We remind the reader that we can consider integral divisors as ideals of O_K and $O_{K,\mathcal{W}}$ as long as $\mathcal{W}$ does not contain any factors of $\mathfrak{I}$.)*

We will next discuss an important notion – that of an "integral basis."

Definition B.1.29. Let M/K be a number field extension. Let $\Omega = \{\omega_1, \ldots, \omega_n\} \subset O_M$ be a basis of M over K. Then Ω is called an integral basis of M over K if for every $x \in O_M$ it is the case that $x = \sum_{i=1}^{n} a_i\omega_i$, where $a_i \in O_K$.

The most important fact for us concerning integral bases is stated below.

Proposition B.1.30. *Every number field K has an integral basis over $\mathbb{Q}$.*

Proof. This follows from Theorem 1 in Appendix B of [37].

We finish this section with two more technical observations concerning extensions of global fields.

Lemma B.1.31. *Let E/F be a Galois extension of fields. Let M/F be any other extension of F. Then ME/M is also a Galois extension.*

Proof. Let α be a generator of E over F. Then all the conjugates of α over F are distinct and are in E. α will also be a generator of ME over M. Further, a conjugate of α over M is also a conjugate of α over F and therefore all the conjugates of α over M are distinct and contained in ME. Hence ME/M is a Galois extension. □

Lemma B.1.32. *Let H be an algebraic function field over a perfect field of constants C and let t be a non-constant element of H. Then the following conditions are equivalent.*

1. *t is not a pth power in H.*
2. *The extension $H/C(t)$ is finite and separable.*

(See Chapter VI of [51].)

B.2 Existence through approximation theorems

In this section we will list various propositions that are consequences of the Strong Approximation Theorem. Before we proceed, we need to say a few words about archimedean valuations. Non-archimedean or discrete valuations give rise to non-archimedean absolute values over number fields. If K is a number field and $\mathfrak{p}$ is its prime then for $x \in K$ we can define

$$|x|_{\mathfrak{p}} = \left(\frac{1}{N\mathfrak{p}}\right)^{\mathrm{ord}_{\mathfrak{p}} x},$$

where $N\mathfrak{p}$ is the size of the residue field of $\mathfrak{p}$. It is not hard to verify that $|\ |_{\mathfrak{p}}$ is in fact an absolute value. Not all absolute values of number fields are generated by its primes. Some absolute values are extensions of the usual absolute value on $\mathbb{Q}$. These are the "archimedean" valuations. Each number field has a finite number of archimedean valuations equal to the number of distinct non-conjugate embeddings of the field into $\mathbb{C}$. For more information on archimedean valuations see, for example, Section 4, Chapter II of [37]. Finally, we note that function fields do not have archimedean valuations.

We are now ready to state the Strong Approximation Theorem.

Theorem B.2.1. *Let K be a global field. Let $\mathcal{M}_K$ be the set of all the absolute values of K. Let $\mathcal{F}_K \subset \mathcal{M}_K$ be a non-empty finite subset. Let $\mathcal{F}_K = \{|\ |_1, \ldots, |\ |_l\}$. Let $a_1, \ldots, a_{l-1} \in K$. Then for any $\varepsilon > 0$ there exists $x \in K$ such that the following conditions are satisfied.*

1. *For $i = 1, \ldots, l-1$ we have that $|x - a_i|_i < \varepsilon$.*
2. *For any $|\ | \notin \mathcal{F}_K$ we have that $|x| \leq 1$.*

A proof of this theorem can be found in the following: Theorem 33:11 Part I, Chapter III of [64], Proposition 3.3.1, Chapter 3 of [33], Proposition 3.6.4, Chapter 3 of [45].

In the following two propositions we will use the Strong Approximation Theorem to show the existence of elements in a global field necessary for the proofs in Chapter 4 on the definability of integrality at finitely many primes.

Proposition B.2.2. *Let K be a number field. Then there exists $g \in O_K$ satisfying the conditions of Notation 4.2.1.*

Proof. Let the divisor of aq^3 be of the form $\prod \mathfrak{T}_j^{m_j}$, where each $\mathfrak{T}_j$ is a prime of K and each m_i is a positive integer. For each i, let $\alpha_i \in K$ be such that $\mathrm{ord}_{\mathfrak{P}_i} \alpha_i = 1$ and let $\gamma \in K$ be such that $\mathrm{ord}_{\mathfrak{C}} \gamma = 1$. (The existence of these elements follows from the fact that valuation rings are local PIDs.) Then, by the strong approximation theorem, there exists $g_0 \in K$ such that

- $\mathrm{ord}_{\mathfrak{P}_i}(g_0 - \alpha_i^{-1}) > 1$,
- $\mathrm{ord}_{\mathfrak{C}}(g_0 - \gamma) > 1$,
- $\mathrm{ord}_{\mathfrak{T}_j}(g_0 - 1) > 3m_j$,
- g_0 is integral at all the other primes.

Note that

$$\begin{aligned}\mathrm{ord}_{\mathfrak{P}_i} g_0 &= \mathrm{ord}_{\mathfrak{P}_i}\left((g_0 - \alpha_i^{-1}) + \alpha_i^{-1}\right) = \min\left(\mathrm{ord}_{\mathfrak{P}_i}(g_0 - \alpha_i^{-1}), \mathrm{ord}_{\mathfrak{P}_i} \alpha_i^{-1}\right)\\ &= \mathrm{ord}_{\mathfrak{P}_i} \alpha_i^{-1} = -1\end{aligned}$$

and, similarly,

$$\begin{aligned}\mathrm{ord}_{\mathfrak{C}} g_0 &= \mathrm{ord}_{\mathfrak{C}}((g - \gamma) + \gamma) = \min(\mathrm{ord}_{\mathfrak{C}_i}(g_0 - \gamma), \mathrm{ord}_{\mathfrak{C}_i} \gamma)\\ &= \mathrm{ord}_{\mathfrak{C}} \gamma = 1.\end{aligned}$$

Finally we note that for all i we have that $\mathrm{ord}_{\mathfrak{T}_i} g_0 = 0$. Now let $g = g_0^{-1}$. Then we need to check only one condition: that

$$\mathrm{ord}_{\mathfrak{T}_i}(g - 1) = \mathrm{ord}_{\mathfrak{T}_i} \frac{1 - g_0}{g_0} = \mathrm{ord}_{\mathfrak{T}_i}(1 - g_0) > 3m_i.$$

□

Lemma B.2.3. *Let K be a global function field. Let $\mathfrak{B}$ be an integral divisor of K and let $\mathfrak{a}$ be a prime divisor of K not dividing $\mathfrak{B}$. Then, for any sufficiently large s, there exists $u \in K$ with a pole of order s at $\mathfrak{a}$, and such that $u \equiv 1 \bmod \mathfrak{B}$ and u is integral at all the other primes of K.*

Proof. By the strong approximation theorem, there exists $u_1 \in K$ such that $u \equiv 1 \bmod \mathfrak{B}$ and u is integral at all the primes of K except for $\mathfrak{a}$. Next, by a consequence of the Riemann–Roch theorem (Corollary 5.5, Chapter II of [113]), we can show that for any natural number s with $s > 2g - 2 + \deg \mathfrak{B}$, where g is the genus of K, there exists $u_2 \in K$ such that $\mathrm{ord}_{\mathfrak{a}} u_2 = -s$, $u_2 \equiv 0 \bmod \mathfrak{B}$, and u_2 is integral at all the other primes of K. Indeed, for any $s > 2g - 1 + \deg \mathfrak{B}$ the dimension of the space of functions with a pole of order at most s and zero modulo $\mathfrak{B}$ is equal to $s \deg \mathfrak{a} - \deg \mathfrak{B} - g + 1$ as a vector space over the field C_K of constants of K. Similarly, the dimension of the space of functions with a pole of order at most $s - 1$ and zero modulo $\mathfrak{B}$ is equal to $(s - 1)\deg \mathfrak{a} - \deg\mathfrak{B} - g + 1$ as a vector space over the field C_K. Thus, the number of functions with a pole of order s at $\mathfrak{a}$ and zero modulo $\mathfrak{B}$ is equal to

$$|C_K|^{(s-1)\deg \mathfrak{a} - \deg \mathfrak{B} - g + 1}\left(|C_K|^{\deg \mathfrak{a}} - 1\right) \neq 0.$$

Now assume that the order of the pole of u_1 is less than s and let $u = u_1 + u_2$. Observe now that u satisfies all the requirements. □

The next lemma plays a role in our construction of the recursive presentations of number fields.

Proposition B.2.4. *Let K be a number field. Let $\Omega = \{\omega_1, \dots, \omega_n\}$ be an integral basis of K over $\mathbb{Q}$. Let p be a rational prime and let $\mathfrak{p}$ be a factor of p in K. Then there exists $\alpha \in K$ satisfying the following requirements:*

1. $\alpha = \sum_{i=1}^{n} a_i \alpha_i,\ a_i \in \mathbb{Z},\ |a_i| < p^2$;
2. $\operatorname{ord}_{\mathfrak{p}} \alpha = 1$;
3. *For all conjugates* $\mathfrak{q} \neq \mathfrak{p}$ *of* $\mathfrak{p}$ *over* $\mathbb{Q}$, $\operatorname{ord}_{\mathfrak{q}} \alpha = 0$.

Proof. By the strong approximation theorem, there exists $\gamma \in O_K$ such that $\operatorname{ord}_{\mathfrak{p}} \gamma = 1$ and, for all conjugates $\mathfrak{q} \neq \mathfrak{p}$ of $\mathfrak{p}$ over $\mathbb{Q}$, $\operatorname{ord}_{\mathfrak{q}} \gamma = 0$. Let $\gamma = \sum_{i=1}^{n} b_i \omega_i, b_i \in \mathbb{Z}$. For each b_i there exists $a_i \in \mathbb{Z}$ such that $|a_i| < p^2$ and $a_i \equiv b_i \bmod p^2$. Let $\alpha = \sum_{i=1}^{n} a_i \omega_i$. Then $p^2 \mid (\alpha - \gamma)$ and $\operatorname{ord}_{\mathfrak{q}} \alpha = \operatorname{ord}_{\mathfrak{q}} \gamma$ for every factor $\mathfrak{q}$ of p in K. □

B.3 Linearly disjoint fields

In this section we explore some properties of linearly disjoint fields. Galois groups of products of linearly disjoint Galois extensions can be decomposed into the products of constituent Galois groups, allowing for a relatively easy analysis of prime splitting. This property makes products of linearly disjoint fields very attractive to us, and we use these fields extensively throughout the book.

A general discussion of linearly disjoint fields together with the definition of linear disjointness can be found in Section 2.5 of [33].

Lemma B.3.1. *Suppose that M/F and L/F are finite field extensions. Then M and L are linearly disjoint over F if and only if $[LM : M] = [L : F]$. Further, $[LM : M] = [L : F]$ if and only if $[LM : L] = [M : F]$.* (See Lemma 2.5.1 and Corollary 2.5.2 of [33].)

The following corollary is obvious but useful.

Corollary B.3.2. *Let M and L be two fields linearly disjoint over a common subfield F. Let L_1, M_1 be fields such that $F \subset M_1 \subset M$ and $F \subset L_1 \subset L$. Then M_1 and L_1 are also linearly disjoint over F.*

Lemma B.3.3. *Suppose that M/F and L/F are finite field extensions, L/F being a Galois extension. Then M and L are linearly disjoint over F if and only if $M \cap L = F$.*

Proof. First of all, it is clear that if $M \cap L \neq F$ then M and L are not linearly disjoint. Suppose now that M and L are not linearly disjoint. Let α be a generator of L over F. Then by Lemma B.3.1, the monic irreducible polynomial $H(T)$ of α over F will factor over M. Let $H_1(T)$ be a factor of $H(T)$ in M. Then the coefficients of $H_1(T)$ are on the one hand elements of M and on the other hand symmetric functions of some conjugates of α over F and thus contained in L, the splitting field of $H(T)$. Hence the coefficients of $H_1(T)$ are contained in $M \cap L$. However, since $H(T)$ does not factor in F, at least one coefficient of $H_1(T)$ is not in F. Thus $M \cap L \neq F$. □

Lemma B.3.4. *Let L be a function field over a finite field of constants C. Let U be a finite extension of L such that the constant fields of U and L are the same. Let C'/C be a finite extension. Then $C'L$ and U are linearly disjoint over L.*

Proof. By Lemma B.3.1 it is enough to show that $[C'U : U] = [C'L : L]$. Let $\alpha \in C'$ generate C' over C. Then it is enough to note that by Theorem 11, Section 3, Chapter XV of [1] we have that $[C(\alpha) : C] = [L(\alpha) : L] = [U(\alpha) : U]$.

□

Lemma B.3.5. *Let M, L, F be as in Lemma B.3.3. Then* $\mathrm{Gal}(ML/M) \cong \mathrm{Gal}(L/F)$ *and this isomorphism is realized by restriction.*

Proof. Let α be a generator of L over F. Then α will generate ML over M. Further, since M and L are linearly disjoint over F, α will have the same number of conjugates over M as over F. Let $\alpha = \alpha_1, \ldots, \alpha_l$, be all the conjugates of α over M and F, where $l = [L : F]$. Since $\{\alpha_i, i = 1, \ldots, l\} \subset L$, we can conclude that $\{\alpha_i, i = 1, \ldots, l\} \subset ML$. Thus ML/M is Galois and of the same degree as L/F. Next consider a map from $\mathrm{Gal}(ML/M)$ to $\mathrm{Gal}(L/F)$ implemented by restriction. Let $\sigma, \tau \in \mathrm{Gal}(LM/M)$ restrict to the same map over F. Then σ and τ send α to the same conjugate. Thus $\sigma = \tau$, since each element of $\mathrm{Gal}(ML/M)$ is determined by its action on the generator of the extension. Hence the restriction sends different maps to different maps. Since both groups are of the same size, the image of the restriction contains all $\mathrm{Gal}(L/F)$. Finally, restriction is certainly a group homomorphism. Thus the groups are isomorphic.

□

The next two lemmas will generalize Lemma B.3.5 to the situations where we consider products of linearly disjoint fields.

Lemma B.3.6. *Let N_1/F, N_2/F be two Galois extensions, linearly disjoint over F. Then*

$$\mathrm{Gal}(N_1N_2/N_1) \cong \mathrm{Gal}(N_2/F),$$
$$\mathrm{Gal}(N_1N_2/N_2) \cong \mathrm{Gal}(N_1/F),$$

and

$$\mathrm{Gal}(N_1N_2/F) \cong \mathrm{Gal}(N_1N_2/N_1) \times \mathrm{Gal}(N_2N_1/N_2).$$

Proof. For $i = 1, 2$, let α_i be a generator of N_i over F. Let $n_i = [N_i : F]$. Then by the definition of linear disjointness, α_i has n_i conjugates over N_j for $j \neq i$ and $i, j \in \{1, 2\}$. Let $\alpha_{i,1} = \alpha_i, \ldots, \alpha_{i,n_i}$ be all the conjugates of α_i over F and N_j, for $j \neq i$ and $i, j \in \{1, 2\}$. Let $\sigma_{i,j} : N_1N_2 \to N_1N_2$ be defined by $\sigma_{i,j}(\alpha_i) = \alpha_{i,j}, \sigma_{i,j}(\alpha_k) = \alpha_k$ for $k \neq i, j \in \{1, \ldots, n_i\}$, and $i, k \in \{1, 2\}$. Then $\sigma_{i,j} \in \mathrm{Gal}(N_1N_2/N_k) \subset \mathrm{Gal}(N_1N_2/F)$. Further, $\sigma_{1,k}\sigma_{2,j} = \sigma_{1,k}\sigma_{2,j}$ since both mappings send α_1 to $\alpha_{1,k}$ and α_2 to $\alpha_{2,j}$. Consider now the set of all products $\{\sigma_{1,k}\sigma_{2,j},\ i = 1, \ldots, n_1,\ j = 1, \ldots, n_2\} \subseteq \mathrm{Gal}(N_1N_2/F)$. This set contains n_1n_2 distinct elements of $\mathrm{Gal}(N_1N_2/F)$, since each automorphism of N_1N_2 over F is determined by its action on α_1 and α_2. But $|\mathrm{Gal}(N_1N_2/F)| = n_1n_2$ and therefore $\{\sigma_{1,k}\sigma_{2,j},\ i = 1, \ldots, n_1,\ j = 1, \ldots, n_2\} = \mathrm{Gal}(N_1N_2/F)$. Finally consider

$$\mathrm{Gal}(N_1N_2/N_i) \to \mathrm{Gal}(N_j/F),$$

where $i \neq j$ and $i, j \in \{1, 2\}$ and the mapping between the groups is realized by restriction. This is an isomorphism by Lemma B.3.5. □

Lemma B.3.7. *Let $N_1/F, \ldots, N_m/F$ be Galois field extensions. Assume further that*

$$G_i = \prod_{j=1,\ldots,m, j\neq i} N_j$$

is linearly disjoint from N_i over F for all $i = 1, \ldots, n$. Then

$$\mathrm{Gal}(N_1 \cdots N_m/F) \cong \mathrm{Gal}(N_1/F) \oplus \cdots \oplus \mathrm{Gal}(N_m/F),$$
$$\mathrm{Gal}\left(\prod_{j=1}^{m} N_j/G_i\right) \cong \mathrm{Gal}(N_i/F).$$

Proof. To prove the first assertion of the lemma, we proceed by induction on m. For $m = 2$ the result follows from Lemma B.3.6. Assume that the statement holds for $m = k$. Then, also by Lemma B.3.6, since N_{k+1} and $\prod_{j=1}^{k} N_j$ are linearly disjoint over F as subfields of linearly disjoint fields (see Corollary B.3.2), we have that

$$\begin{aligned}\mathrm{Gal}\left(N_{k+1}\prod_{j=1}^{k} N_j/F\right) &\cong \mathrm{Gal}(N_{k+1}/F) \oplus (\mathrm{Gal}(N_1/F) \oplus \cdots \oplus \mathrm{Gal}(N_k/F)) \\ &\cong \mathrm{Gal}(N_{k+1}/F) \oplus \mathrm{Gal}(N_1/F) \oplus \cdots \oplus \mathrm{Gal}(N_k/F),\end{aligned}$$

with $\mathrm{Gal}(\prod_{j=1}^{k+1} N_j / \prod_{i=1}^{k} N_i) \cong \mathrm{Gal}(N_{k+1}/F)$.

By renumbering the fields N_i if necessary and setting $k = m - 1$, we also conclude that in general for $i = 1, \ldots, m$ we have that

$$\mathrm{Gal}\left(\prod_{j=1}^{m} N_j/G_i\right) \cong \mathrm{Gal}(N_i/F).$$

□

Lemma B.3.8. *Let G/F, H/F be Galois extensions of number fields, where G, F are totally real and H is a totally complex number field such that $[H : F] = 2$. Then G and H are linearly disjoint from each other over F.*

Proof. By Lemma B.3.3 it is enough to show that $G \cap H = F$. Since $[H : F]$ is a prime number, $G \cap H = F$ or $G \cap H = H$. Since H is totally complex, $H \not\subseteq G$. Thus the first alternative holds. □

Proposition B.3.9. *Let K be any number field. Let q be a prime number. Let n be a positive integer. Then in the algebraic closure of K there exist $\beta_1, \ldots, \beta_n$ such that for all $i = 1, \ldots, n$ we have that $K(\beta_i)/K$ is a cyclic extension of degree q, $K(\beta_i)$ is linearly disjoint from $\prod_{j \neq i} K(\beta_j)$, and $\beta_1, \ldots, \beta_n$ are totally real, i.e. all their conjugates over $\mathbb{Q}$ are real.*

Proof. Let p be a prime equivalent to 1 mod q if q is odd and let $p \equiv 1 \bmod 4$ otherwise. (Such a prime exists by a theorem on primes in arithmetic progressions. See Theorem 9.4.1 of [90].) Let ξ_p be a primitive pth root of unity. Then by Theorem 9.1 and Exercise 7, Section 9, Chapter I of [37], we have that $\mathbb{Q}(\xi_p)/\mathbb{Q}$ is a cyclic extension of degree $p - 1$, where p is ramified completely. Further, $\mathbb{Q}(\cos(2\pi/p)) \subset \mathbb{Q}(\xi_p)$ is a totally real subextension of degree 2. Let $H \lhd \mathrm{Gal}\big(\mathbb{Q}(\cos(2\pi/p))/\mathbb{Q}\big)$ be a normal subgroup of order $(p - 1)/2q$. Let F_H be the fixed field of H. We claim that $F_H/\mathbb{Q}$ is a cyclic extension of degree q. Indeed, since $\mathrm{Gal}\big(\mathbb{Q}(\cos(2\pi/p))/\mathbb{Q}\big)$ is cyclic, every quotient group of $\mathrm{Gal}\big(\mathbb{Q}(\cos(2\pi/p))/\mathbb{Q}\big)$ is cyclic and $[F_H : \mathbb{Q}] = (p - 1/2)/((p - 1)/2q) = q$.

Using the fact that there are infinitely many primes in arithmetic progression we can find a sequence $p_1, \ldots, p_n$ of prime numbers such that for all $i = 1, \ldots, n$ it is the case that p_i is not ramified in the extension $K/\mathbb{Q}$ and that $q \mid (p_i - 1)$ if q is odd and $4 \mid (p_i - 1)$ otherwise. Let F_i be a totally real subfield of $\mathbb{Q}(\xi_{p_i})$ of degree q. Observe that in the extension $\left(K \prod_{j \neq i} F_j\right)/\mathbb{Q}$ we know that p_i is not ramified, and in the extension $\left(K \prod_{j=1}^{n} F_j\right)/\mathbb{Q}$ we have that p_i is ramified with ramification degree q. Thus

$$\left[K \prod_{j=1}^{n} F_j : K \prod_{j \neq i} F_j\right] = q.$$

By Lemma B.3.1, $K \prod_{j \neq i} F_j$ and F_i are linearly disjoint over $\mathbb{Q}$. □

Lemma B.3.10. *Let p, q be distinct prime numbers. Let ξ_p and ξ_q be the pth and qth primitive roots of unity respectively. Then $\mathbb{Q}(\xi_q)$ and $\mathbb{Q}(\xi_p)$ are linearly disjoint over $\mathbb{Q}$.*

Proof. The proof follows immediately from Lemma B.3.3 if we take into account the fact that p and only p is ramified completely in the first extension and q and only q is ramified completely in the second. □

Corollary B.3.11. *Let K be a number field. Then for all but finitely many p, we have that K and $\mathbb{Q}(\xi_p)$ are linearly disjoint over $\mathbb{Q}$.*

Proof. Let p be a prime not ramified in K and consider $\mathbb{Q}(\xi_p)$. Now, by Lemma B.3.3 we have that $\mathbb{Q}(\xi_p) \cap K$ is linearly disjoint if and only if $K \cap \mathbb{Q}(\xi_p) = \mathbb{Q}$. But since p is ramified completely in $\mathbb{Q}(\xi_p)$ and not at all in K, the intersection of these fields must $\mathbb{Q}$. Since only finitely many primes can ramify in a number field (see Theorem 7.3 of [37]), the conclusion of the corollary follows. □

B.4 Divisors, prime and composite, under extensions

In this section we will examine how prime and composite divisors behave under finite extensions of global fields. We start with basic results.

Lemma B.4.1. *Let M/L be a Galois extension of global fields. Let $\mathfrak{p}$ be a prime of L and let $\mathfrak{P}$ be a prime of M above it. Let $R(\mathfrak{P})$, $R(\mathfrak{p})$ be the residue fields of $\mathfrak{P}$ and $\mathfrak{p}$ respectively. Let*

$$G(\mathfrak{P}) = \{\sigma \in \mathrm{Gal}(M/L) : \sigma(\mathfrak{P}) = \mathfrak{P}\}.$$

Then $G(\beta)$ *is called the decomposition group of* $\mathfrak{P}$. *Let*

$$T(\mathfrak{P}) = \{\sigma \in \mathrm{Gal}(M/L) : \sigma(x) \equiv x \bmod \mathfrak{P} \text{ for all } x \in M\}.$$

(In other words $T(\mathfrak{P})$ fixes all the classes modulo $\mathfrak{P}$.) *Then* $T(\mathfrak{P})$ *is called the inertia group of* $\mathfrak{P}$. *The following statements are true.*

- $|T(\mathfrak{P})| = e(\mathfrak{P}/\mathfrak{p})$.
- $|G(\mathfrak{P})| = e(\mathfrak{P}/\mathfrak{p})f(\mathfrak{P}/\mathfrak{p})$.
- *The number of factors of* $\mathfrak{p}$ *in* M *is equal to* $[M : L]/|G(\mathfrak{P})|$, *so that* $\mathfrak{p}$ *splits completely if and only if* $|G(\mathfrak{P})|$ *is trivial.*
- $G(\mathfrak{P}) \cong \mathrm{Gal}(M_{\mathfrak{P}}/L_{\mathfrak{p}})$. *(Here* $M_{\mathfrak{P}}$ *and* $L_{\mathfrak{p}}$ *are the completions of* M *and* L *under* $\mathfrak{P}$ *and* $\mathfrak{p}$ *respectively.)*
- *There exists a natural surjective homomorphism from* $G(\mathfrak{P})$ *to*

  $$\mathrm{Gal}(R(\mathfrak{P})/R(\mathfrak{p})).$$

  The kernel of this homomorphism is $T(\mathfrak{P})$. $T(\mathfrak{P})$ *is normal in* $G(\mathfrak{P})$ *and*

  $$G(\mathfrak{P})/T(\mathfrak{P}) \cong \mathrm{Gal}(R(\mathfrak{P})/R(\mathfrak{p}))$$

  is cyclic as a Galois group of a finite field extension.
- $\mathfrak{P}$ *is totally ramified in the extension* $M/M^{T(\mathfrak{P})}$, *where* $M^{T(\mathfrak{P})}$ *is the fixed field of* $T(\mathfrak{P})$ *and* $M^{T(\mathfrak{P})}/L$ *is unramified with respect to* $\mathfrak{p}$.

  If $\mathfrak{p}$ *is unramified in the extension* M/L *then the generator of* $G(\mathfrak{P})$ *which is the inverse image of the Frobenius automorphism of* $R(\mathfrak{P})/R(\mathfrak{p})$ *is called the Frobenius automorphism of* $\mathfrak{P}$.

See Section 6.2 of [33] for more details.

The next result relates the divisor of a norm under a finite extension to the norm of the divisor.

Proposition B.4.2. Let M/K *be a global field extension of degree* n. *Let* $y \in M$. *Then* $\mathfrak{D}_K(\mathbf{N}_{M/K}y) = \mathbf{N}_{M/K}(\mathfrak{D}_M(y))$. (Here $\mathfrak{D}_K()$ denotes the K-divisor of an element of K, while $\mathfrak{D}_M()$ denotes the M-divisor of an element of M.)

Proof. This follows from Proposition 8.1 of [37]. □

The lemma below addresses the issue of the order of a norm at a prime that does not split in an extension. This plays an important role in the definition of integrality at finitely many primes.

Lemma B.4.3. *Let* M/K *be an extension of degree* n *of global fields. Let* $\mathfrak{q}_K$ *be a prime of* K. *Suppose that* $\mathfrak{q}_K$ *does not split in the extension* M/K. *Let* $x \in K$ *be such that* x *is a* K*-norm of some element* $y \in M$. *Then* $\mathrm{ord}_{\mathfrak{q}}\, x \equiv 0 \bmod n$.

Proof. Let $y \in M$ be such that $\mathbf{N}_{M/K}(y) = x$. Let $\mathfrak{D}_M(y)$ be the divisor of y in M. Then by Proposition B.4.2 and Corollary 8.5, Chapter I of [37], we deduce the following:

$$\mathbf{N}_{M/K}(\mathfrak{D}_M(y)) = \prod_{\mathfrak{p}_M \in \mathcal{M}_M} \mathbf{N}_{M/K}\left(\mathfrak{p}_M^{\mathrm{ord}_{\mathfrak{p}_M} y}\right)$$

$$= \prod_{\mathfrak{p}_K \in \mathcal{M}_K} \mathfrak{p}_K^{f(\mathfrak{p}_M/\mathfrak{p}_K)\mathrm{ord}_{\mathfrak{p}_M} y} = \mathfrak{D}_K(x),$$

where $\mathfrak{p}_K$ is the K-prime below $\mathfrak{p}_M$ in K, and $f(\mathfrak{p}_M/\mathfrak{p}_K)$ is the relative degree of $\mathfrak{p}_M$ over K. If a prime $\mathfrak{p}_K$ does not split in the extension M/K, it means that there is only one factor $\mathfrak{p}_M$ above it in M. Therefore, by Theorem 1, Section 1, Chapter IV of [3] for the function field case and by Corollary 6.7, Chapter I of [37] for the number field case, $f(\mathfrak{p}_M/\mathfrak{p}_K) = n$. Thus for this prime $\mathfrak{P}_K$ we have that $\mathrm{ord}_{\mathfrak{p}_K} x = \mathrm{ord}_{\mathfrak{p}_K} \mathfrak{D}_K(x) = n \, \mathrm{ord}_{\mathfrak{p}_M} y$. □

The sequence of five lemmas below takes a close look at primes splitting and not splitting in products of linearly disjoint fields.

Lemma B.4.4. *Let F be a global field. Let N_1 be a cyclic extension of F and let N_2 be a Galois extension of F, linearly disjoint from N_1 over F. Then there are infinitely many primes $\mathfrak{p}$ of F that do not split in N_1 but split completely in N_2.*

Proof. Consider the extension N_1N_2/F. This extension is Galois. Further, linear disjointness guarantees by Lemma B.3.6 that

$$\mathrm{Gal}(N_1N_2/F) \cong \mathrm{Gal}(N_1/F) \times \mathrm{Gal}(N_2/F)$$

and

$$\mathrm{Gal}(N_1N_2/N_1) \cong \mathrm{Gal}(N_2/F),$$
$$\mathrm{Gal}(N_1N_2/N_2) \cong \mathrm{Gal}(N_1/F),$$

where the last two isomorphisms are realized by restriction. Let σ be a generator of $\mathrm{Gal}(N_1/F)$ and consider a N_1N_2-prime $\mathfrak{P}$ whose Frobenius automorphism is $(\sigma, \mathrm{id}_{N_2})$, where id_{N_2} is the identity element of $\mathrm{Gal}(N_2/F)$. Then on the one hand σ is an element of the decomposition group of $\mathfrak{P}_1 = \mathfrak{P} \cap N_1$ over F. Hence this decomposition group is all of $\mathrm{Gal}(N_1/F)$. Thus, $\mathfrak{p} = \mathfrak{P} \cap F$ does not split in the extension N_1/F. On the other hand, the decomposition group of $\mathfrak{P}$ over N_2 is $\langle(\sigma, \mathrm{id}_{N_2})\rangle \cap \mathrm{Gal}(N_1N_2/N_2) = \mathrm{Gal}(N_1N_2/N_2)$. Then, by looking at the quotient of the decomposition group of $\mathfrak{P}$ over F and over N_2, we conclude that the decomposition group of $\mathfrak{P}_2 = \mathfrak{P} \cap N_2$ over F is trivial and therefore that

$\mathfrak{p} = \mathfrak{P} \cap F$ splits completely in the extension N_2/F. Now the result follows by the Chebotarev density theorem (see Theorem 10.4, Chapter V of [37]). □

Lemma B.4.5. *Let $K \subset L \subset M$ be a finite extension of global fields, $M/K, L/K$ being Galois extensions. Let $\mathfrak{P}_K$ be a prime of K. Let $\mathfrak{P}_{L,1}, \mathfrak{P}_{L,2}$ be any two factors of $\mathfrak{P}_K$ in L. Then in M it is the case that $\mathfrak{P}_{L,1}, \mathfrak{P}_{L,2}$ have the same number of factors.*

Proof. By Proposition 11, Section 5, Chapter 1 of [46], there exists a $\bar{\sigma} \in \mathrm{Gal}(L/K)$ such that $\bar{\sigma}(\mathfrak{P}_{L,1}) = \mathfrak{P}_{L,2}$. Let $\sigma \in \mathrm{Gal}(M/K)$ be an extension of $\bar{\sigma}$ to M. Then every factor of $\mathfrak{P}_{L,1}$ has to be mapped to a factor of $\mathfrak{P}_{L,2}$ by σ. Thus $\mathfrak{P}_{L,1}, \mathfrak{P}_{L,2}$ must have the same number of factors in M. □

The next four lemmas consider when splitting "below" implies splitting "above" and vice versa.

Lemma B.4.6. *Let K/E be a separable extension of global fields. Let K_G be the Galois closure of K over E. Then an E-prime $\mathfrak{P}_E$ splits completely in K if and only if it splits completely in K_G.* (See Proposition 6.3.2 of [45].)

Lemma B.4.7. *Let N_1/F, N_2/F be two separable global field extensions, with N_1, N_2 linearly disjoint over F and N_1/F Galois. Let $\mathfrak{p}$ be a prime of F splitting completely into distinct factors in the extension N_1/F. Let $\mathfrak{P}_2$ be a prime above $\mathfrak{p}$ in N_2. Then $\mathfrak{P}_2$ splits completely into distinct factors in the extension N_1N_2/N_2.*

Proof. First of all we observe that as in Lemma B.3.5 we have that $\mathrm{Gal}(N_1N_2/N_2) \cong \mathrm{Gal}(N_1/F)$, the isomorphism being realized by restriction. Let $\mathfrak{P}_{1,2}$ lie above $\mathfrak{P}_2$ in N_1N_2 and let σ be any element of its decomposition group over F. Then $\sigma_{|N_1}$ is an element of the decomposition group of $\mathfrak{P}_1 = \mathfrak{P}_{1,2} \cap N_1$. However, since $\mathfrak{p}$ splits completely in N_1 into distinct factors, the decomposition group of $\mathfrak{P}_1$ is trivial and therefore $\sigma_{|N_1} = \mathrm{id}_{N_1}$. Let $\alpha \in N_1$ generate N_1 over F and N_1N_2 over N_2. Therefore, $\alpha = \sigma_{|N_1}(\alpha) = \sigma(\alpha)$. Thus the decomposition group of $\mathfrak{P}_{1,2}$ over N_2 is trivial and therefore $\mathfrak{P}_2$ splits completely in the extension N_1N_2/N_1. □

Lemma B.4.8. *Let N_1/F, N_2/F be two separable global field extensions, with N_1, N_2 Galois over F and $[N_1 : F]$ relatively prime to $[N_2 : F]$. Let $\mathfrak{p}$ be a prime of F not splitting in N_1. Let $\mathfrak{P}_2$ be above $\mathfrak{p}$ in N_2. Then $\mathfrak{P}_2$ does not split in the extension N_1N_2/N_2.*

Proof. First, consider the diagram below.

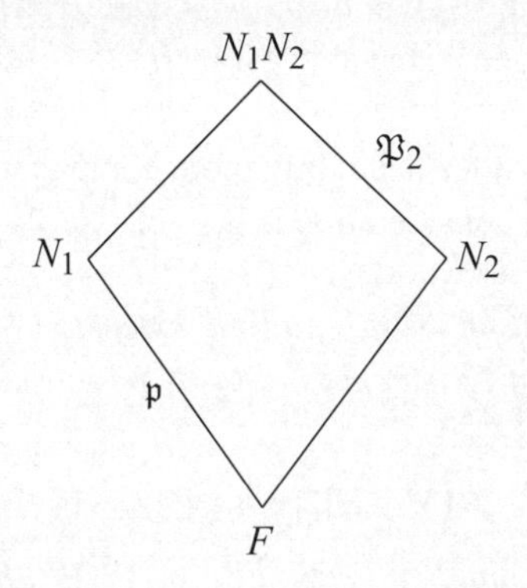

Now note that the extension $F \subset N_1 \subset N_1N_2$ is a Galois extension by Lemma B.3.6. The number of factors of $\mathfrak{p}$ in N_1N_2 must be a divisor of $[N_1N_2 : N_1] = [N_2 : F]$. Suppose that $\mathfrak{P}_2$ does not remain prime in the extension $N_2 \subset N_2N_1$. Then the number of factors of $\mathfrak{P}_2$ in N_1N_2 is a non-trivial divisor of $[N_1N_2 : N_1]$. Since N_1N_2/F is Galois, every factor of $\mathfrak{p}$ in N_2 has the same number of factors in N_1N_2 by Lemma B.4.5, and therefore the number of factors of $\mathfrak{p}$ in N_1N_2 must have a non-trivial common factor with $[N_1 : F]$. But this contradicts the fact that this number must be a divisor of $[N_2 : F]$. Thus $\mathfrak{P}_2$ does not split in the extension N_1N_2/N_2. □

Lemma B.4.9. *Let F/E and H/E be Galois extensions of global fields with F and H linearly disjoint over E. Let $\mathfrak{p}_F$ be a prime of F not splitting in the extension HF/F. Let $\mathfrak{p}_E$ be the prime below it in E. Then there is only one prime above $\mathfrak{p}_E$ in H.*

Proof. Consider the following diagram.

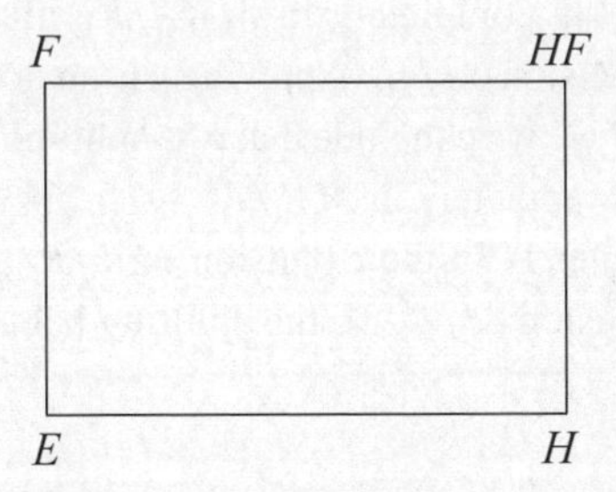

By Lemma B.3.5, $\mathrm{Gal}(FH/F) \cong \mathrm{Gal}(H/E)$ and the isomorphism is realized by restriction. Let $\mathfrak{p}_{FH}$ be the single factor of $\mathfrak{p}_F$ in FH. Then its Frobenius automorphism σ is a generator of Gal(FH/F) and restriction of σ to H will generate Gal(H/E). But $\sigma|_H$ will fix $\mathfrak{p}_H = \mathfrak{p}_{FH} \cap H$, a prime of H lying above $\mathfrak{p}_E$. Thus, $\mathfrak{p}_E$ has only one prime above it in H. □

The next lemma considers a slightly more complicated situation where one of the extensions under consideration is not necessarily Galois.

Lemma B.4.10. *Let K/E be a separable extension of global fields. Let K_G be the Galois closure of K over E. Let M/E be a cyclic extension of E such that*

$$([M : E], [K_G : E]) = 1$$

and $[M : E]$ is a prime number. Let $\mathfrak{p}_1, \ldots, \mathfrak{p}_{[K:E]}$ be all the factors in K of an E-prime $\mathfrak{P}_E$, splitting completely in the extension K/E. Then Parts 1 and 2 below are true.

1. *The following statements are equivalent:*
   (a) *for some i we have that $\mathfrak{p}_i$ splits completely in the extension MK/K;*
   (b) *$\mathfrak{P}_E$ splits completely in the extension M/E;*
   (c) *some factor of $\mathfrak{P}_E$ in K_G splits completely in the extension MK_G/K_G.*
2. *Either all $\mathfrak{p}_1, \ldots, \mathfrak{p}_{[K:E]}$ split completely in MK_G/K or none does.*

Proof. By Lemma B.4.6, $\mathfrak{P}_E$ splits completely in K_G/E if and only if $\mathfrak{P}_E$ splits completely in K/E. Thus, we can conclude that for all i we have that $\mathfrak{p}_i$ splits completely in the extension K_G/K. Therefore, $\mathfrak{p}_i$ splits completely in MK_G/K if and only if every factor $\mathfrak{p}_{K_G}$ of $\mathfrak{p}_i$ in K_G splits completely in the extension MK_G/K_G. By Lemma B.3.3, $([M : E], [K_G : E]) = 1$ implies that M and K_G are linearly disjoint over E and consequently that $[MK_G : K_G] = [M : E]$ is a prime number. Thus, either $\mathfrak{p}_{K_G}$ splits completely or it is inert in the extension MK_G/K_G. Therefore $([M : E], [K_G : E]) = 1$ also implies by Lemmas B.4.8 and B.4.7 that $\mathfrak{p}_{K_G}$ splits completely in MK_G/K_G if and only if $\mathfrak{P}_E$ splits completely in the extension M/E. Applying an analogous argument to the Galois extension K_GM/K we conclude that $\mathfrak{p}_{K_G}$ splits completely in MK_G/K_G if and only if $\mathfrak{p}_i$ splits completely in MK/K.

Thus $\mathfrak{p}_i$ splits completely in the extension MK/K if and only if $\mathfrak{p}_E$ splits completely in the extension M/E. So the splitting behavior is uniform across $\mathfrak{P}_E$-factors in K. □

The next lemma deals with cyclic extensions of prime degree. They play an important role in implementation of the norm method for defining integrality at finitely many primes.

Lemma B.4.11. *Let q be a rational prime. Let G be any field of characteristic different from q. Let $a \in G$ and let α be a root of the polynomial $T^q - a$. Then the following statements are true.*

1. *If $a \in G$ is not a qth power then the polynomial $X^q - a$ is irreducible over G.*
2. *Assume that G has all the qth roots of unity. Then $[G(\alpha) : G] = q$ or $[G(\alpha) : G] = 1$. Further, in the first case the extension is cyclic.*
3. *If G is a global field and $\mathfrak{P}$ is a prime of G such that $\mathrm{ord}_{\mathfrak{P}}\, q = 0$ and $\mathrm{ord}_{\mathfrak{P}}\, a = 0$ then $\mathfrak{P}$ is not ramified in the extension $G(\alpha)/G$, and $\mathfrak{P}$ has a relative-degree-1 factor in this extension if and only if a is equivalent to a qth power modulo $\mathfrak{P}$. If G has a primitive qth root of unity, then we can substitute "will split completely" for "will have a relative-degree-1 factor."*
4. *If G is a global field, $\mathrm{ord}_{\mathfrak{P}}\, a \not\equiv 0 \bmod q$, and the extension $G(\alpha)/G$ is not trivial then $\mathfrak{P}$ is ramified completely in this extension.*

Proof. Part of the lemma follows from Theorem 16 p. 221 of [47]. The rest is left as an exercise for the reader. □

The next proposition describes a generalization of sorts for the notion of an integral basis. This result allows us to make sure that the norm equations widely used in this book have the solutions we require.

Lemma B.4.12. *Let K/F be a finite separable global field extension. Let K_G be the Galois closure of K over F. Let $\Omega = \{\omega_1, \ldots, \omega_m\} \subset K$ be a basis of K over F. Let $\mathcal{D}$ be the determinant of the matrix $(\omega_{i,j})$, where $\omega_{i,1} = \omega_i, \ldots, \omega_{i,m}$ are all the conjugates of ω_i over F. (Note that $\mathcal{D}^2$ is the discriminant of Ω.) Let $\mathfrak{p}$ be a prime of F such that all the elements of Ω are integral with respect to $\mathfrak{p}$. Let $\mathfrak{p} = \prod_{i=1}^{s} \mathfrak{P}_i^{e}$ be the factorization of $\mathfrak{p}$ in K_G. Let $h \in K$ be integral at $\mathfrak{p}$ and let $h = \sum_{i=1}^{m} a_i\omega_i$, where $a_i \in F$. Then for $i = 1, \ldots, m$ we have that $\mathrm{ord}_{\mathfrak{p}}\, a_i \geq e^{-1}\mathrm{ord}_{\mathfrak{P}_j}\, \mathcal{D}^{-1}$, for all $i = 1, \ldots, s$. In particular, if for all $j = 1, \ldots, s$ we have that $\mathrm{ord}_{\mathfrak{P}_j}\, \mathcal{D} = 0$ then for all $i = 1, \ldots, m$ we also have that the a_i are integral at $\mathfrak{p}$ and that Ω is an integral basis with respect to $\mathfrak{p}$. Finally, in all cases $\mathcal{D}^2 a_i$ is integral at $\mathfrak{p}$.*

Proof. Consider the following linear system:

$$\sum_{i=1}^{m} a_i\omega_{i,r} = h_r,$$

where $r = 1, \ldots, m$ and $h_1 = h, \ldots, h_m$ are all the conjugates of h over F. The determinant of this system is $\mathcal{D}$. Solving this system by Cramer's rule, we conclude that

$$a_i = \frac{\mathcal{D}_i}{\mathcal{D}},$$

where $\mathcal{D}_i \in K_G$ is integral with respect to $\mathfrak{p}$. □

The following proposition explains how to determine whether a particular prime splits in a finite extension.

Lemma B.4.13. *Let K be a global field. Let $\mathfrak{p}$ be a prime of K. Let E/K be a finite separable extension of K generated by an element $\gamma \in E$, integral at $\mathfrak{p}$. Let $g(t)$ be the monic irreducible polynomial of γ over K. Assume further that the discriminant of γ is a unit at $\mathfrak{p}$ and $g(t) = \prod_{i=1}^{l} g_i(t) \bmod \mathfrak{p}$, where $g_i(t)$ is a monic polynomial irreducible in the residue field of $\mathfrak{p}$ of degree f_i. Then in E we have that $\mathfrak{p} = \prod_{i=1}^{k} \mathfrak{p}_i$, where $\mathfrak{p}_i$ is a prime of E of relative degree f_i over $\mathfrak{p}$.*

Proof. See Proposition 25, Section 8, Chapter I of [46] and Lemma B.4.12. □

The lemma below tells us that for each prime we can find an extension where it splits. This issue comes up when we consider the definability of the set of non-zero elements of local subrings of number fields.

Lemma B.4.14. *Let K be a global field. Let $\mathfrak{p}$ be any prime of K. Then there exists an extension M of K where $\mathfrak{p}$ has two or more distinct factors.*

Proof. Let $\mathfrak{q} \neq \mathfrak{p}$ be any other prime of K. By the strong approximation theorem there exists $a \in K$ such that $\mathrm{ord}_{\mathfrak{q}}\, a = 1$ and $a \equiv 1 \bmod \mathfrak{p}$. Let α be a root of $T^p - a$, where p is a rational prime distinct from the characteristic of the field and relatively prime to $\mathfrak{p}$ in the case of a number field. Then $M = K(\alpha)$ is an extension of degree p, $\mathfrak{p}$ does not divide the discriminant of α, and $T^p - a$ are factors modulo $\mathfrak{p}$. Thus, by Lemmas B.4.13 and B.4.12, $\mathfrak{p}$ splits in the extension M/K into distinct factors. □

When dealing with function fields, it is often useful to assume that the primes under consideration are of degree 1. The next lemma tells us what needs to be done to achieve this.

Lemma B.4.15. *Let M be a function field over a finite field of constants. Let $\mathfrak{p}$ be a prime of M. Let $C_\mathfrak{p}$ be the residue field of $\mathfrak{p}$. Then in the extension $C_\mathfrak{p}M/M$, we have that $\mathfrak{p}$ will split completely into factors of degree 1.*

Proof. Let $\gamma \in C_\mathfrak{p}$ be a generator of $C_\mathfrak{p}$ over C_M, the constant field of M. Then γ will also generate the extension $C_\mathfrak{p}M/M$. Let $f(t)$ be the monic irreducible polynomial of γ over C_M and M (it must be the same by Theorem 11, Section 3, Chapter XV of [1]). Then it splits completely mod $\mathfrak{p}$, since all finite extensions of finite fields are Galois. Further, since α is a constant its discriminant over M is not divisible by $\mathfrak{p}$. Thus, by Lemma B.4.13, $\mathfrak{p}$ will split into $[C_\mathfrak{p}M : M]$ factors of relative degree 1 over $\mathfrak{p}$. Hence the residue field of every factor of $\mathfrak{p}$ will be $C_\mathfrak{p}$. But by Theorem 13, Section 3, Chapter XV of [1], the constant field of $C_\mathfrak{p}M$ is $C_\mathfrak{p}$ and therefore every factor of $\mathfrak{p}$ will be of degree 1. □

The result below describes a useful property of degree-1 primes.

Lemma B.4.16. *Let M be a function field over a finite field of constants. Let $\mathfrak{p}$ be a prime of M of degree 1. Let E/M be a constant field extension. Then $\mathfrak{p}$ will not split in the extension E/M and will retain degree 1.*

Proof. Let $\gamma \in C_E$, the constant field of E, generate the extension. Let $f(t)$ be the monic irreducible polynomial of γ over M. Then $f(t)$ must also be the monic irreducible polynomial of γ over C_M by Theorem 11, Section 3, Chapter XV of [1]. Thus $f(t)$ will remain prime modulo $\mathfrak{p}$. Therefore, by Lemma B.4.13 we have that $\mathfrak{p}$ will remain prime in E. The single factor of $\mathfrak{p}$ in E will therefore have a relative degree $[E : M]$ over $\mathfrak{p}$ and thus the residue field of the factor must be of degree $[E : M]$ over the residue field of $\mathfrak{p}$ equal to C_M. But by Theorem 13, Section 3, Chapter XV of [1], the constant field of E is equal to $C_M(\gamma)$ and therefore is also of degree $[E : M]$ over C_M. Since the residue field of the factor must be an extension of the constant field of E, these fields must be equal. □

The following proposition explains why constant field extensions are intrinsically simpler than non-constant ones.

Lemma B.4.17. *Let M be a function field over a finite field of constants. Let E/M be a constant field extension. Then no prime of M ramifies in this extension.*

Proof. Since E/M is a constant field extension, it is generated by a constant element of E. The discriminant of the power basis of this constant element is not divisible by any prime. Therefore, by Proposition 8, Section 2, Chapter III of [46], no prime ramifies in this extension. □

The next two lemmas tell us how to generate large sets of elements without zeros at any prime belonging to a fixed infinite set of primes.

Lemma B.4.18. *Let M/K be a separable global field extension generated by an element α of M. Let $G(T)$ be the monic irreducible polynomial of α over K. Let $\mathfrak{P}_K$ be a prime of K such that $\mathfrak{P}_K$ does not have a relative degree-1 factor in M, does not divide the discriminant of $G(T)$, and is not a pole of any coefficient of $G(T)$. Then for all $x \in K$ we have that $\text{ord}_{\mathfrak{P}_K} G(x) \leq 0$.*

Proof. By Lemma B.4.12 the power basis of α is an integral basis with respect to $\mathfrak{P}_K$. Therefore, by Lemma B.4.13, $G(T)$ has no roots modulo $\mathfrak{P}_K$ if and only if $\mathfrak{P}_K$ has no relative-degree-1 factors in M. Thus if $\text{ord}_{\mathfrak{P}_K} x \geq 0$ we conclude that $\text{ord}_{\mathfrak{P}_K} G(x) = 0$. However, if $\text{ord}_{\mathfrak{P}_K} x < 0$ then $\text{ord}_{\mathfrak{P}_K} G(x) < 0$, since $G(T)$ is monic and all the coefficients are integral at $\mathfrak{P}_K$. □

Lemma B.4.19. *Let K be a global field. Let $\mathcal{W}$ be a set of non-archimedean primes of K such that in some finite extension M of K only finitely many primes of $\mathcal{W}$ have a factor of relative degree 1. Then there exists a polynomial $F(X) \in K[X]$ such that for all $x \in K$ and for all $\mathfrak{p} \in \mathcal{W}$ we have that $\text{ord}_{\mathfrak{p}} F(x) \leq 0$.*

Proof. From Lemma B.4.18 we know that there exists a monic $G(X) \in K[X]$ such that for all but finitely many $\mathfrak{p} \in \mathcal{W}$ and for all $x \in K$ we have that $\text{ord}_{\mathfrak{p}} G(x) \leq 0$. Let $\mathfrak{p}_1, \ldots, \mathfrak{p}_l$ be all the "exceptions." Let $G(X) = \sum_{i=0}^{k} A_i X^i$. Let $m = \max(1, |\text{ord}_{\mathfrak{p}_i} A_j| : i = 1, \ldots, l,\ j = 1, \ldots, k)$. Let $a \in K$ be such that for all $i = 1, \ldots, l$ we have that $\text{ord}_{\mathfrak{p}_i} a = -2m$ (such an a exists by the Strong Approximation Theorem) and consider $F(X) = G(X^{3m} + a)$. Observe that, by our assumptions on $G(X)$, for all $\mathfrak{p} \in \mathcal{W} \setminus \{\mathfrak{p}_1, \ldots, \mathfrak{p}_l\}$ and for all $x \in K$ it is the case that $\text{ord}_{\mathfrak{p}} F(x) \leq 0$. Next let $x \in K$ and consider $\text{ord}_{\mathfrak{p}_i} F(x)$. Suppose that $\text{ord}_{\mathfrak{p}_i} x \geq 0$. Then $\text{ord}_{\mathfrak{p}_i}(x^{3m} + a) = \text{ord}_{\mathfrak{p}_i} a = -2m$. Thus

$$\begin{aligned}\text{ord}_{\mathfrak{p}_i} F(x) &= \text{ord}_{\mathfrak{p}_i}\left(\sum_{j=0}^{k} A_j (x^{3m} + a)^j\right)\\ &= \min\left(\text{ord}_{\mathfrak{p}_i} A_i + \text{ord}_{\mathfrak{p}_i}(a + x^{3m})^i,\ i = 0, \ldots, k\right)\\ &= \min\left(\text{ord}_{\mathfrak{p}_i} A_i - 2im,\ i = 0, \ldots, k\right) = -2km < 0,\end{aligned}$$

since for $i = 0, \ldots, k-1$ we have that

$$\begin{aligned}\mathrm{ord}_{\mathfrak{p}_i} A_i - 2im &\geq -2im - m \geq -2(k-1)m - m \\ &= -2km + m > -2km = \mathrm{ord}_{\mathfrak{p}_i}(a + x^{3m})^k.\end{aligned}$$

Suppose now that $\mathrm{ord}_{\mathfrak{p}_i} x < 0$. In this case, $\mathrm{ord}_{\mathfrak{p}_i}(x^{3m} + a) = 3m\,\mathrm{ord}_{\mathfrak{p}_i} x < -3m$. Further, as above,

$$\begin{aligned}\mathrm{ord}_{\mathfrak{p}_i} F(x) &= \mathrm{ord}_{\mathfrak{p}_i}\left(\sum_{j=0}^{l} A_j(x^{3m} + a)^j\right) \\ &= \min\left(\mathrm{ord}_{\mathfrak{p}_i} A_i + \mathrm{ord}_{\mathfrak{p}_i}(a + x^{3m})^i,\ i = 0, \ldots, k\right) \\ &\leq -3km < 0,\end{aligned}$$

since for $i = 0, \ldots, k-1$ we have that

$$\begin{aligned}\mathrm{ord}_{\mathfrak{p}_i} A_i + \mathrm{ord}_{\mathfrak{p}_i}(a + x^{3m})^i &\geq 3im\,\mathrm{ord}_{\mathfrak{p}_i} x - m \geq 3(k-1)m\,\mathrm{ord}_{\mathfrak{p}_i} x - m \\ &= 3km\,\mathrm{ord}_{\mathfrak{p}_i} x - (3m\,\mathrm{ord}_{\mathfrak{p}_i} x + m) \\ &> 3km\,\mathrm{ord}_{\mathfrak{p}_i} x = \mathrm{ord}_{\mathfrak{p}_i}(a + x^{3m})^k.\end{aligned}$$

Thus for all $\mathfrak{p} \in \mathcal{W}$ and for all $x \in K$ we have that $\mathrm{ord}_{\mathfrak{p}} F(x) \leq 0$. □

The next lemma explains how to determine the intersection of a ring of $\mathcal{W}$-integers with a subextension, knowing the pattern of prime splitting for primes below primes in $\mathcal{W}$.

Lemma B.4.20. *Let K/L be a finite extension of global fields. Let $\mathcal{W}_K$ be a set of K-primes. Let $\mathcal{V}_L$ be the set of all the primes $\mathfrak{p}_L$ of L such that every K-factor of $\mathfrak{p}_L$ is in $\mathcal{W}_K$. Then $O_{K,\mathcal{W}_K} \cap L = O_{L,\mathcal{V}_L}$.*

Proof. Suppose that $x \in O_{K,\mathcal{W}_K} \cap L$. Then for all $\mathfrak{q}_K \notin \mathcal{W}_K$ we have that $\mathrm{ord}_{\mathfrak{q}_K} x \geq 0$. Let $\mathfrak{t}_L$ be a prime of L not in $\mathcal{V}_L$. Then there exists at least one K-factor $\mathfrak{T}_K$ of $\mathfrak{t}_L$ such that $\mathfrak{T}_K \notin \mathcal{W}_K$. Therefore, $\mathrm{ord}_{\mathfrak{T}_K} x \geq 0$. Next consider the divisor of x in L

$$\mathfrak{t}_L^a \prod \mathfrak{t}_{L,i}^{a_i}, \tag{B.4.1}$$

where $a, a_i \in \mathbb{Z}$, only finitely many a_i are not zero, for all i, we have that $\mathfrak{t}_{L,i} \neq \mathfrak{t}_L$, and all $\mathfrak{t}_{L,i}$ are distinct. Hence the divisor of x in K has the form

$$\prod_j \mathfrak{T}_{K,j}^{ae_j} \prod_{j=1}^{m} \prod_j \mathfrak{T}_{K,i,j}^{a_i e_{i,j}}, \tag{B.4.2}$$

where $\mathfrak{t}_L = \prod_j \mathfrak{T}_{K,j}^{e_j}$ and $\mathfrak{t}_{L,i} = \prod_j \mathfrak{T}_{K,i,j}^{e_{i,j}}$. Thus in K we have that

$$\operatorname{ord}_{\mathfrak{T}_{K,j}} x = ae_j.$$

But for some j this order must be non-negative. Since the ramification degree is always positive we must conclude that $a \geq 0$. Thus, $\operatorname{ord}_{\mathfrak{t}_L} x \geq 0$ for all $\mathfrak{t}_L \notin \mathcal{V}_L$ and consequently $x \in O_{L,\mathcal{V}_L}$.

Conversely, suppose that $x \in O_{L,\mathcal{V}_L}$. Let $\mathfrak{T}_K \notin \mathcal{W}_K$. Let $\mathfrak{t}_L$ be the prime below $\mathfrak{T}_K$ in L. Then by the definition of $\mathcal{V}_L$ we have that $\mathfrak{t}_L \notin \mathcal{V}_L$ and $\operatorname{ord}_{\mathfrak{t}_L} x \geq 0$. Next, as before let (B.4.1) be the L-divisor of x and let (B.4.2) be the K-divisor of x. Then $a \geq 0$ and, for some j, we have that $\operatorname{ord}_{\mathfrak{T}_K} x = ae_j \geq 0$ for all $\mathfrak{T}_K \notin \mathcal{W}_K$. Hence $x \in O_{K,\mathcal{W}_K}$. Since by assumption $x \in L$, we can conclude that $x \in O_{K,\mathcal{W}_K} \cap L$. □

The following proposition tells us that only non-constant extensions change the degree of divisors.

Lemma B.4.21. *Let H/L be a finite separable extension of function fields and let C_H be the constant field of H. Let $\mathfrak{u}$ be an integral divisor of L. Then* $\deg_H \mathfrak{u} = [H : C_H L] \deg_L \mathfrak{u}$. (See Theorems 9 and 14, Section 3, Chapter XV of [1].)

The next lemma describes the behavior of some primes under constant field extensions.

Lemma B.4.22. *Let H be a function field over a finite field of constants C. Let $\mathfrak{p}$ be a prime of H. Let $\hat{C}$ be a finite extension of C such that $[\hat{C} : C]$ is prime to the degree of $\mathfrak{p}$. Then $\mathfrak{p}$ remains prime in $\hat{C}H$.*

Proof. Since $\hat{C}/C$ is a separable extension, by Theorem 14, Section 3, Chapter XV of [1] we have that $\hat{C}$ is the constant field of $\hat{C}H$. Let $\mathfrak{P}$ be a $\hat{C}H$-prime above $\mathfrak{p}$, let $R_\mathfrak{p}$ and $R_\mathfrak{P}$ be the residue fields of $\mathfrak{p}$ and $\mathfrak{P}$ respectively, and consider the following diagram:

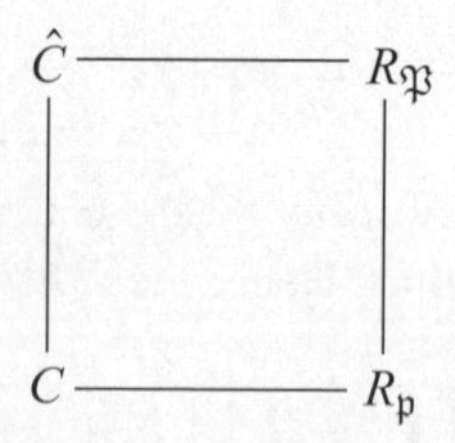

From the diagram we conclude that

$$[R_{\mathfrak{P}} : C] = [R_{\mathfrak{P}} : R_{\mathfrak{p}}][R_{\mathfrak{p}} : C] = [R_{\mathfrak{P}} : \hat{C}][\hat{C} : C]$$

or, in other words,

$$f(\mathfrak{P}/\mathfrak{p}) \deg \mathfrak{p} = \deg \mathfrak{P}\, [\hat{C} : C].$$

Thus, since $([\hat{C} : C], \deg \mathfrak{p}) = 1$, we must conclude that $\deg \mathfrak{p}$ divides $\deg \mathfrak{P}$. Hence, $\deg \mathfrak{P}$ is at least as big as the degree of the divisor $\mathfrak{p}$ in $\hat{C}H$ because by Lemma B.4.21 the degree of a divisor stays the same under a separable constant field extension. However, since $\mathfrak{P} \mid \mathfrak{p}$, $\deg \mathfrak{P}$ is less than or equal to the degree of $\mathfrak{p}$ as a divisor of $\hat{C}H$. Thus we must conclude that $\deg \mathfrak{p}$ as a divisor of $\hat{C}H$ is equal to $\deg \mathfrak{P}$ and so $\mathfrak{P}$ is the only prime of $\hat{C}H$ above $\mathfrak{p}$. □

The following two propositions are necessary ingredients of the proof of the fact that if under a separable extension of function fields there is one prime which does not split then there are infinitely many such primes. This assertion plays an important role in the proof of the Diophantine undecidability of global function fields.

Lemma B.4.23. *Let M be a Galois extension of a function field L over a finite field of constants, and assume that U is a function field such that $L \subset U \subset M$ and U is not necessarily Galois over L. Further, let $\mathfrak{p}_L$ be a prime of L which does not split in U. Let $\mathfrak{p}_U$ be the prime above $\mathfrak{p}_L$ in U, let $\mathfrak{p}_M$ be a prime of M above $\mathfrak{p}_U$, let $G(\mathfrak{p}_M)$ be the decomposition group of $\mathfrak{p}_M$ over L, and let $\sigma \in G(\mathfrak{p}_M)$ be such that its coset modulo the inertia group of $\mathfrak{p}_M$ induces the Frobenius automorphism $\phi_{\mathfrak{p}_M}$ on the residue field of $\mathfrak{p}_M$ over the residue field of $\mathfrak{p}_L$. (In other words, $\phi_{\mathfrak{p}_M}(c) = c^{\mathbf{N}\mathfrak{p}_L}$ for all c in the residue field of $\mathfrak{p}_M$.) The following diagram corresponds to the data in the lemma, $T(\mathfrak{p}_M)$ being the inertia group of $\mathfrak{p}_M$ and $M^{T(\mathfrak{p}_M)}$ the fixed field of $T(\mathfrak{p}_M)$:*

$$\mathfrak{p}_L \in L \;\bullet\!\!-\!\!-\!\!-\!\!\bullet\; \mathfrak{p}_U \in U \;-\!\!-\!\!\bullet\; M^{T(\mathfrak{p}_M)} \;-\!\!-\!\!-\!\!\bullet\; \mathfrak{p}_M \in M$$

Then $\sigma^{f(\mathfrak{p}_U/\mathfrak{p}_L)} \in$ Gal(M/U), and $f(\mathfrak{p}_U/\mathfrak{p}_L) = [U : L]$ is the smallest positive exponent such that the corresponding power of σ is in Gal(M/U).

Proof. First of all observe that, since $\mathfrak{p}_L$ does not split in U, $\mathfrak{p}_U$ is not ramified over L and therefore indeed $U \subseteq M^{T(\mathfrak{p}_M)}$ by Lemma B.4.1. Further, $T(\mathfrak{p}_M) \subset \text{Gal}(M/U)$. Next observe that since the equivalence class of σ generates $G(\mathfrak{p}_M)/T(\mathfrak{p}_M)$, every element $\phi \in G(\mathfrak{p}_M)$ can be written as $\sigma^i \psi$, $\psi \in$

$T(\mathfrak{p}_M)$, $i = 0, \ldots, f(\mathfrak{p}_M/\mathfrak{p}_L) - 1$. Let $H(\mathfrak{p}_M)$ be the decomposition group of $\mathfrak{p}_M$ with respect to U and note that $T(\mathfrak{p}_M) \subseteq H(\mathfrak{p}_M)$. Next we observe that

$$f(\mathfrak{p}_U/\mathfrak{p}_L) = \frac{f(\mathfrak{p}_M/\mathfrak{p}_L)}{f(\mathfrak{p}_M/\mathfrak{p}_U)} = \frac{|G(\mathfrak{p}_M)/T(\mathfrak{p}_M)|}{|H(\mathfrak{p}_M)/T(\mathfrak{p}_M)|} = \frac{|G(\mathfrak{p}_M)|}{|H(\mathfrak{p}_M)|}.$$

Further, observe that

$$\begin{aligned} H(\mathfrak{p}_M) &= G(\mathfrak{p}_M) \cap \mathrm{Gal}(M/U) \\ &= \left\{\sigma^i\psi,\ \psi \in T(\mathfrak{p}_M),\ i = 0, \ldots, f(\mathfrak{p}_M/\mathfrak{p}_L) - 1\right\} \cap \mathrm{Gal}(M/U). \end{aligned}$$

However, for $\psi \in T(\mathfrak{p}_M)$ we have that $\sigma^i\psi \in H(\mathfrak{p}_M)$ if and only if $\sigma^i \in H(\mathfrak{p}_M)$, since $T(\mathfrak{p}_M) \subseteq H(\mathfrak{p}_M)$. Let r be the smallest positive exponent such that $\sigma^r \in H(\mathfrak{p}_M)$. Then every element of $H(\mathfrak{p}_M)$ can be written as $\sigma^{rm}\psi$, where $\psi \in T(\mathfrak{p}_M)$ and $m \in \mathbb{N}$. Thus

$$f(\mathfrak{p}_U/\mathfrak{p}_L) = \frac{|G(\mathfrak{p}_M)|}{|H(\mathfrak{p}_M)|} = r. \qquad \square$$

Corollary B.4.24. *Let M, L, U, σ be as in Lemma B.4.23. Let $\mathfrak{q}_M$ be a prime of M whose Frobenius automorphism over L is σ. Let $\mathfrak{q}_L = \mathfrak{q}_M \cap L$. Then $\mathfrak{q}_L$ does not split in the extension U/L.*

Proof. By assumption, $\mathfrak{q}_L$ is not ramified in the extension U/L and $G(\mathfrak{q}_M)$, the decomposition group of $\mathfrak{q}_M$ over L, is a cyclic group generated by σ. Similarly $H(\mathfrak{q}_M)$, the decomposition group of $\mathfrak{q}_M$ over U, is also a cyclic group generated by σ. Further, the size of the quotient group $|G(\mathfrak{q}_M)/H(\mathfrak{q}_M)| = r$, since r is the smallest positive integer such that $\sigma^r \in \mathrm{Gal}(M/U) \cap G(\mathfrak{q}_M) = H(\mathfrak{q}_M)$. Therefore, by Lemma B.4.23,

$$f(\mathfrak{q}_U/\mathfrak{q}_L) = \frac{f(\mathfrak{q}_M/\mathfrak{q}_L)}{f(\mathfrak{q}_M/\mathfrak{q}_U)} = \frac{|G(\mathfrak{q}_M)|}{|H(\mathfrak{p}_M)|} = r = [U : L],$$

where $\mathfrak{q}_U = \mathfrak{q}_M \cap U$. Therefore the assertion of the corollary is true. $\square$

The next nine lemmas play an important role in the proofs of results concerning the Diophantine definability over holomorphy rings of function fields. Their main role is to ensure the existence of "sufficiently" many degree-1 primes not splitting in the given cyclic extensions. (See Section 10.2 for more details.)

Lemma B.4.25. *Let L be a function field over a finite field of constants C. Let $\mathfrak{q}$ be a prime of L. Let $F(T) \in L[T]$ be monic and irreducible with coefficients integral with respect to $\mathfrak{q}$. Assume that $F(T)$ remains irreducible modulo $\mathfrak{q}$. Let $R(\mathfrak{q})$ be the residue field of $\mathfrak{q}$. Let $E = R(\mathfrak{q})L$ and let $\mathfrak{t}$ be a prime above $\mathfrak{q}$ in E. Then $F(T)$ is irreducible modulo $\mathfrak{t}$.*

Proof. It is enough to show that the residue fields of $\mathfrak{q}$ and $\mathfrak{t}$ are the same. By Lemma B.4.14 we know that $\mathfrak{q}$ will split into factors of degree 1 in E. Thus, the residue field of $\mathfrak{t}$ is the constant field of E. Since constant field extensions of global fields are separable, by Theorem 13, Section 3, Chapter XV of [1], the constant field of E is actually $R(\mathfrak{q})$. □

Lemma B.4.26. *Let U/L be a finite separable extension of function fields over a finite field of constants. Assume that the constant fields of U and L are the same and denote this finite field by C. Let $\mathfrak{q}_L$ be a prime of L not splitting in U. Let C'/C be a finite extension where $[C' : C] = \deg \mathfrak{q}_L$. Let $\mathfrak{q}'_L$ lie above $\mathfrak{q}_L$ in $L' = C'L$. Then $\mathfrak{q}'_L$ does not split in the extension U'/L', where $U' = C'U$.*

Proof. By Lemma B.3.4, $[U' : L'] = [U : L]$. Let $\alpha \in U$ be an element integral at $\mathfrak{q}_L$ and such that its residue class modulo $\mathfrak{q}_U$ generates the residue field of $\mathfrak{q}_U$ over the residue field of $\mathfrak{q}_L$. (Here $\mathfrak{q}_U$ is the U-prime above $\mathfrak{q}_L$.) Then α also generates U over L, the discriminant of the monic irreducible polynomial of α over L is a unit at $\mathfrak{q}_L$, and by Lemma B.4.13 this polynomial does not factor modulo $\mathfrak{q}$. Since a finite field has only one extension of every degree, by Lemma B.4.15 we have that $\mathfrak{q}$ will split completely into degree-1 factors in the extension U'/U. Thus $\mathfrak{q}'$ is of degree 1 and its residue field is C', which is also the residue field of $\mathfrak{q}$. Since $[U' : L'] = [U : L]$, the monic irreducible polynomial of α over L is the same as over L' and, by Lemma B.4.25, the monic irreducible polynomial of α over L' will be irreducible modulo $\mathfrak{q}'$. Therefore, again by Lemma B.4.13, $\mathfrak{q}'$ will not split in the extension U'/L'. □

Proposition B.4.27. *Let M be a Galois extension of an algebraic function field L over a finite field of constants, let C_L be the constant field of L, let C_M be the constant field of M, and let t be a non-constant element of L. Let $\sigma \in \mathrm{Gal}(M/L)$ and let*

$$\mathcal{C} = \{\tau\sigma\tau^{-1} \mid \tau \in \mathrm{Gal}(M/L)\}.$$

Let $\mathcal{P}(L)$, as before, be the set of all primes of L. Then, for a positive integer k, let

$$\mathcal{P}_k(L) = \left\{\mathfrak{p}_L \mid \mathfrak{p}_L \in \mathcal{P}(L) \wedge \deg \mathfrak{p}_L = k \wedge \mathfrak{p}_L \text{ is unramified over } C_L(t)\right\}.$$

Further, let p^r be the size of C_L, let $\phi = \phi_{C_L}$ be the generator of $\mathrm{Gal}(C_M/C_L)$ sending each element $c \in C_M$ to c^{p^r}, and assume that for every $\psi \in \mathcal{C}$ we have that $\psi_{|C_M} = \phi^a$ for some natural integer a different from zero. Then if k is a positive integer such that $k \equiv a$ modulo $[C_M : C_L]$, $m = [M : C_M L]$,

$d = [L : C_L(t)]$, and

$$C_k(M/L, \mathcal{C}) = \left\{ \mathfrak{p}_L \mid \mathfrak{p}_L \in \mathcal{P}_k(L) \wedge \exists \mathfrak{P}_M \in \mathcal{P}(M) : \mathfrak{P}_M \mid \mathfrak{p}_L \wedge \left(\frac{M/L}{\mathfrak{P}_M} \right) \in \mathcal{C} \right\}$$

then

$$\left| \left| C_k \left(\frac{M}{L}, \mathcal{C} \right) \right| - \frac{|\mathcal{C}|}{km} p^{rk} \right| < \frac{|\mathcal{C}|}{km} \left((m + 2g_M) p^{rk/2} + m(3g_L + 1) p^{kr/4} + 2(g_M + dm) \right), \tag{B.4.3}$$

where g_M, g_L are the genus of M and L respectively.

The proof of this proposition can be found in Proposition 13.4 of [34].

Corollary B.4.28. *Let M/L be a Galois extension of function fields over the same finite field of constants C. Let t, d, m, r, g_L, g_M be as in Proposition B.4.27. Let $\sigma \in \mathrm{Gal}(M/L)$. Then, for any sufficiently large positive integer k, there exists an L-prime $\mathfrak{p}_L$ of degree k such that σ is the Frobenius automorphism of a factor of $\mathfrak{p}_L$ in M.*

Proof. Since we have assumed that M and L have the same field of constants, inequality (B.4.3) holds for any k and implies that

$$\left| C_k \left(\frac{M}{L}, \mathcal{C} \right) \right| > \frac{|\mathcal{C}|}{km} p^{rk} \left(1 - \frac{m + 2g_M}{p^{rk/2}} - \frac{m(3g_L + 1)}{p^{3kr/4}} - \frac{2(g_M + dm)}{p^{rk}} \right). \tag{B.4.4}$$

Thus, for all sufficiently large k we have that $|C_k(M/L, \mathcal{C})| > 1$. □

Corollary B.4.29. *Let M/L be a Galois extension of function fields over the same finite field of constants C. Let t, d, m, r, g_L, g_M be as in Proposition B.4.27. Let $\sigma \in \mathrm{Gal}(M/L)$, let $\mathcal{C}$ be the conjugacy class of σ, and assume that*

$$\frac{1}{4}|C| > (m + 4g_M + 3mg_L + 2dm)^2.$$

Then

$$\left| C_1 \left(\frac{M}{L}, \mathcal{C} \right) \right| > \frac{|C|}{2m}.$$

Proof. As for Corollary B.4.28 we start with inequality (B.4.4). Substituting 1 for k and keeping in mind that $|C| = p^r$, we obtain

$$\left|C_1\left(\frac{M}{L}, \mathcal{C}\right)\right| > \frac{|C|}{m} p^r \left(1 - \frac{m + 2g_M}{p^{r/2}} - \frac{m(3g_L + 1)}{p^{3r/4}} - \frac{2(g_M + dm)}{p^r}\right), \tag{B.4.5}$$

$$\left|C_1\left(\frac{M}{L}, \mathcal{C}\right)\right| > \frac{|C|}{m}\left(1 - \frac{m + 3g_L m + 4g_M + 2dm}{\sqrt{|C|}}\right) > \frac{|C|}{2m}. \tag{B.4.6}$$

□

Corollary B.4.30. *Let G/F be a cyclic extension of function fields over the same finite field of constants $|C| = p^r$. Let $t \in F$ be such that it is not a pth power in F and let $d = [F : C(t)]$. Let $m = [G : F]$. Let g_G, g_F be the genuses of G and F respectively. Let $l > 2\log_p 2(m + 4g_G + 3mg_F + 1 + 2dm)$, $l \equiv 0 \bmod r$. Let C_l be the splitting field of the polynomial $X^{p^l} - X$ over the field of characteristic p. Then there are more than $p^l/2m$ primes of C_lF of degree 1 not splitting in the extension C_lG/C_lF.*

Proof. By Lemma B.3.5, C_lG/C_lF is also a cyclic extension. Let σ be a generator of the Galois group and let $\mathcal{C} = \{\sigma\}$. We want to estimate the lower bound on $|C_1(C_lG/C_lF, \mathcal{C})|$. First of all, we note that the constant field of C_lF and C_lG is indeed C_l, and $|C_l| > (2(m + 4g_G + 3mg_F + 1 + 2dm))^2$. Thus we can apply Corollary B.4.29 to conclude that

$$|C_1(C_lG/C_lF, \mathcal{C})| > |C_l|/2m > p^l/2m.$$

However, none of the primes in $C_1(C_lG/C_lF, \mathcal{C})$ split in the extension C_lG/C_lF. □

Proposition B.4.31. *Let G/F be a cyclic extension of function fields over the same field of constants C of size p^r for some positive $r \in \mathbb{N}$. Let $m, l, d, t, g_G, g_F, C_l$ be as in Corollary B.4.30. Let C_{qr}, C_{ql} be finite extensions of C_r and C_l respectively, of degree q where $(l, q) = 1$ and $(m, q) = 1$. Then there are at least $p^l/2m$ primes of C_lF of degree 1 not splitting in the extension $C_{ql}G/C_lF$. Further, if $\mathfrak{p}_{C_lF}$ is a degree-1 prime of C_lF not splitting in the extension C_lG/C_lF and $\mathfrak{p}_{C_lG}$ is a prime above it in C_lG then $\mathfrak{p}_G$, the prime below $\mathfrak{p}_{C_lG}$ in G, does not split in the extension $C_{qr}G/G$.*

Proof. Consider the following two-dimensional diagram.

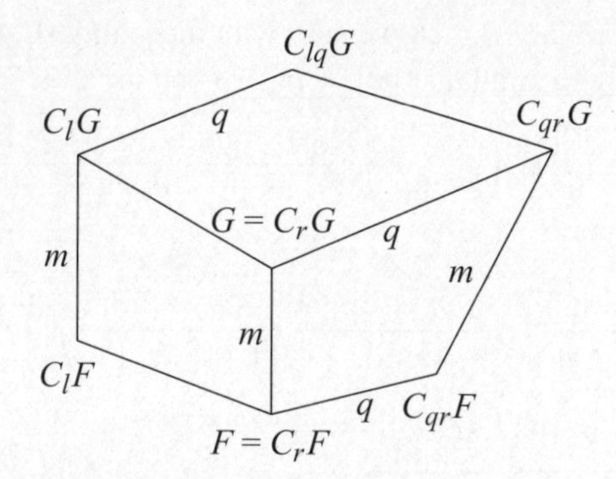

First of all we observe that C_l and C_{qr} are linearly disjoint over C_r, since

$$[C_l : C_r] \mid l, \qquad [C_{qr} : C_r] = q, \qquad (q, l) = 1,$$

and therefore

$$C_l \cap Cqr = C_r.$$

Consequently $C_l C_q = C_{lq}$ and $[C_{lq} : C_l] = q$. Further,

$$[GC_{lq} : GC_l] = [C_{lq} : C_q] = q.$$

Next, directly from Corollary B.4.30 we conclude that $C_l F$ has the requisite number of degree-1 primes not splitting in the extension $C_l G/C_l F$. We show that these degree-1 primes of $C_l F$ actually continue to be inert in the extension $C_{lq}G/C_l F$. So, let $\mathfrak{p}_{C_l F}$ be a degree-1 prime of $C_l F$ inert in the extension $C_l G/C_l F$. Let $\mathfrak{p}_{C_l G}$ lie above $\mathfrak{p}_{C_l F}$ in $C_l G$. Then, since the constant fields of $C_l F$ and $C_l G$ are the same,

$$\deg \mathfrak{p}_{C_l G} = m = [G : F] = [C_l G : C_l F]$$

by Lemma B.4.21. Thus, by Lemma B.4.22, $\mathfrak{p}_{C_l G}$ will not split in the extension $C_{ql}G/C_l G$ because the degree of the extension q is relatively prime to the degree of this prime m. This proves the first assertion of the proposition.

Next we note that $C_l G$ and $C_{qr}G$ are linearly disjoint Galois extensions of G, by Lemma B.3.3, since $C_l G \cap C_{qr}G = (C_l \cap C_{qr})G = C_r G = G$. Now we can conclude that $\mathfrak{p}_G$ does not split in the extension $C_{qr}G/G$ by Lemma B.4.9, since the prime $\mathfrak{p}_{C_l G}$ above it does not split in the extension $C_{ql}G/C_l G$. □

Lemma B.4.32. *Let C be a finite field of positive characteristic p of size p^r. Let t be transcendental over C and let $q \neq p$ be a rational prime. Let β be an element of the algebraic closure of C of degree q over C. Let $l \in \mathbb{N}$ be such* *that $(l, qr) = 1$. Let $P(t)$ be a polynomial in t over C such that $P(t)$ is a prime factor of*

$$t^{p^l} - t \tag{B.4.7}$$

in $C[t]$ and $C(t)$-prime $\mathfrak{P}$ corresponds to the polynomial $P(t)$. Then neither $\mathfrak{P}$ nor $P(t)$ split in the extension $C(t, \beta)/C(t)$.

Proof. Let α be a root of $P(t)$ in the algebraic closure of C. Then $P(t)$ is the monic irreducible polynomial of α over C and $\alpha^{p^l} - \alpha = 0$. Therefore $\alpha \in C_l$, a finite field of p^l elements with $[C_l : \mathbb{F}_p] = l$. Let $n_\alpha = [\mathbb{F}_p(\alpha) : \mathbb{F}_p]$. Then $n_\alpha \mid l$ and therefore $(n_\alpha, qr) = 1$. Consequently $\mathbb{F}_p(\alpha)$ and C are linearly disjoint over $\mathbb{F}_p$ and therefore the monic irreducible polynomial $P(t)$ of α over C is also of degree n_α. Further, $P(t) \in \mathbb{F}_p[t]$ and will not factor under the extension of degree q. Finally, the prime corresponding to $P(t)$ is also of degree $n(\alpha)$ and thus it will not factor in the extension $C(t, \beta)/C$ by Lemma B.4.22. □

The next proposition bounds the number of ramified primes in a series of extensions of global function fields.

Proposition B.4.33. *Let M/K be a finite separable extension of function fields. Let $\mathcal{E}_{M/K}$ be the set of all the primes of M ramified in the extension M/K. Let*

$$\mathfrak{E}_{M/K} = \prod_{\mathfrak{e} \in \mathcal{E}_{M/K}} \mathfrak{e}.$$

Let C_K be the constant field of K and let C/C_K be a finite separable extension. Then the number of CM primes ramified in the extension CM/CK is bounded by the degree of the divisor $\mathfrak{E}_{M/K}$ in M.

Proof. Consider the following diagram.

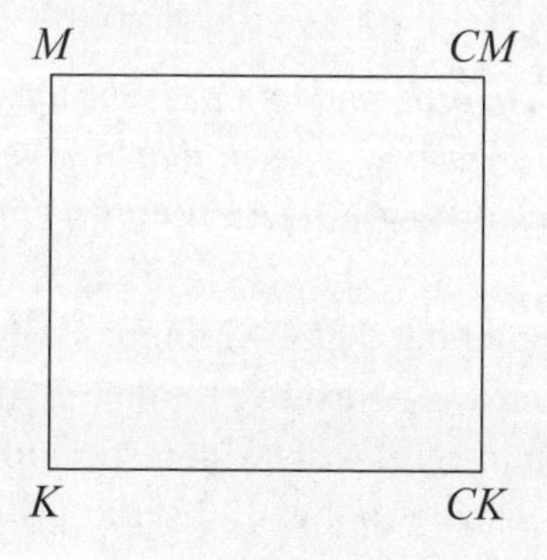

Observe the following. By Lemma B.4.17 no prime ramifies in the extensions CM/M and CK/K. Let $\mathfrak{P}_{CM}$ be a prime of CM. Let $\mathfrak{P}_M$, $\mathfrak{P}_{CK}$, and $\mathfrak{P}_K$ be the primes below $\mathfrak{P}_{CM}$ in M, CK, and K respectively. Then

$$e(\mathfrak{P}_{CM}/\mathfrak{P}_K) = e(\mathfrak{P}_{CM}/\mathfrak{P}_{CK})e(\mathfrak{P}_{CK}/\mathfrak{P}_K) = e(\mathfrak{P}_{CM}/\mathfrak{P}_M)e(\mathfrak{P}_M/\mathfrak{P}_K).$$

Thus, since $e(\mathfrak{P}_{CM}/\mathfrak{P}_M) = e(\mathfrak{P}_{CK}/\mathfrak{P}_K) = 1$, we conclude that $e(\mathfrak{P}_{CM}/\mathfrak{P}_{CK}) = e(\mathfrak{P}_M/\mathfrak{P}_K)$. Therefore, a prime of CM is ramified in the extension CM/CK if and only if the prime below it in M is ramified in the extension M/K. Let $\mathcal{E}'$ be the set of all the primes of CM ramified in the extension CM/CK. Let $\mathfrak{E}' = \prod_{\mathfrak{e}\in\mathcal{E}'} \mathfrak{e}$. Then if we consider $\mathfrak{E}_{M/K}$ as a divisor of CM, we conclude that $\mathfrak{E}_{M/K} = \mathfrak{E}'$. Further, by Lemma B.4.21,

$$\deg_M \mathfrak{E}_{M/K} = \deg_{CM} \mathfrak{E}_{M/K} = \deg_{CM} \mathfrak{E}'.$$

Therefore, the number of primes in $\mathcal{E}'$ cannot exceed $\deg_M \mathfrak{E}$. □

The lemma below bounds a number of "inconvenient" primes that can occur in a series of finite extensions of global function fields.

Proposition B.4.34. *Let M/K be a finite separable extension of function fields of degree k over the same field of constants. Let $\alpha \in M$ be a generator of M over K. Let $A = \{1, \dots, \alpha^{k-1}\}$ be the power basis of α. Let $\alpha_1 = \alpha, \alpha_2, \dots, \alpha_{k-1}$ be all the conjugates of α over K. Let*

$$\mathcal{D}(\alpha) = \prod_{i<j} (\alpha_i - \alpha_j)^2,$$

let

$$\mathfrak{n}(\alpha) = \prod_{\mathfrak{q}\in\mathcal{P}(K),\ \mathrm{ord}_{\mathfrak{q}} \mathcal{D}(\alpha)>0} \mathfrak{q},$$

and let

$$\mathfrak{d}(\alpha) = \prod_{\mathfrak{q}\in\mathcal{P}(K), \alpha \text{ is not integral at } \mathfrak{q}} \mathfrak{q}.$$

Let C_K be the constant field of K and let C/C_K be a finite separable extension. Then the number of CK primes $\mathfrak{q}_{CK}$ such that A is not an integral basis with respect to $\mathfrak{q}_{CK}$ is bounded by $\deg(\mathfrak{n}(\alpha)\mathfrak{d}(\alpha)) = n(\alpha)$ *in K.*

Proof. First of all, we point out that, by Lemma B.3.4, we have that A is also a power basis of CM over CK. Next, by Lemma B.4.12, if $\mathfrak{q}_{CK}$ is a prime of CK such that A is not an integral basis with respect to $\mathfrak{q}_{CK}$ then either α is not integral at $\mathfrak{q}_{CK}$ or $\mathrm{ord}_{\mathfrak{q}_{CK}} \mathcal{D}(\alpha) > 0$. Let $\mathfrak{q}_K$ be a K-prime below $\mathfrak{q}_{CK}$. Then α is not integral at $\mathfrak{q}_K$ or $\mathfrak{q}_K$ is a zero of $\mathcal{D}(\alpha)$. Therefore, $\mathrm{ord}_{\mathfrak{q}_K}(\mathfrak{n}(\alpha)\mathfrak{d}(\alpha)) > 0$ and consequently

$$\mathrm{ord}_{\mathfrak{q}_{CK}}(\mathfrak{n}(\alpha)\mathfrak{d}(\alpha)) > 0 \tag{B.4.8}$$

in CK. The number of primes in CK satisfying (B.4.8) is bounded by the CK-degree of $\mathfrak{n}(\alpha)\mathfrak{d}(\alpha)$. However, by Lemma B.4.21, $\deg_{CK}(\mathfrak{n}(\alpha)\mathfrak{d}(\alpha)) = \deg_K(\mathfrak{n}(\alpha)\mathfrak{d}(\alpha))$ and the assertion of the lemma holds. □

Lemma B.4.35. *Let H/K be a separable extension of global fields. Let $\mathfrak{p}$ be a prime of K remaining prime in H. Then there exists an element $\delta \in H$, generating H over K, such that the power basis of δ is an integral basis with respect to $\mathfrak{p}$ and the monic irreducible polynomial of δ over K remains irreducible modulo $\mathfrak{p}$.*

Proof. Let $\mathfrak{P}$ be the single factor of $\mathfrak{p}$ in K. Let $\delta \in H$ be an element integral at $\mathfrak{p}$ and such that its residue class generates the residue field of $\mathfrak{P}$ over the residue field of $\mathfrak{p}$. Let $f(T)$ be the monic irreducible polynomial of δ over K. Let $\bar{f}(T)$ be the polynomial obtained from $f(T)$ by reducing its coefficients modulo $\mathfrak{p}$. Then $\bar{f}$ is irreducible over the residue field of $\mathfrak{p}$ and therefore cannot have multiple roots. Thus $\mathfrak{p}$ does not divide the discriminant of $f(T)$ and so, by Lemma B.4.12, the power basis of δ is integral with respect to $\mathfrak{p}$. □

Lemma B.4.36. *Let G/F be a finite extension of function fields over the same field of constants C. Let $\mathfrak{p}$ be a degree-1 prime of G. Then it lies above a degree-1 prime of F.*

Proof. Let $\mathfrak{P}$ be a prime below $\mathfrak{p}$ in F. Then the residue field of $\mathfrak{P}$ is a finite extension of C which must be contained in the residue field of $\mathfrak{p}$, which is C. Thus the residue field of $\mathfrak{P}$ is C. □

B.5 Density of prime sets

In this section we will discuss the density of prime sets in global fields. One could think of the density of an infinite prime set as a way to describe its size. There are two commonly used kinds of density: Dirichlet and natural. We start by defining the Dirichlet density.

Definition B.5.1. Let K be a global field. Let $\mathcal{P}$ be the set of all the non-archimedean primes of K. Let $\mathcal{A} \subseteq \mathcal{P}$. Then the Dirichlet density of $\mathcal{A}$ (denoted by $\delta(\mathcal{A})$) is the following limit, assuming it exists:

$$\lim_{Re(s)>1} \frac{\sum_{\mathfrak{p}\in\mathcal{A}} 1/N\mathfrak{p}^s}{\sum_{\mathfrak{p}\in\mathcal{P}} 1/N\mathfrak{p}^s},$$

where $N\mathfrak{p}$ is the norm of prime $\mathfrak{p}$, i.e. the number of elements in the residue field of $\mathfrak{p}$.

In the case of number fields there is an alternative (but equivalent) definition of the Dirichlet density, which can be found in Section 6.5 of [33]:

$$\delta(\mathcal{A}) = \lim_{Re(s)>1} \frac{\sum_{\mathfrak{p}\in\mathcal{A}} 1/N\mathfrak{p}^s}{-\log(s-1)}.$$

The advantage of this form is that the denominator does not depend on the field. This simplifies comparisons of the densities of prime sets of different fields. In the function field case there is a similar formula but it depends on the size of the constant field and so should be used for interfield comparisons only when the constant fields are the same:

$$\delta(\mathcal{A}) = \lim_{Re(s)>1} \frac{\sum_{\mathfrak{p}\in\mathcal{A}} 1/N\mathfrak{p}^s}{-\log(1-q^{1-s})},$$

where q is the size of the constant field of K. (See Lemma 6.4.10 of [33].)

One of the main tools for determining the density of a set is the Chebotarev density theorem. It can be found for the number field case in Theorem 10.4, Chapter V of [37] and for the function field case in Chapter 6 of [33]. An immediate consequence of this theorem is the following lemma.

Lemma B.5.2. *Let E/K be a Galois extension of global fields. Let $\mathcal{W}_K$ be the set of all the primes of K unramified and splitting completely in E/K. Then the Dirichlet density of $\mathcal{W}_K$ is equal to* $1/[E:K]$.

Proof. A prime $\mathfrak{p}_K$ splits completely into distinct factors if and only if one of its factors has the identity as its Frobenius automorphism. The conjugacy class of the identity contains just one element. Thus, the density of unramified K-primes splitting completely in E is $1/[K:E]$ by the Chebotarev density theorem. □

The next two lemmas describes important properties of the Dirichlet density.

Lemma B.5.3. *Let K be a global field. Let $\mathcal{P}(K)$ denote the set of all the primes of K. For $\mathfrak{p}\in\mathcal{P}(K)$ let $N\mathfrak{p}$ denote the norm of $\mathfrak{p}$, that is the size of the residue field. Then for any s such that $\Re s > 1$, we have that $\sum_{\mathfrak{p}\in\mathcal{P}(K)} N\mathfrak{p}^{-s} < \infty$, i.e. $\left|\sum_{\mathfrak{p}\in\mathcal{P}(K)} N\mathfrak{p}^{-s}\right|$ is bounded, and* $\lim_{s\to 1^+}\sum_{\mathfrak{p}\in\mathcal{P}(K)} N\mathfrak{p}^{-s} = \infty$.

Proof. For the function field case the lemma follows from the fact that as $s \to 1^+$ we have that $\sum_{\mathfrak{p}\in\mathcal{P}(K)} N\mathfrak{p}^{-s} = -\log(1-q^{1-s}) + O(1)$, where q is the size of the constant field of K. (See Lemma 6.4.10 of [33].)

For the number field case the lemma follows from the fact that

$$\sum_{\mathfrak{p}\in\mathcal{P}(K)} N\mathfrak{p}^{-s} = -\log(s-1) + O(1).$$

(See Section 6.5 of [33].) □

Lemma B.5.4. *Let K/E be a finite separable extension of global fields (not necessarily Galois). Let $\mathcal{U}_K$ be a set of primes of K containing all primes of K of relative degree 1 over E. Then the Dirichlet density of $\mathcal{U}_K$ is 1.*

Proof. It is sufficient to show that the density of the complement of $\mathcal{U}_K$ is 0. Let $\mathcal{H}_K$ be the complement of $\mathcal{U}_K$. Then, by Lemma B.5.3,

$$0 \le \left| \lim_{s\to 1^+} \frac{\sum_{\mathfrak{p}\in\mathcal{H}_K} N\mathfrak{p}^{-s}}{\sum_{\mathfrak{p}\in\mathcal{P}(K)} N\mathfrak{p}^{-s}} \right| \le \lim_{s\to 1^+} [K:E] \frac{\sum_{\mathfrak{p}\in\mathcal{P}(E)} |N\mathfrak{p}^{-2s}|}{\left|\sum_{\mathfrak{p}\in\mathcal{P}(K)} N\mathfrak{p}^{-s}\right|}$$

$$\le \lim_{s\to 1^+} [K:E] \frac{\sum_{\mathfrak{p}\in\mathcal{P}(E)} |N\mathfrak{p}^{-2}|}{\left|\sum_{\mathfrak{p}\in\mathcal{P}(K)} N\mathfrak{p}^{-s}\right|} = 0.$$

Thus $\delta(\mathcal{H}_K) = 0$. □

The next two lemmas contain the details of the density calculations for the sets that we used for our definability results over number and function fields. We treat the number field case first.

Lemma B.5.5. *Let $F \subset K \subset K_G$ be a finite extension of number fields, K_G being the Galois closure of K over F. Let M be a cyclic extension of F of prime degree q not dividing $[K_G : F]$. Let $\mathcal{W}_K$ be a set of primes of K formed in the following fashion from the set $\mathcal{P}(K)$ of all primes of K. From each set of primes of K lying above the same prime of F, remove the prime of the highest relative degree over F. Next remove all the primes splitting in the extension MK/K. The remaining primes will form the set $\mathcal{W}_K$. Then the Dirichlet density of $\mathcal{W}_K$ exists and is greater than $1-[K:F]^{-1}-1/q$.*

Proof. Let $\mathcal{V}_K = \mathcal{P}(K) \setminus \mathcal{W}_K$. We would like to estimate the density of $\mathcal{V}_K$. Let $\mathcal{V}_1$ be the set of all primes of $\mathcal{V}_K$ which were removed in the first step. Let $\mathcal{K}$ be all the primes of K which split in the extension MK/K. Let $\mathcal{K}_1 = \mathcal{V}_1 \cap \mathcal{K}$.

We want to estimate the Dirichlet density of $\mathcal{K}_1$. By Lemma B.5.4, we can disregard all the primes of $\mathcal{K}_1$ of relative degree higher than 1 over F. Let

$$\tilde{\mathcal{K}}_1 = \mathcal{K}_1 \cap \{\text{the primes of relative degree 1 over } F\}.$$

Let

$$\mathcal{K}_F = \{\mathfrak{P}_F \in \mathcal{P}(F) : \mathfrak{P}_F \text{ splits completely in the extension } MK_G/F\}.$$

Let $\mathcal{K}_K$ be the set of all primes of K above $\mathcal{K}_F$. We claim that $\tilde{\mathcal{K}}_1$ contains exactly one prime per set of F-conjugates in $\mathcal{K}_K$ and no other primes. Indeed, suppose that $\mathfrak{p} \in \tilde{\mathcal{K}}_1$. Then it is a prime of relative degree 1 over F and also is of the highest relative degree among its conjugates over F. Thus, all the conjugates of $\mathfrak{p}$ are of relative degree 1 over F and $\mathfrak{p}$ lies above a prime $\mathfrak{p}_F$ of F splitting completely in the extension K/F and therefore, by Proposition 6.3.2 of [45], splitting completely in the extension K_G/F. Further, we have that $\mathfrak{p}$ splits completely in the extension KM/K. By Lemma B.4.10 this can happen if and only if all the F-conjugates of $\mathfrak{p}$ split completely in the extension MK/K and all the conjugates of $\mathfrak{p}_F$ in K_G split completely in the extension MK_G/K_G. Thus $\mathfrak{p}_F$ splits completely in the extension MK_G/F. Conversely, let $\mathfrak{p}$ be a prime lying above a prime of $\mathfrak{p}_F$ splitting completely in the extension MK_G/F. Then all the F-conjugates of $\mathfrak{p}$ are of relative degree 1 over F and, by Lemma B.4.10 again, $\mathfrak{p}$ and all its F-conjugates split completely in the extension MK/K.

Now using the alternative formula for the Dirichlet density (Definition B.5.1) and the fact that all the F-conjugates of elements in $\tilde{\mathcal{K}}_1$ are of relative degree 1 we can conclude that

$$\delta(\tilde{\mathcal{K}}_1) = \delta(\mathcal{K}_F) = \frac{1}{[K_G : F]q},$$

where the last equality follows from Lemmas B.5.2 and B.3.6.

Next consider the density of $\mathcal{V}_1$. As above, we need to consider the primes of relative degree 1 only. Thus it is enough to look at the set of primes of K containing exactly one representative for every set of conjugates lying above a prime of F splitting completely in the extension K/F, which is the same set as the set of F-primes splitting completely in the extension K_G/F by Proposition 6.3.2 of [45] again. Thus $\delta(\mathcal{V}_1) = 1/[K_G : F]$. Further, by similar arguments the Dirichlet density of $\mathcal{K}$ exists and is equal to $1/q$.

Since $\mathcal{V}_1$, $\mathcal{K}$, and $\mathcal{V}_1 \cap \mathcal{K}$ have a Dirichlet density, we conclude that $\mathcal{V}_1 \cup \mathcal{K}$ also has such a density and, by Proposition 4.6 of [37], this density is less than or equal to the density of $\mathcal{V}_1$ plus the density of $\mathcal{K}$. Thus, the density of $\mathcal{V}$ is less then or equal to the density of $\mathcal{V}_1$ plus the density of $\mathcal{K}$, and consequently the density of $\mathcal{V}$ is less or equal to $1/[K : F] + 1/q$. Thus, the Dirichlet density of $\mathcal{W}_K$ is greater than $1 - 1/[K_G : F] - 1/q$. □

We now prove the analogous result for the function field case.

Proposition B.5.6. *Let K/E be a separable extension of function fields over the same finite field of constants C. Let K_G be the Galois closure of K over E. Let M/E be a constant field extension of E of prime degree such that $([M : E], [K_G : E]) = 1$. Then the following statements are true.*

1. *Let $\mathcal{V}_{K,1}$ be the set of primes of K splitting completely in the extension MK_G/K. Then*
$$\delta(\mathcal{V}_{K,1}) = \frac{1}{[K_G : K][MK : K]},$$
where δ denotes the Dirichlet density.
2. *Let $\mathcal{W}_{K,1}$ be the set of all the K-primes lying above E-primes splitting completely in the extension K/E. Let $\mathcal{Z}_{K,1} = \mathcal{W}_{K,1} \setminus \mathcal{V}_{K,1}$. Then*
$$\delta(\mathcal{Z}_{K,1}) = \frac{[MK : K] - 1}{[MK_G : K]}.$$
3. *Let $\mathcal{Z}_{E,1}$ be the set of E-primes below the primes $\mathcal{Z}_{K,1}$. Then all the primes in $\mathcal{Z}_{E,1}$ split completely in K_G/E. No prime above $\mathcal{Z}_{E,1}$ is in $\mathcal{V}_{K,1}$.*
4. *Let $\mathcal{G}_{K,1}$ be a set of K-primes such that it contains exactly one prime above each prime in $\mathcal{Z}_{E,1}$. Then $\mathcal{G}_{K,1} \subset \mathcal{Z}_{K,1}$ and*
$$\delta(\mathcal{G}_{K,1}) = \frac{\delta(\mathcal{Z}_{K,1})}{[K : E]} = \frac{[MK : K] - 1}{[MK_G : E]}.$$
5. *Let $\mathcal{V}_{K,2}$ be the set of all the primes of K of relative degrees greater than or equal to 2 over E. Let $\mathcal{V}_K = \mathcal{V}_{K,1} \cup \mathcal{G}_{K,1} \cup \mathcal{V}_{K,2}$. Let $\mathcal{W}_K = \mathcal{P}(K) \setminus \mathcal{V}_K$. Then*
$$\delta(\mathcal{W}_K) = 1 - \frac{[MK : K] - 1}{[MK_G : E]} - \frac{1}{[K_G : K][MK : K]}$$
$$> 1 - \frac{1}{[K : E]} - \frac{1}{[MK : K]},$$
no prime of $\mathcal{W}_K$ has a relative-degree-1 factor in the extension MK_G/K and no prime of E has all its K-conjugates in $\mathcal{W}_K$.

The diagram below illustrates the extensions under discussion.

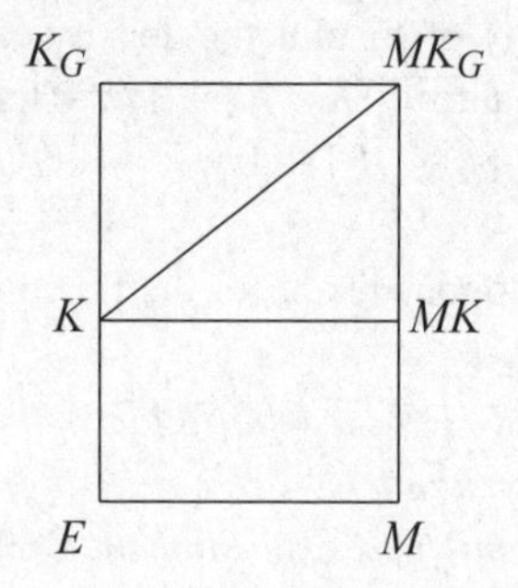

Proof.
1. First of all we note that given that $([M : E], [K_G : E]) = 1$, it is the case that $M \cap K = E$, $MK \cap K_G = K$, and thus $[MK_G : K] = [K_G : K][MK : K]$, by Lemma B.3.3. Further, the extension MK_G/K is Galois. Thus the first assertion follows by Lemma B.5.2.
2. Let $\mathfrak{p} \in \mathcal{Z}_{K,1}$ and assume that $\mathfrak{p}$ lies above an E-prime without MK_G-factors ramified in the extension MK_G/E. Then $\mathfrak{p}$ lies above a prime $\mathfrak{P}_E$ of E splitting completely in the extension K/E. Therefore, by Lemma B.4.6, $\mathfrak{p}$ splits completely in the extension K_G/K and thus the number of factors of $\mathfrak{p}$ in MK_G is divisible by $[K_G : K]$ by Proposition B.1.11.

However, since $\mathfrak{p}$ does not split completely in MK_G, the number of factors of $\mathfrak{p}$ in MK_G is not equal to $[MK_G : K] = [MK : K][K_G : K]$ but must be a divisor of this number, by Proposition B.1.11 again. Since $[MK : K]$ is a prime number, we must conclude that $\mathfrak{p}$ has exactly $[K_G : K]$ factors in MK_G. Thus, the Frobenius automorphism of any factor of $\mathfrak{p}$ in the extension MK_G/K must be of order $[MK : K]$. Similarly, the Frobenius automorphism of any factor of $\mathfrak{P}_E$ in $\mathrm{Gal}(MK_G/E)$ must also be of order $[MK : K] = [M : E]$.

Conversely, if a K-prime $\mathfrak{p}$ with no factors ramified in the extension MK_G/K lies above an E-prime $\mathfrak{P}_E$ not ramified in the extension MK_G/E and whose factors in MK_G have Frobenius automorphisms of order $[MK : K]$, then $\mathfrak{p}$ lies above an E-prime splitting completely in K and $\mathfrak{p}$ will have $[K_G : K]$ factors in MK_G, thus not splitting completely. Therefore, a K-prime $\mathfrak{p}$ with no ramified factors in the extension MK_G/E is in $\mathcal{Z}_{K,1}$ if and only if it lies above an E-prime not ramified in the extension MK_G/E and whose factors in MK_G have Frobenius automorphisms over E of order $[MK : K] = [M : E]$.

Further, $\mathrm{Gal}(MK_G/E) = \mathrm{Gal}(K_G/E) \times \mathrm{Gal}(M/E)$, by Lemmas B.3.3 and B.3.6, and for $\sigma \in \mathrm{Gal}(M/E)$, $\tau \in \mathrm{Gal}(K_G/E)$ we have that $\sigma\tau = \tau\sigma$. Thus the only elements of order $[M : E]$ in $\mathrm{Gal}(MK_G/E)$ are non-trivial elements of $\mathrm{Gal}(M/E)$. Since there are $[M : E] - 1 = [MK : K] - 1$ such elements, by the Chebotarev density theorem the density of the set of E-primes whose factors have Frobenius automorphisms of order $[M : E] = [MK : K]$ is

$$\frac{[MK : K] - 1}{[MK_G : E]}.$$

However, for each such E-prime there are $[K : E]$ primes in $\mathcal{Z}_{K,1}$ and the norms of primes in $\mathcal{Z}_{K,1}$ are the same as the norms of the corresponding E-primes below. Therefore using the alternative definition of the Dirichlet density for function fields (see Definition B.5.1) and the fact that E and K have the same field of constants, we can conclude that the density of primes in $\mathcal{Z}_{K,1}$ is

$$[K : E]\frac{[MK : K] - 1}{[MK_G : E]} = \frac{[MK : K] - 1}{[MK_G : K]}.$$

3. For each prime of $\mathcal{Z}_{E,1}$, there exists at least one factor $\mathfrak{p}$ in K not splitting completely in the extension MK_G/K. Thus by Lemma B.4.10 no conjugate of $\mathfrak{p}$ over E splits completely in the extension MK_G/K, and thus no conjugate of $\mathfrak{p}$ is in $\mathcal{V}_{K,1}$.
4. From the discussion above it follows that $\mathcal{Z}_{K,1}$ is closed under conjugation over E and that each set of conjugates contains exactly $[K : E]$ primes. Hence $G_{K,1}$ will contain one prime for each $\mathcal{Z}_{K,1}$-set of conjugates over E. Thus

$$\delta(\mathcal{G}_{K,1}) = \frac{\delta(\mathcal{Z}_{K,1})}{[K : E]} = \frac{[MK : K] - 1}{[MK_G : E]}.$$

5. First of all we observe that by Lemma B.5.3, used in the same manner as in the proof of Lemma B.5.4, it is the case that $\delta(\mathcal{V}_{K,2}) = 0$. Note also that the density of the set $\tilde{\mathcal{V}}_{K,2} = \mathcal{V}_{K,2} \setminus (\mathcal{V}_{K,1} \cup \mathcal{G}_{K,1})$ is zero by the same lemma. Now, the sets $\tilde{\mathcal{V}}_{K,2}$, $\mathcal{G}_{K,1}$, $\mathcal{V}_{K,1}$ are all disjoint and have Dirichlet densities; thus the density of their union is just the sum of the densities of the three sets. Therefore the density calculation for $\mathcal{W}_K$ follows from the density calculations for $\tilde{\mathcal{V}}_{K,2}$, $\mathcal{V}_{K,1}$, and $\mathcal{G}_{K,1}$.

Next we consider a prime $\mathfrak{p}$ in $\mathcal{W}_K$. By construction $\mathfrak{p} \notin \mathcal{V}_{K,1}$ and thus does not split completely in the extension MK_G/K. Therefore, $\mathfrak{p}$ has no factors of relative degree 1 in the extension MK_G/K. Finally, let $\mathfrak{P}_E$ be a prime of E. If $\mathfrak{P}_E$ splits completely in the extension MK_G/E then none of its factors is in $\mathcal{W}_K$. If $\mathfrak{P}_E$ does not split completely in the extension MK_G/E but splits completely in the extension K/E (and therefore by Lemma B.4.6 in the extension K_G/E), then $\mathfrak{P}_E \in \mathcal{Z}_{E,1}$ and by construction $\mathfrak{P}_E$ will be missing a factor in $\mathcal{W}_K$. Finally, if $\mathfrak{P}_E$ does not split completely in K then it has a factor of relative degree at least 2 in K, and all such primes were removed from $\mathcal{W}_K$. □

Lemma B.5.7. *Let K be a global field. Let $F_1/K, \ldots, F_n/K$ be cyclic extensions of prime degree q linearly disjoint over K. Then the Dirichlet density of the set of K primes splitting in every extension F_i/K is $1/q^n$.*

Proof. Let $F = \prod_{i=1}^{n} F_i$ and let $\mathfrak{P}$ be a prime in the product field. Then

$$\mathrm{Gal}(F/K) = \mathrm{Gal}(F_1/K) \oplus \cdots \oplus \mathrm{Gal}(F_n/K)$$

by Lemma B.3.7. Let $\mathfrak{P}_i$ be the prime below $\mathfrak{P}$ in F_i. Let $(\sigma_1, \ldots, \sigma_n)$, $\sigma_i \in \mathrm{Gal}(F_i/K)$ be the Frobenius automorphism of $\mathfrak{P}$ over K. Then σ_i is an element of the decomposition group of $\mathfrak{P}_i$ over K. Since the $[F_i : K] = q$, a prime number, unless σ_i is equal to the identity element it will generate $\mathrm{Gal}(F_i/K)$. Let $\mathfrak{P}_K$ be the prime below $\mathfrak{P}$ in K. Then $\mathfrak{P}_K$ splits in every F_i if and only if $\sigma_i = \mathrm{id}_i$ for all i. (Here id_i is the identity element of $\mathrm{Gal}(F_i/K)$.) Thus by the Chebotarev density theorem the Dirichlet density of the set of K-primes splitting in every F_i is equal to $1/q^n$. □

Lemma B.5.8. *Let K be a function field over a finite field of constants. Let $K_0 = K \subset K_1 \subset \cdots \subset K_n$ be a tower of constant field extensions with $[K_{i+1} : K_i] = q$, where q is a prime number. Let $\mathcal{W}$ be the set of all the primes $\mathfrak{p}$ of K such that, for some i, none of the factors of $\mathfrak{p}$ in K_i splits in the extension K_{i+1}/K_i. Then the Dirichlet density of $\mathcal{W}$ is $(q^n - 1)/q^n$.*

Proof. First of all we note the following. Let $\mathfrak{p}$ be a K-prime. Let $\mathfrak{p}_{i,1}$ and $\mathfrak{p}_{i,2}$ be two factors of $\mathfrak{p}$ in K_i. Then either both these factors split in K_{i+1} or neither does. (This follows from the fact that K_{i+1}/K is a Galois extension and by Lemma B.4.5.) Suppose now that $\mathfrak{p} \notin \mathcal{W}$. Let $\mathfrak{p}_n$ be a factor of $\mathfrak{p}$ in K_n and let $\mathfrak{p}_i = \mathfrak{p}_n \cap K_i$. Then since all the extensions in the tower are of prime degree and $\mathfrak{p} \notin \mathcal{W}$ we have that $f(\mathfrak{p}_{i+1}/\mathfrak{p}_i) = 1$. (Otherwise some factor of $\mathfrak{p}$ in K_i does not split in K_{i+1}, implying that none of the factors of $\mathfrak{p}$ in K_i splits in K_{i+1} and $\mathfrak{p} \in \mathcal{W}$.) Thus, $f(\mathfrak{p}_n/\mathfrak{p}) = 1$ and $\mathfrak{p}$ splits completely in this extension. By the Chebotarev density theorem the Dirichlet density of this set is $1/q^n$ and the assertion of the lemma follows. □

We now proceed to a brief discussion of natural density.

Definition B.5.9. Let K be a number field. Let $\mathcal{P}(K)$ be the set of all primes of K. Let $\mathcal{T}(K) \subset \mathcal{P}(K)$. Then the natural density of $\mathcal{T}(K)$ is defined to be the following limit if it exists:

$$\lim_{X \to \infty} \frac{\{\#\mathfrak{P} \in \mathcal{T}(K) : N\mathfrak{P} \le X\}}{\{\#\mathfrak{P} \in \mathcal{P}(K) : N\mathfrak{P} \le X\}}$$

If the limit above does not exist, we can consider lim sup or lim inf in place of lim. If lim sup replaces the limit then the resulting number is called the upper natural density. Similarly, if we use lim inf then we will obtain the lower natural density.

Proposition B.5.10. Let $K, \mathcal{T}(K), \mathcal{P}(K)$ be as in Definition B.5.9. Assume that the natural density of $\mathcal{T}(K)$ exists and denote it by d. Then $\delta(\mathcal{T}(K))$, the Dirichlet density of $\mathcal{T}(K)$, also exists and is equal to the natural density of $\mathcal{T}(K)$.

Proof. Unfortunately, a detailed proof of this proposition would be rather lengthy and somewhat tedious. Since the proposition has no direct bearing on any other proposition in the book and is included here for the sake of completeness only, we will restrict ourselves to outlining the main ideas of a proof suggested by Bjorn Poonen.

The starting point for the proof is the following estimates. The first is a version of the prime number theorem:

$$\#\{\mathfrak{p} \in \mathcal{P}(K), N\mathfrak{p} \leq x\} = \frac{x}{\log x} + \varepsilon_1(x)\frac{x}{\log x},$$

where $\varepsilon_1(x) \to 0$ as $x \to \infty$. (See Theorem 4, Section 5, Chapter XV of [46].) The second is really the alternative definition of the Dirichlet density introduced at the beginning of this section,

$$\lim_{s\to 1^+} \log(s-1) \sum_{\mathfrak{p}\in\mathcal{P}(K)} N\mathfrak{p}^{-s} = -1. \tag{B.5.1}$$

The third and fourth inequalities, below, are of an elementary nature. Let $s = \sigma + it$, where $\sigma, t \in \mathbb{R}$. Then there exists a positive real constant C such that for all s sufficiently close to 1 while $\sigma > 1$ we have that

$$\left|\frac{s-1}{\sigma-1}\right| < C. \tag{B.5.2}$$

(See Section 2, Chapter IV of [37].) Further, there exists a positive real constant ε such that for any real h with $|h-1| < \varepsilon$ and any complex s with $\Re s > 1$ and $|s-1| < \varepsilon$, we have that

$$\Re\, h^s > 0. \tag{B.5.3}$$

Finally we need the following familiar expansion for $0 < x < 1$:

$$-\log(1-x) = \sum_{i=1}^{\infty} \frac{x^i}{i}.$$

Now we are ready to proceed. Throughout our discussion below we will assume that r and $s = \sigma + it, \sigma, t \in \mathbb{R}$, are close enough to 1 that (B.5.2) and (B.5.3) with $h = r$ hold. From the definition of the natural density we can conclude that

$$\#\{\mathfrak{p} \in \mathcal{T}(K), N\mathfrak{p} \leq x\} = d\frac{x}{\log x} + \varepsilon_2(x)\frac{x}{\log x},$$

where $\varepsilon_2(x) \to 0$ as $x \to \infty$.

Next let $r > 1$ be a real transcendental number and note that

$$\mu(n) = \#\{\mathfrak{p} \in \mathcal{P}(K),\ r^n < N\mathfrak{p} < r^{n+1}\} = \frac{r-1}{\log r}\frac{r^n}{n} + \varepsilon_3(r^n)\frac{r^n}{n\log r},$$

where $\varepsilon_3 \to 0$ as $r^n \to \infty$. Similarly,

$$\mu_{\mathcal{T}}(n) = \#\{\mathfrak{p} \in \mathcal{T}(K),\ r^n < N\mathfrak{p} < r^{n+1}\} = d\frac{r-1}{\log r}\frac{r^n}{n} + \varepsilon_4(r^n)\frac{r^n}{n\log r},$$

where $\varepsilon_4(r^n) \to 0$ as $r^n \to \infty$. Next let $s \in \mathbb{C}$ with $\Re s > 1$ and write

$$\begin{aligned}\sum_{\mathfrak{p}\in\mathcal{T}(K)} N\mathfrak{p}^{-s} &= \sum_{n=1}^{\infty} \sum_{\mathfrak{p}\in\mathcal{T}(K), r^n<N\mathfrak{p}<r^{n+1}} r^{-(n+1)s} \\ &\quad + \sum_{n=1}^{\infty} \sum_{\mathfrak{p}\in\mathcal{T}(K), r^n<N\mathfrak{p}<r^{n+1}} r^{-(n+1)s}\delta(s,r,n,\mathfrak{p}),\end{aligned}$$

where $\delta(s,r,n,\mathfrak{p}) = (r^{(n+1)s}N\mathfrak{p}^{-s} - 1)$ while $r^n < N\mathfrak{p} < r^{n+1}$. Thus

$$\begin{aligned}\left|\sum_{\mathfrak{p}\in\mathcal{T}(K)} N\mathfrak{p}^{-s}\right| &= \left|\sum_{n=1}^{\infty} \mu_{\mathcal{T}}(n) r^{-(n+1)s}\right. \\ &\quad \left. + \sum_{n=0}^{\infty} \sum_{\mathfrak{p}\in\mathcal{T}(K), r^n<N\mathfrak{p}<r^{n+1}} r^{-(n+1)s}\delta(s,r,n,\mathfrak{p})\right| \\ &\leq \left|\sum_{n=1}^{\infty} d\frac{r-1}{\log r}\frac{r^n}{n} r^{-(n+1)s}\right| + \left|\sum_{n=0}^{\infty} \varepsilon_4(r^n)\frac{r^n}{n\log r} r^{-(n+1)s}\right| \\ &\quad + \sum_{n=1}^{\infty}\left| d\frac{r-1}{\log r}\frac{r^n}{n} r^{-(n+1)s}\right| \delta(s,r,n) \\ &\quad + \sum_{n=0}^{\infty}\left|\varepsilon_4(r^n)\frac{r^n}{n\log r} r^{-(n+1)s}\right| \delta(s,r,n),\end{aligned}$$

where $\varepsilon_4(r^n) \to 0$ as $r^n \to \infty$ and $\delta(s,r,n) = \max_{r^n<N\mathfrak{p}<r^{n+1}} |\delta(s,r,n,\mathfrak{p})|$. Similarly, we obtain a lower bound for $\left|\sum_{\mathfrak{p}\in\mathcal{T}(K)} N^{-s}\right|$ by subtracting "error" terms. The analogous inequalities hold for $\mathcal{P}(K)$ too, of course, but with $d = 1$.

Now noting that $|\mathbf{N}\mathfrak{p}^{-s}| = \mathbf{N}\mathfrak{p}^{-\sigma}$, we will proceed to get an upper bound for $\delta(s, r, n)$. In order to do this we observe that if $r^n < \mathbf{N}\mathfrak{p} < r^{n+1}$ then

$$1 < \left|\frac{r^{n+1}}{\mathbf{N}\mathfrak{p}}\right| < r,$$

and therefore (B.5.3) will hold with $h = r^{n+1}/\mathbf{N}\mathfrak{p}$. Thus, for s sufficiently close to 1 (given a fixed value of r),

$$\delta(s, r, n) = \max |(r^{(n+1)s}\mathbf{N}\mathfrak{p}^{-s} - 1)| \leq r^{\sigma} - 1 + r^{\sigma}|\sin(t \ln r)| = \delta(r, s). \tag{B.5.4}$$

Observe that we have a bound on $\delta(s, r, n)$ that is independent of n. Observe also that for any $\varepsilon > 0$ we can find an $r > 1$ and arbitrarily close to 1 such that for all s sufficiently close to 1 we have that $\delta(s, r) < \varepsilon$. Now, to finish the proof, it is enough to show that for any $\delta > 0$ for some r and all s sufficiently close to 1 we have that

$$\sum_{n=1}^{\infty} \left|\frac{r-1}{\log r}\frac{r^n}{n}r^{-(n+1)s}\right| \delta(s, r, n) < \delta \log |s - 1| \tag{B.5.5}$$

and

$$\left|\sum_{n=1}^{\infty} \varepsilon(r^n)\frac{r^{-(n+1)s+n}}{n \log r}\right| < \delta \log |s - 1|, \text{ where } \varepsilon(r^n) \to 0 \text{ as } r^n \to \infty \tag{B.5.6}$$

Indeed if we can show that (B.5.5) and (B.5.6) hold then, in view of (B.5.1), we will be able to conclude that the Dirichlet densities of $\mathcal{T}(K)$ and $\mathcal{P}(K)$ "reside" for the most part in the sums

$$\sum_{n=1}^{\infty} d\frac{r-1}{\log r}\frac{r^n}{n}r^{-(n+1)s}$$

and

$$\sum_{n=1}^{\infty} \frac{r-1}{\log r}\frac{r^n}{n}r^{-(n+1)s}$$

respectively. From that point on it will not be hard to obtain the desired ratio. Further, if we let $s = \sigma + it$, $\sigma > 1$, as above then using (B.5.2) we can also replace $|s - 1|$ by $|\sigma - 1|$ in our estimates and simplify our calculations.

Now we will show that (B.5.5) holds. First we observe that

$$\lim_{r \to 1} \frac{r-1}{\log r} = C_1,$$

where C_1 does not depend on r. Further, we observe that

$$\lim_{\sigma\to 1} \frac{1 - r^{1-\sigma}}{\sigma - 1} = C(r),$$

where $C(r)$ depends on r only.

Next consider the following inequalities for some fixed r, s sufficiently close to 1 that (B.5.4) holds:

$$\begin{aligned}
\sum_{n=1}^{\infty} \left| r^{-(n+1)s} \frac{r^n}{n} \frac{r-1}{\log r} \right| \delta(s, r, n) &< r^{-\sigma}\delta(s, r)\frac{r-1}{\log r} \sum_{n=1}^{\infty} \frac{r^{n(1-\sigma)}}{n} \\
&< -r^{-\sigma}\delta(s, r)\frac{r-1}{\log r} \log\left(1 - r^{1-\sigma}\right) \\
&= -r^{-\sigma}\delta(s, r)\frac{r-1}{\log r} \left(\log(\sigma - 1) - \log\left(\frac{1 - r^{1-\sigma}}{\sigma - 1}\right)\right) \\
&= -r^{-\sigma}\delta(s, r)\frac{r-1}{\log r} \log(\sigma - 1) + r^{-\sigma}\delta(s, r)\frac{r-1}{\log r} \log\left(\frac{1 - r^{1-\sigma}}{\sigma - 1}\right) \\
&< -\delta(s, r)\frac{r-1}{\log r} \log(\sigma - 1) + \hat{C}(r),
\end{aligned}$$

where $\hat{C}(r)$ depends on r only. By moving r closer to 1 if necessary and then letting $s \to 1$ we can make sure that (B.5.5) holds.

We now show that (B.5.6) holds. Fix $1 < r < 2$ and let M be a positive integer such that for all $n > M$ we have that $|\varepsilon(r^n)| < (\log r)\varepsilon$, where ε is an arbitrary positive real number. Finally, consider the following inequalities:

$$\begin{aligned}
\left| \sum_{n=1}^{\infty} \varepsilon(r^n) \frac{r^{-(n+1)s+n}}{n \log r} \right| &\leq \sum_{n=1}^{M} |\varepsilon(r^n)| \frac{1}{n \log r} + \sum_{n=M+1}^{\infty} |\varepsilon(r^n)| \frac{r^{-(n+1)\sigma+n}}{n \log r} \\
&\leq \tilde{C}(r) - \varepsilon \sum_{n=M}^{\infty} \log\left(1 - r^{1-\sigma}\right) \\
&< \tilde{C}(r) - \varepsilon \log(\sigma - 1) - \varepsilon \log\left(\frac{1 - r^{1-\sigma}}{\sigma - 1}\right) \\
&< \bar{C}(r) - \varepsilon \log(\sigma - 1)
\end{aligned}$$

where $\tilde{C}(r)$, $\bar{C}(r)$ are constants depending on r only. We leave the rest of the proof to the reader. □

We leave the following easy proposition as another exercise for the reader.

Proposition B.5.11. *Let $\mathcal{A}$, $\mathcal{B}$ be sets of primes of a number field K such that both sets have a natural density. Assume further that the natural density of $\mathcal{A}$ is*

1. *Then the natural density of $\mathcal{B} \cap \mathcal{A}$ exists and is equal to the natural density of $\mathcal{B}$.*

B.6 Elliptic curves

In this section we discuss some general properties of elliptic curves which played a role in the proof of Poonen's results. We will use the following notation.

Notation B.6.1.

- Let K denote a global field.
- Let $\bar{K}$ denote the algebraic closure of K.
- Let M denote a finite extension of K.
- Let E/K denote an elliptic curve defined over K for which we will fix an (affine) Weierstrass equation,

$$y^2 = x^3 + a_2x^2 + a_4x + a_6, \qquad \text{(B.6.1)}$$

  where for $i = 2, 4, 6$ we have that $a_i \in K$.
- Let $E(M)$ be the set of all points of E/K in M, or the set of all solutions to (B.6.1) in M together with the point at infinity, O, the unit element of the group.
- Let $E[m](M) \subset E(M)$ be the set of all points of $E(M)$ of order dividing m.
- Let $\mathcal{P}(K)$ denote the set of all non-archimedean primes of K.
- Let $\mathfrak{p}$ denote a non-archimedean prime of K such that
  (a) for $i = 2, 4, 6$ we have that $\text{ord}_{\mathfrak{p}}\, a_i \geq 0$;
  (b) $\mathfrak{p}$ does not divide the discriminant of the chosen Weierstrass equation and therefore E has a good reduction mod $\mathfrak{p}$ (see Section 2, Chapter VII of [113] for a definition of the reduction of an elliptic curve);
  (c) $\mathfrak{p}$ is not ramified over $\mathbb{Q}$.
- Let $\mathcal{S}_0(K)$ denote the set of all primes $\mathfrak{q}$ of K such that $\mathfrak{q}$ does not satisfy the requirements for $\mathfrak{p}$, listed above (note that $\mathcal{S}_0(K)$ is finite);
- Let $\mathcal{M}_K$ denote the set of all normalized absolute values of K. (See the definition of normalized valuations in Section 5, Chapter VIII of [113].)
- Let $\mathcal{M}_{K,\infty} \subset \mathcal{M}_K$ denote the set of all archimedean absolute values of K.
- Let $\mathcal{M}_{K,0} \subset \mathcal{M}_K$ denote the set of all normalized non-archimedean absolute values of K.
- Let $\mathcal{M}_{K,\mathcal{S}_0(K)} \subset \mathcal{M}_{K,0}$ denote the set of all non-archimedean absolute values of K corresponding to primes of $\mathcal{S}_0$.
- Let k denote the residue field of $\mathfrak{p}$.

- Let $K_\mathfrak{p}$ denote the completion of K under the corresponding $\mathfrak{p}$-adic absolute value.
- Let $E(K_\mathfrak{p})$ be the set of all points of E in $K_\mathfrak{p}$, i.e. the set of all solutions to (B.6.1) in $K_\mathfrak{p}$ together with the point at infinity, O, the unit element of the group.
- Let $R_\mathfrak{p}$ denote the ring of integers of $K_\mathfrak{p}$, i.e. the elements of $K_\mathfrak{p}$ with non-negative order at $\mathfrak{p}$.
- Let $\mathcal{M}$ denote the maximum ideal of $R_\mathfrak{p}$.
- Let $\hat{E}$ denote the formal group over $R_\mathfrak{p}$ associated with E. (See the definition of a formal group in Section 2, Chapter IV of [113] and a description of $\hat{E}$ in Example 2.2.3, Chapter IV of [113].)
- Let $\hat{E}(\mathcal{M})$ denote the group associated with $\hat{E}$.
- Let $E_0(K_\mathfrak{p}) = E(K_\mathfrak{p})$.
- For $n \geq 1$ let $E_n(K_\mathfrak{p})$ be the set of $K_\mathfrak{p}$-points of E whose affine coordinates x and y derived from the fixed Weierstrass equation for E have poles at $\mathfrak{p}$ of order at least $2n$ and $3n$ respectively.
- Let $\pi : E(K) \to E(k)$ be the reduction modulo $\mathfrak{p}$.
- Let $\Pi : E(K_\mathfrak{p}) \to E(k)$ be the reduction modulo $\mathfrak{p}$.
- Given a point $P \in E(K) \setminus \{O\}$ of infinite order and an integer $n \neq 0$, define $x_n(P)$, $y_n(P)$ to be the affine coordinates of $[n]P$ derived from the fixed Weierstrass equation of E.
- For an integer $n \neq 0$ write the divisor of $x_n(P)$ in the form $(\mathfrak{a}_n/\mathfrak{d}_n)\mathfrak{s}_n$, where
  (a) For $n \geq 1$ let $\mathfrak{a}_n(P) = \prod \mathfrak{q}^{a(\mathfrak{q})}$, where the product is taken over all the primes $\mathfrak{q} \in \mathcal{P}(K) \setminus \mathcal{S}_0(K)$ such that $a(\mathfrak{q}) = \operatorname{ord}_\mathfrak{q} x_n > 0$,
  (b) $\mathfrak{d}_n(P) = \prod \mathfrak{q}^{a(\mathfrak{q})}$, where the product is taken over all the primes $\mathfrak{q} \in \mathcal{P}(K) \setminus \mathcal{S}_0(K)$ such that $a(\mathfrak{q}) = \operatorname{ord}_\mathfrak{q} x_n < 0$,
  (c) $\mathfrak{s}_n(P) = \prod \mathfrak{q}^{a(\mathfrak{q})}$, where the product is taken over all the primes $\mathfrak{q} \in \mathcal{S}_0(K)$ such that $a(\mathfrak{q}) = \operatorname{ord}_\mathfrak{q} x_n$.
- For an integer $n \neq 0$ let $d_n \in O_K$ be an element whose divisor is $\mathfrak{d}_n^{h_K}$, where h_K is the class number of K.
- For an integer $n \neq 0$ let $\mathcal{S}_n(P) = \{\mathfrak{q} \in \mathcal{P}(K) | \operatorname{ord}_\mathfrak{q} \mathfrak{d}_n(P) > 0\}$.
- For $u \in O_K \setminus \{0\}$, define

$$h_0(u) = -\sum_{v \in \mathcal{M}_{K,0} \setminus \mathcal{M}_{K,\mathcal{S}_0(K)}} \frac{n_v}{[K:\mathbb{Q}]} \log |u|_v = \sum_{\mathfrak{p} \in \mathcal{P}(K) \setminus \mathcal{S}_0(K)} \frac{\operatorname{ord}_\mathfrak{p} u}{[K:\mathbb{Q}]} \log N\mathfrak{p},$$

where $n_v = [K_v : \mathbb{Q}_v]$ and K_v, $\mathbb{Q}_v$ are completions of K and $\mathbb{Q}$ respectively under v. It is not hard to see by using the product formula that, since $u \in O_K$,

$$h_0(u) = h(u) + \sum_{v \in \mathcal{M}_{K,\infty} \wedge |u|_v < 1} \frac{n_v}{[K:\mathbb{Q}]} \log |u|_v + \sum_{v \in \mathcal{M}_{K,\mathcal{S}_0(K)}} \frac{n_v}{[K:\mathbb{Q}]} \log |u|_v,$$

where $h(u)$ is the usual logarithmic height of u. If $u \in \mathbb{Z}$ then

$$h_0(u) = h(u) + \sum_{v \in \mathcal{M}_{K,\mathcal{S}_0(K)}} \frac{n_v}{[K : \mathbb{Q}]} \log |u|_v$$

and if $d_n \in \mathbb{Z}$ then $h_0(d_n) = h(d_n)$. (See the definitions and notation in Sections 5 and 6, Chapter VIII of [113].)
One way to look at h_0 is to say that h_0 captures the height at the "relevant" primes only.

- For any $P \in K$, let $\hat{h}(P)$ be the canonical height of P. (See the definition in Section 9, Chapter VIII of [113].)

The first proposition asserts the existence of an elliptic curve necessary for carrying out the proof of Poonen's results.

Proposition B.6.2. *Consider the curve E defined by the equation $y^2 = x^3 - 6x^2 + 17x$. Then the following statements are true.*

1. *This equation defines an elliptic curve.*
2. $E(\mathbb{Q}) \cong \mathbb{Z}/2 \oplus \mathbb{Z}$.
3. $E(\mathbb{R}) \cong \mathbb{R}/\mathbb{Z}$ *as topological groups.*
4. *E does not have complex multiplication.*

Proof.

1. This statement follows by Proposition 3.1, Chapter III of [113]. (The discriminant of this equation $\Delta = -2^9 17^2$. See Section 1, Chapter III of [113] for the definition of the discriminant.)
2. The proof of this statement is in Example 4.10, Chapter X of [113].
3. First of all, observe that the discriminant of this curve is negative. Second, by Corollary 2.3.1, Chapter V of [114], an elliptic curve defined over $\mathbb{R}$ and having a negative discriminant is isomorphic as a topological group to $\mathbb{R}/\mathbb{Z}$.
4. By Theorem 6.1, Chapter II of [114], if E has complex multiplication then its j-invariant is an integer. The j-invariant of this curve is

$$\frac{(36 - 24(17/2))^3}{-2^9 17^2} \notin \mathbb{Z}.$$

(See Section 1, Chapter III of [113] for a definition of the j-invariant). □

The following sequence of propositions B.6.3 – B.6.8 describes the behavior of primes in the denominators of multiples of a point of infinite order.

Proposition B.6.3. *The map* $\phi : \hat{E}(\mathcal{M}) \to E_1(K_\mathfrak{p})$ *defined by*

$$z \mapsto \left(\frac{z}{w(z)}, -\frac{1}{w(z)}\right) = (x, y),$$

where $w(z) \in R_\mathfrak{p}[[z]]$ *is defined in Proposition 1.1, Chapter IV of [113], is a group isomorphism.*

See Proposition 2.2, Chapter VII of [113] for a proof.

Proposition B.6.4. *Let* $P \neq O$ *be a point of infinite order on* E. *Let* $n \in \mathbb{Z}_{>0}$. *Then*

$$\{l \in \mathbb{Z} \setminus \{0\} : \mathfrak{p}^n | \mathfrak{d}_l\} \cup \{0\}$$

is a subgroup of $\mathbb{Z}$ *under addition.*

Proof. Let $\phi : \hat{E}(\mathcal{M}) \to E_1(K_\mathfrak{p})$ be the map defined in Proposition B.6.3. Then from the definition of $w(z)$ it follows that for all $n \in \mathbb{Z}_{>0}$ we have that

$$\operatorname{ord}_\mathfrak{p} z = n \Leftrightarrow \operatorname{ord}_\mathfrak{p} x = -2n \wedge \operatorname{ord}_\mathfrak{p} y = -3n.$$

In other words, $\phi : \hat{E}(\mathcal{M}^n) \to E_n(K_\mathfrak{p})$ is also an isomorphism. Therefore $E_n(K_\mathfrak{p})$ is a group for all $n \in \mathbb{Z}_{>0}$. Finally, we observe that $E_n(K) = E(K) \cap E_n(K_p)$ and the assertion of the lemma holds. □

From this lemma we derive the following corollary.

Corollary B.6.5. *Let* $\mathfrak{b}$ *be any integral divisor of* K. *Then* $\{l \in \mathbb{Z} \setminus \{0\} : \mathfrak{b} | \mathfrak{d}_l\} \cup \{0\}$ *is a subgroup of* $\mathbb{Z}$ *under addition.*

Proposition B.6.6. *Let* $P \in E_1(K_\mathfrak{p})$ *(so that* $\operatorname{ord}_\mathfrak{p} x(P) < 0$*). Then*

1. *if* $n \in R_\mathfrak{p} \setminus \mathcal{M}$, $\operatorname{ord}_\mathfrak{p} x_n(P) = \operatorname{ord}_\mathfrak{p} x_1(P)$ *and* $\operatorname{ord}_\mathfrak{p} y_n(P) = \operatorname{ord}_\mathfrak{p} y_1(P)$;
2. *if* p *is a rational prime below* $\mathfrak{p}$, $\operatorname{ord}_\mathfrak{p} x_p(P) = \operatorname{ord}_\mathfrak{p} x_1(P) + 2$ *and* $\operatorname{ord}_\mathfrak{p} y_p(P) = \operatorname{ord}_\mathfrak{p} y_1(P) + 3$.

Proof. By the assumption and Proposition 2.2, Chapter VII of [113], for some $z \in \mathcal{M}$,

$$(x_1(P), y_1(P)) = \left(\frac{z}{w(z)}, -\frac{1}{w(z)}\right).$$

Further, $[p]z = pf(z) + g(z^p)$, by Proposition 2.3, Chapter IV and Corollary 4.4, Chapter IV of [113], where $[p]$ denotes multiplication by p in the formal group, $f(T), g(T)$ are power series in T with $g(0) = 0$, and $f(z) = z +$ higher-order terms. Thus, since $\mathfrak{p} \mid p$, $x_1([p]P)$ will have a pole of order exactly 2 greater than that of $x_1(P)$ and similarly $y_1([p]P)$ will have a pole of order exactly 3 greater than that of $y_1(P)$. If we replace p with a rational prime $q \neq p$ with $\mathfrak{p} \nmid q$, however, then $x_1([q]P)$ and $y_1([q]P)$ will have poles of the same order at $\mathfrak{p}$ as do $x_1(P)$ and $y_1(P)$ respectively. □

From this lemma and the definition of $\mathfrak{d}_i(P)$ for a point $P \in E(K)$ of infinite order, we derive the following corollary.

Corollary B.6.7. *Let $m, k \in \mathbb{Z}_{>0}$, $m \mid k$. Then $\mathfrak{d}_m(P) \mid \mathfrak{d}_k(P)$.*

Lemma B.6.8. *There exists $c \in \mathbb{R}$ such that $h_0(d_n(P)) = (c - o(1))n^2$ as $n \to \infty$.*

Proof. By Theorem 9.3, Chapter VIII of [113], we have that

$$h(x_n(P)) + O(1) = \hat{h}(nP) = n^2\hat{h}(P).$$

Thus, for some positive constant $\bar{c}$ it is the case that $h(x_n(P)) = n^2\bar{c} + O(1)$ and $h_K h(x_n(P)) = h_K n^2\bar{c} + O(1)$. Next, for $i \in \mathbb{N}$ we can write

$$x_i^{h_K}(P) = \frac{a_i(P)}{d_i(P)} s_i(P),$$

where $a_i(P)$ has the divisor $\mathfrak{a}_i^{h_K}(P)$, $d_i(P)$ has the divisor $\mathfrak{d}_i^{h_K}(P)$, and only primes of $\mathcal{S}_0(K)$ occur in the divisor of $s_i(P)$. By the definition of $h(x_n(P))$ (see the definitions of regular and logarithmic height and the height notation in Section 5, Chapter VIII of [113]) and the product formula,

$$\begin{aligned}
h(x_n(P)) &= [K:\mathbb{Q}]^{-1}\left(\sum_{v\in\mathcal{M}_K} n_v \log\max(|x_n(P)|_v, 1)\right) \\
&= [K:\mathbb{Q}]^{-1}\left(\sum_{v\in\mathcal{M}_{K,0}} n_v \log\max(|x_n(P)|_v, 1)\right. \\
&\quad \left. + \sum_{v\in\mathcal{M}_{K,\infty}} n_v \log\max(|x_n(P)|_v, 1)\right) \\
&= [K:\mathbb{Q}]^{-1}\left(-\sum_{v|\mathfrak{d}_n(P)} h_K^{-1} n_v \log|d_n(P)|_v\right. \\
&\quad \left. + \sum_{v\in\mathcal{M}_{K,\infty}\cup\mathcal{S}_0} n_v \log\max(|x_n(P)|_v, 1)\right) \\
&= h_K^{-1} h_0(d_n(P)) + [K:\mathbb{Q}]^{-1}\left(\sum_{v\in\mathcal{M}_{K,\infty}\cup\mathcal{S}_0} n_v \log\max(|x_n(P)|_v, 1)\right).
\end{aligned}$$

Thus, by the Theorem on p. 101 of [89],

$$\begin{aligned} h_0(d_n(P)) &= h_K h(x_n(P)) + o(h(x_n(P))) \\ &= h_K n^2 \bar{c} + O(1) + o(h(x_n(P))) = (c - o(1))n^2. \end{aligned}$$ □

From the definition of h_0 it is clear that the following lemma is true.

Lemma B.6.9. *Let $u, v \in O_K \setminus \{0\}$. Then*

1. $h_0(uv) = h_0(u) + h_0(v)$.
2. $u|v \Rightarrow h_0(u) \leq h_0(v)$.

The next sequence of results, B.6.10–B.6.14, deals with automorphisms of torsion elements of elliptic curves.

Proposition B.6.10. *Let $\bar{F}$ be any algebraically closed field. Let $m \in \mathbb{N}$. If* char $\bar{F} = 0$ *or if m is prime to* char $\bar{F}$, *then* $E(\bar{F})[m] = (\mathbb{Z}/m) \times (\mathbb{Z}/m)$. (See Corollary 6.4, Chapter III of [113].)

Proposition B.6.11. *Let $m \geq 1$ be prime to $\mathfrak{p}$. Then π restricted to $E(K)[m]$ is injective.*

Proof. By Proposition 3.1(b), Chapter VII of [113], Π is injective when restricted to $E(K_{\mathfrak{p}})[m]$. Since π is the restriction of Π to K, the assertion follows. □

Corollary B.6.12. *Let $m \in \mathbb{N}$, $m \geq 1$ be prime to $\mathfrak{p}$. Suppose that $E(K)[m] \cong (\mathbb{Z}/m) \times (\mathbb{Z}/m)$. Then $E(K)[m] \cong E(\bar{k})[m]$ as groups and reduction modulo $\mathfrak{p}$ induces an isomorphism. (Here $\bar{k}$ is the algebraic closure of k.)*

Proof. $\pi(E(K)[m]) \subseteq E(k)[m] = (\mathbb{Z}/m) \times (\mathbb{Z}/m)$, since m is prime to $\mathfrak{p}$. But π is injective on $E(K)[m]$. Thus $|\pi(E(K)[m])| = |E(\bar{k})[m]|$ and the assertion follows. □

Proposition B.6.13. *Let l be a rational prime number. Then the group* Aut$(E(\bar{K})[l])$ *of automorphisms of $E(\bar{K})[l]$ is isomorphic to $GL_2(\mathbb{Z}/l)$. Further, let*

$$n_l = \frac{\#\{\sigma \in \mathrm{Aut}(E(\bar{K})[l]) : \exists P \in E(\bar{K})[l] \setminus \{O\}, \sigma(P) = P\}}{\#\{\sigma \in \mathrm{Aut}(E(\bar{K})[l])\}}. \quad \text{(B.6.2)}$$

Then as $l \to \infty$ we have that $n_l = 1/l + 1/l^2 + o(1/l^2)$.

Proof. Since $E(\bar{K})[l] = (\mathbb{Z}/l) \times (\mathbb{Z}/l)$, a two-dimensional vector space over $\mathbb{Z}/l$, $\mathrm{Aut}(E(\bar{K})[l])$ is isomorphic to the space of 2×2 invertible matrices, otherwise known as $\mathrm{GL}_2(\mathbb{Z}/l)$. Thus we need to determine the size of $\mathrm{GL}_2(\mathbb{Z}/l)$ as well as the number of matrices with a fixed non-zero vector. Let

$$\begin{pmatrix} a & b \\ c & d \end{pmatrix} \in \mathrm{GL}_2(\mathbb{Z}/l). \tag{B.6.3}$$

We need to determine the number of distinct 4-tuples $(a, b, c, d) \in (\mathbb{Z}/l)^4$ such that $ac - bd = 0$ in $\mathbb{Z}/l$. If $a \neq 0$ then c is uniquely determined for any value of b and d. Thus we get $(l-1)l^2$ singular matrices for this case. If $a = 0$ then for any values of c we can have any of the $2l - 1$ values of the pair (b, d) where one entry is 0. Thus, for this case we have $l(2l-1)$ singular matrices. So altogether there are $(l-1)l^2 + l(2l-1) = l^3 - l^2 + 2l^2 - l = l^3 + l^2 - l$ singular 2×2 matrices over $\mathbb{Z}/l$. Hence there are $l^4 - l^3 - l^2 + l = (l^2 - l)(l^2 - 1)$ matrices in $GL_2(\mathbb{Z}/l)$.

Next we calculate the number of non-singular matrices with a fixed non-zero vector. To this effect we first determine the number of matrices of the form (B.6.3) satisfying

$$\begin{cases} (a-1)(d-1) - bc = 0 \\ \qquad\qquad ad - bc = 0. \end{cases} \tag{B.6.4}$$

This system of equations implies that $(a-1)(d-1) = ad$ or $1 - a = d$. Thus we have that $a(1-a) = bc$. Using the same strategy as above, we conclude that if $b \neq 0$ it is the case that c is determined by the value of a, and therefore we have $l(l-1)$ solutions in this case. If $b = 0$ then $a = 1, 0$ and c can take any value. Therefore for this case we have $2l$ solutions. So the total number of matrices satisfying (B.6.4) is $l(l-1) + 2l = l^2 + l$. Consequently the number of non-singular matrices satisfying $(a-1)(d-1) - bc = 0$ is $l^3 + l^2 - l - l^2 - l = l^3 - 2l$. Hence,

$$n_l = \frac{l^3 - 2l}{l^4 - l^3 - l^2 + l} = \frac{1}{l} + \frac{1}{l^2} + o\left(\frac{1}{l^2}\right)$$

as $l \to \infty$. □

Proposition B.6.14. *Let l be a rational prime number. Then the following statements are true.*

1. $\mathrm{Gal}(\bar{K}/K)$ *acts on* $E(\bar{K})[l]$ *for all* l, *or, in other words, there is a homomorphism*

$$\Lambda_l : \mathrm{Gal}(\bar{K}/K) \to \mathrm{Aut}(E(\bar{K})[l]) \cong GL_2(\mathbb{Z}/l).$$

(See Section 7, Chapter 3 of [113].)

2. *There exists a finite set of rational primes $\mathcal{S}(K)$ such that for all rational primes $l \notin \mathcal{S}(K)$ we have that Λ_l is onto.* (See (7) on page 260 of [87].)

B.7 Coordinate polynomials

In this section we introduce coordinate polynomials, a computational device designed to simplify translation from polynomials with variables ranging in a finite extension of a field to polynomials with variables ranging in the field defined below. This definition of coordinate polynomials was suggested to the author by Laurent Moret-Bailly. We start with a definition of a map.

Definition B.7.1. Let F be a field and let M be a finite extension of F. Let $\Omega = \{\omega_1, \dots, \omega_k\}$ be a basis of M over K. Then let $\pi : F^k \to M$ be defined by

$$(b_1, \dots, b_k) \mapsto \sum_{i=1}^{k} b_i \omega_i.$$

Let

$$\bar{\sigma} = (\sigma_1, \dots, \sigma_k) : M \to F^k$$

be a linear F-section of π defined by $\sigma_j(\sum_{i=1}^{k} b_i \omega_i) = b_j$. Next we extend $\bar{\sigma}$ to polynomials over M. Let $Q(z_1, \dots, z_m) \in M[z_1, \dots, z_m]$ and suppose that

$$Q(z_1, \dots, z_m) = \sum_{i_1,\dots,i_m} A_{i_1,\dots,i_m} z_1^{i_1} \cdots z_m^{i_m}$$

where $A_{i_1,\dots,i_m} \in M$. Then define

$$\sigma_j(Q(z_1, \dots, z_m)) = \sum_{i_1,\dots,i_m} \sigma_j(A_{i_1,\dots,i_m}) z_1^{i_1} \cdots z_m^{i_m},$$

so that

$$\bar{\sigma}(Q(z_1, \dots, z_m)) = \big(\sigma_1(Q(z_1, \dots, z_m)), \dots, \sigma_k(Q(z_1, \dots, z_m))\big).$$

We are now ready to define the coordinate polynomials.

Definition B.7.2. Let M, F, Ω be as above. Let

$$P(X_1, \dots, X_m, w_1, \dots, w_u) \in M[X_1, \dots, X_m, w_1, \dots, w_u].$$

Then define the k-tuple of polynomials

$$\bar{\sigma}\left(P\left(\sum_{i=1}^{k} x_{1,i}\omega_i, \dots, \sum_{i=1}^{k} x_{m,i}\omega_i\right), w_1, \dots, w_u\right)$$

$$= \big(P_1^{\Omega}(x_{1,1}, \dots, x_{m,k}, w_1, \dots, w_u), \dots, P_k^{\Omega}(x_{1,1}, \dots, x_{m,k}, w_1, \dots, w_u)\big)$$

to be the coordinate polynomials of $P(X_1, \ldots, X_m, w_1, \ldots, w_u)$ with respect to Ω and $\bar{X} = (X_1, \ldots, X_m)$.

The most important properties of coordinate polynomials are described in the following lemma, whose proof we leave to the reader.

Lemma B.7.3. *In the above notation, the following statements are true.*

1. *For $j = 1, \ldots, k$, we have that*
2. $$P_j^{\Omega}(x_{1,1}, \ldots, x_{k,m}, w_1, \ldots, w_u) \in F[x_{1,1}, \ldots, x_{m,k}, w_1, \ldots, w_u].$$
$$P\left(\sum_{j=1}^{k} x_{1,j}\omega_j, \ldots, \sum_{j=1}^{k} x_{m,j}\omega_j, w_1, \ldots, w_u\right)$$
$$= \sum_{j=1}^{m} P_j^{\Omega}(x_{1,1}, \ldots, x_{m,k}, w_1, \ldots, w_u)\omega_j.$$
3. $$\exists X_1, \ldots, X_m \in M, w_1, \ldots, w_u \in F : P(X_1, \ldots, X_m, w_1, \ldots, w_u) = 0$$
$$\Updownarrow$$
$$\exists x_{1,1}, \ldots, x_{m,k}, w_1, \ldots, w_u \in F : \bigwedge_{i=1}^{k} P_i^{\Omega}(x_{1,1}, \ldots, x_{m,k}, w_1, \ldots, w_u) = 0.$$
4. *Let $A_{ijr} = \sigma_r(\omega_i\omega_j)$. Then the coefficients of coordinate polynomials are themselves polynomials in $\{A_{i,j,r}, i, j, r \in \{1, \ldots, k\}\}$. Furthermore, these polynomials depend on $P(X_1, \ldots, X_m, w_1, \ldots, w_u)$ only and can be computed effectively and uniformly in $P(X_1, \ldots, X_m, w_1, \ldots, w_u)$ from coefficients of $P(X_1, \ldots, X_m, w_1, \ldots, w_u)$. (Here by a procedure "uniform in P" we mean an effective procedure which takes the coefficients of P as inputs and produces the coefficients of P_j^{Ω}.)*

Remark B.7.4. If $P[X_1, \ldots, X_m, w_1, \ldots, w_u] \in F[X_1, \ldots, X_m, w_1, \ldots, w_u]$ then the procedure described in Part 4 of Lemma B.7.3 can be carried out even if the given set $\{\omega_1, \ldots, \omega_k\}$ is not a basis of some finite extension M over F but is simply a set of elements in the algebraic closure of F subject to the condition $\omega_i\omega_j = \sum_{r=1}^{k} A_{i,\,j,r}\omega_r$ for all pairs $(i, j) \in \{1, \ldots, k\}^2$ and for some set

$$A = \{A_{i,\,j,r}, i, j, r \in \{1, \ldots, k\}\} \subset F.$$

Note that once we have a set A as above then inductively we can rewrite any finite product of elements of Ω as a linear combination of elements of Ω with coefficients in F.

If Ω is not a basis of some finite extension M of F (or we are not sure whether Ω is a basis, as is the case below), we will call the polynomials generated by this procedure "pseudo-coordinate" polynomials and use the notation $P^{\Omega,A}$ for these polynomials. Note that Parts 1 and 2 of Lemma B.7.3 remain true for pseudo-coordinate polynomials also, while Part 3 fails if Ω is not a basis.

A situation which occurs frequently can be described as follows. Let F be a field as above, let $a_0, \dots, a_m$ represent variables ranging over F, and let t be a variable ranging over the algebraic closure of F. Assume that the variables satisfy the following equation:

$$t^m + a_{m-1}t^{m-1} + \cdots + a_0 = 0. \tag{B.7.1}$$

Let $\Omega = \{1, t, \dots, t^{m-1}\}$; so, for this Ω we have that $\omega_i = t^{i-1}$. Then from (B.7.1) we conclude that for $i + j < m$ we have that $A_{i,j,i+j} = a_{i+j}$ and $A_{i,j,r} = 0$ for $r \neq i + j$. Next we consider the case $i + j = m$. Here we have $A_{i,j,r} = a_r$. For the case $i + j > m$ we proceed inductively. Assume that for $m < i + j < k$ we have that $A_{i,\,j,r} = Q_{i+j,r}(a_0, \dots, a_{m-1}) \in F[a_0, \dots, a_{m-1}]$. Then observe that

$$\begin{aligned} t^k &= -a_{m-1}t^{k-1} - \cdots - a_0 t^{k-m} \\ &= -a_{m-1}\sum_{r=0}^{m-1} Q_{k-1,r}(\bar{a})t^r - \cdots - \cdots a_0 \sum_{r=0}^{m-1} Q_{k-m,r}(\bar{a})t^r \\ &= \sum_{r=0}^{m-1} Q_{k,r}(\bar{a})t^r, \end{aligned}$$

where $\bar{a} = (a_0, \dots, a_{m-1})$. Thus we conclude that, assuming that $\bar{a} = (a_{m-1}, \dots, a_0)$ takes values in F^m, we have that $A_{i,\,j,r} = Q_{i+j,r}(\bar{a}) \in F[\bar{a}]$ for all $i, j, r \in \{0, \dots, m-1\}$ and we can construct pseudo-coordinate polynomials $P^{\Omega,A}$ whose coefficients will be polynomials in $a_0, \dots, a_{m-1}$. Note further that these polynomial coefficients depend on m only and can be constructed effectively given an m.

For those values of $\bar{a}$ for which the polynomial $T^m + a_{m-1}T^{m-1} + \cdots + a_0$ is irreducible over F, the pseudo-coordinate polynomials will actually be the real thing. But it is important to note that pseudo-coordinate polynomials are defined unconditionally.

With the discussion above in mind we prove the following lemma.

Lemma B.7.5. *Let L be a field. Let $G(T, Z_1, \dots, Z_l) \in F[T, Z_1, \dots, Z_l]$ be a monic polynomial in T of degree d in T. Let*

$$P(X_1, \dots, X_m, w_1, \dots, w_u) \in F[X_1, \dots, X_m, w_1, \dots, w_u].$$

For $k = 1, \ldots, d-1$ let $\Omega_k = \{1, t, \ldots, t^{k-1}\}$, where t is a root of the equation $G(T, z_1, \ldots, z_l) = 0$ for some $z_1, \ldots, z_l \in F$. Further, let $A^{(k)} = \{A^{(k)}_{i,j,r}\}$ be generated formally as above from the equation $t^k + a_{k,k-1}t^{k-1} + \cdots + a_{k,0} = 0$. Let $P_i^{\Omega_k, A^{(k)}}$, $k < d$, as above be the ith *pseudo-coordinate polynomial corresponding to Ω_k and $A^{(k)}$. The set $A^{(d)}$ should be generated using the coefficients of $G(T, z_1, \ldots, z_l)$. Then for any $z_1, \ldots, z_l \in F$ and for any t in the algebraic closure of F satisfying $G(t, z_1, \ldots, z_l) = 0$,*

$$\exists X_1, \ldots, X_m \in F(t),\ w_1, \ldots, w_u \in F : P(X_1, \ldots, X_m, w_1, \ldots, w_u) = 0 \tag{B.7.2}$$

$$\Updownarrow$$

$$\exists a_{1,0}, \ldots, a_{d-1,d-2}, b_{1,0}, \ldots, b_{d-1,d-2}, x_{1,0}, \ldots, x_{m,d-1}, w_1, \ldots, w_u \in F :$$

$$\bigvee_{k=1}^{d-1} \begin{cases} t^k + a_{k,k-1}t^{k-1} + a_{k,0} = 0, \\ \left(T^k + a_{k,k-1}T^{k-1} + a_{k,0}\right)\left(T^{d-k} + b_{k,d-k-1}T^{d-k-1} + \cdots + b_{k,0}\right) \\ \qquad = G(T, z_1, \ldots, z_r), \\ \bigwedge_{j=1}^{k} P_j^{\Omega_k, A^{(k)}}(x_{1,0}, \ldots, x_{m,k-1}, w_1, \ldots, w_u) = 0 \end{cases}$$

$$\bigvee$$

$$\bigwedge_{j=0}^{d-1} P_j^{\Omega_d, A^{(d)}}(x_{1,0}, \ldots, x_{m,d-1}, w_1, \ldots, w_u) = 0. \tag{B.7.3}$$

Here the equality

$$\left(T^i + a_{i,i-1}T^{i-1} + a_{i,0}\right)\left(T^{d-i} + b_{i,d-i-1}T^{d-i-1} + \ldots + b_{i,0}\right) = G(T, z_1, \ldots, z_r)$$

should be read as an equality of polynomials in T which can be rewritten as a system of polynomial equations in the variables $a_{i,j}$, $b_{k,l}$, and z_s. Also, for $1 \le k < d$ and $1 \le j \le k$ we have that

$$P_j^{\Omega_k, A^{(k)}}(x_{1,0}, \ldots, x_{m,k-1}, w_1, \ldots, w_u)$$

$$= R_{j,k}(a_{k,0}, \ldots, a_{k,k-1}, x_{1,0}, \ldots, x_{m,k-1}, w_1, \ldots, w_u),$$

where $R_{j,k}(a_{k,0}, \ldots, a_{k,k-1}, x_{1,0}, \ldots, x_{m,i-1}, w_1, \ldots, w_u)$ is a polynomial over F whose coefficients depend on P, j, and k only and can be constructed effectively from these data, while

$$P_j^{\Omega_d, A^{(d)}}(x_{1,0}, \ldots, x_{m,d-1}, w_1, \ldots, w_u) = R_{j,d}(x_{1,0}, \ldots, x_{m,d-1}, w_1, \ldots, w_u),$$

where $R_{j,d}(x_{1,0}, \ldots, x_{m,d-1}, w_1, \ldots, w_u)$, is a polynomial over F whose coefficients depend on G, P, and j only and can be constructed effectively from these data.

Proof. Let $t, z_1, \ldots, z_l$ be given. Let $1 \leq k \leq d$ be the degree of t over F. Let $T^k + c_{k-1}T^{k-1} + \cdots + c_0$ be the monic irreducible polynomial of t over F. Then either $k = d$ and $G(T, z_1, \ldots, z_l)$ is irreducible, or $k < d$ and for $a_{k,i} = c_i$ and some F-values of the $b_{k,j}$ the polynomial-in-T equation in (B.7.3) holds. Further, for this value of k we also have that the $P_i^{\Omega_k, A^{(k)}} = P_i^{\Omega_k}, i = 1, \ldots, k$, are coordinate polynomials. Now the conclusion of the lemma follows from Lemma B.7.3 and Remark B.7.4. □

B.8 Basic facts about local fields

In this section we define local fields and list those of their properties that are important for our purposes. We start by defining notation.

Notation B.8.1. Let K be a global field and let $\mathfrak{p}$ be a non-archimedean prime of K. Let $K_{\mathfrak{p}}$ denote the completion of K under the non-archimedean absolute value corresponding to $\mathfrak{p}$.

B.8.2. Hensel's lemma *Let $f(X) \in K_{\mathfrak{p}}[X]$ be such that all the coefficients of f are integral at $\mathfrak{p}$. Suppose that there exists an integral $K_{\mathfrak{p}}$-element α_0 such that* $\operatorname{ord}_{\mathfrak{p}} f(\alpha_0) > 2 \operatorname{ord}_{\mathfrak{p}} f'(\alpha_0)$. *Then $f(x)$ has a root α in $K_{\mathfrak{p}}$.* (See Proposition 2, Section 2, Chapter II of [46].)

Lemma B.8.3. *Let q be a rational prime relatively prime to the characteristic $p > 0$ of the residue field of $K_{\mathfrak{p}}$. Let $a \in K_{\mathfrak{p}}$ be a unit such that a is a qth power modulo $\mathfrak{p}$. Then a is a qth power in $K_{\mathfrak{p}}$.*

Proof. First of all observe the following. If K is a function field then q is automatically prime to $\mathfrak{p}$. If K is a number field then $\mathfrak{p}$ is a factor of p. Thus if p is prime to q, $\mathfrak{p}$ is prime to q. Next consider the polynomial $f(X) = X^q - a$ with $f'(X) = qX^{p-1}$. Let $b^q \equiv a \bmod \mathfrak{p}$. Note that

$$2 \operatorname{ord}_{\mathfrak{p}} f'(b) = 2 \operatorname{ord}_{\mathfrak{p}} q + 2(p-1) \operatorname{ord}_{\mathfrak{p}} b = 0 < \operatorname{ord}_{\mathfrak{p}} f(b),$$

by assumption. Therefore, by Hensel's Lemma B.8.2, $f(X)$ has a root α in $K_{\mathfrak{p}}$ or, in other words, for some $\alpha \in K_{\mathfrak{p}}$ we have that $\alpha^p - a = 0$. □

Lemma B.8.4. *Let K be a number field. Let $p > 0$ be the characteristic of the residue field of $\mathfrak{p}$. Let a be a unit of $K_{\mathfrak{p}}$ such that for some $\varepsilon \in K$ we have that $\varepsilon^p \equiv a \bmod \mathfrak{p}^{2e(\mathfrak{p}/p)+1}$. Then a is a pth power in $K_{\mathfrak{p}}$.*

Proof. We apply Hensel's lemma again. Let $f(X) = X^p - a$ as before and note that

$$2\,\mathrm{ord}_{\mathfrak{p}} f'(\varepsilon) = 2\,\mathrm{ord}_{\mathfrak{p}}\, p + 2(p-1)\,\mathrm{ord}_{\mathfrak{p}}\, \varepsilon = 2e(\mathfrak{p}/p) < 2e(\mathfrak{p}/p) + 1$$
$$= \mathrm{ord}_{\mathfrak{p}} f(\varepsilon).$$

Thus f has a root in $K_{\mathfrak{p}}$ and a is a pth power. □

Lemma B.8.5. *Let M be a finite separable extension of K. Let $\mathfrak{P}$ be a factor of $\mathfrak{p}$ in M. Finally, let $M_{\mathfrak{P}}$ be the completion of M under $\mathfrak{P}$. Then if $\mathfrak{P}/\mathfrak{p}$ is unramified, every $K_{\mathfrak{p}}$-unit is a norm of some element of $M_{\mathfrak{P}}$.* (See Theorem 2, Section 2, Chapter 7 of [1] or the corollary to Proposition 6, Section 2, Chapter XII of [119].)

B.9 Derivations

In this section we discuss derivations over function fields. They are used to ensure that zeros and poles of certain functions are simple. To introduce the notion of derivation we will need the proposition stated below.

Proposition B.9.1. *Let K be a function field over a finite field of constants C. Let*

$$K_0 = \{w^p, w \in K\}$$

and let $t \in K$ be such that $K/C(t)$ is a separable extension. Then $K = K_0(t)$ and $[K : K_0] = p$.

See Section 2, Chapter III of [21], for a proof.

Definition B.9.2. Let t, K, K_0 be as in Proposition B.9.1. Let $x \in K, x = \sum_{i=0}^{p-1} u_i t^i$, $u_i \in K_0$. Then define the global derivation $dx/dt = \sum_{i=0}^{p-2} i u_i t^{i-1}$.

Proposition B.9.1 assures us that we have defined dx/dt for all x. We leave the proof of the next proposition to the reader.

Proposition B.9.3. *The derivation in Definition B.9.2 satisfies the usual differentiation rules concerning the derivatives of the sum and product of functions as well as the chain rule.*

Next we state the main proposition of this section.

Proposition B.9.4. *Let K be a function field over a finite field of constants C_K. Let $\mathfrak{p}$ be a prime of K and let $t \in K$ be such that $C/K(t)$ is separable and* $\mathrm{ord}_{\mathfrak{p}}\, t = 1$. *Then, for any $x \in K$, if* $\mathrm{ord}_{\mathfrak{p}}\, x \geq 0$ *then* $\mathrm{ord}_{\mathfrak{p}}\, dx/dt \geq \mathrm{ord}_{\mathfrak{p}}\, x - 1$.

Proof. We present a proof of this proposition that was suggested to the author by Laurent Moret-Bailly. First of all we observe that by Proposition B.9.1 the derivation with respect to t is well defined. Next write $x = \sum_{i=0}^{p-1} u_i t^i$, where $u_i \in K_0$ as in Definition B.9.2. Observe that $\mathrm{ord}_{\mathfrak{p}}\, u_i t^i \equiv i \bmod p$ and, therefore, for $i \neq j$ we have that $\mathrm{ord}_{\mathfrak{p}}\, u_i t^i \neq \mathrm{ord}_{\mathfrak{p}}\, u_j t^j$. Thus

$$0 \leq \mathrm{ord}_{\mathfrak{p}}\, x = \min_{i=0,\ldots,p-1} \mathrm{ord}_{\mathfrak{p}}\, u_i t^i$$

and $\mathrm{ord}_{\mathfrak{p}}\, u_i \geq 0$ for all i. Now using the definition of derivation we consider two cases. In the first case $\mathrm{ord}_{\mathfrak{p}}\, x \equiv 0 \bmod p$ and therefore

$$\mathrm{ord}_{\mathfrak{p}}\, x = \mathrm{ord}_{\mathfrak{p}} u_0 \leq \left(\min_{1 \leq j \leq p-1} \mathrm{ord}_{\mathfrak{p}} t^j u_j \right) - 1.$$

Thus

$$\begin{aligned} \mathrm{ord}_{\mathfrak{p}} \frac{dx}{dt} &= \min \left(\mathrm{ord}_{\mathfrak{p}}\, u_1, \mathrm{ord}_{\mathfrak{p}}\, 2u_2 t, \ldots, \mathrm{ord}_{\mathfrak{p}}\, (p-1) u_{p-1} t^{p-2} \right) \\ &= \left(\min_{1 \leq j \leq p-1} \mathrm{ord}_{\mathfrak{p}}\, t^j u_j \right) - 1 \geq \mathrm{ord}_{\mathfrak{p}}\, x. \end{aligned}$$

In the second case we have that $\mathrm{ord}_{\mathfrak{p}}\, x \not\equiv 0 \bmod p$. Then

$$\min_{i=0,\ldots,p-1} \mathrm{ord}_{\mathfrak{p}} u_i t^i = \mathrm{ord}_{\mathfrak{p}} u_j t^j,$$

where $1 \leq j \leq p-1$. But in this case

$$\mathrm{ord}_{\mathfrak{p}}\, j u_j t^{j-1} = \min \left(\mathrm{ord}_{\mathfrak{p}}\, u_1, \mathrm{ord}_{\mathfrak{p}}\, 2u_2 t, \ldots, \mathrm{ord}_{\mathfrak{p}}\, (p-1) u_{p-1} t^{p-2} \right),$$

so that $\mathrm{ord}_{\mathfrak{p}}\, dx/dt = \mathrm{ord}_{\mathfrak{p}}\, x - 1$. □

We can strengthen the result of the lemma in the following fashion.

Corollary B.9.5. *Let $\mathfrak{q}$ be a prime of a global function field K of characteristic p and let $w \in K$ be such that* $\mathrm{ord}_{\mathfrak{q}}\, w = 1$. *Let $t \in K$ be, as before, such that $K/C(t)$ is separable. Assume further that* $\mathrm{ord}_{\mathfrak{q}}\, dw/dt \geq 0$. *Then for any $x \in K$ integral at $\mathfrak{q}$, we have that* $\mathrm{ord}_{\mathfrak{q}}\, dx/dt \geq \mathrm{ord}_{\mathfrak{q}}\, x - 1$.

Proof. Since w has order 1 at a prime, it is not a pth power and therefore the derivation with respect to w is defined. Now we use the chain rule:

$$\operatorname{ord}_{\mathfrak{q}} \frac{dx}{dt} = \operatorname{ord}_{\mathfrak{q}} \frac{dx}{dw} + \operatorname{ord}_{\mathfrak{q}} \frac{dw}{dt} \geq \operatorname{ord}_{\mathfrak{q}} \frac{dx}{dw} \geq \operatorname{ord}_{\mathfrak{q}} x - 1,$$

where the last inequality holds by Proposition B.9.4. □

Remark B.9.6. Using the product rule, it is not hard to show that the corollary above holds even if x is not integral at p. We leave the details to the reader.

Corollary B.9.7. *Let K be a function field over a finite field of constants C. Let $t \in K$ be such that t is not a pth power in K. Let $\mathfrak{p}$ be a prime of K such that it is not ramified in the extension $K/C(t)$ and is not a pole of t. Then, for any $x \in K$, if $\operatorname{ord}_{\mathfrak{p}} x > 1$ then $\operatorname{ord}_{\mathfrak{p}} dx/dt > 0$.*

Proof. Let $P(t)$ be a monic irreducible polynomial corresponding to the prime of $C(t)$ lying below $\mathfrak{p}$. (We know that such a polynomial exists because $\mathfrak{p}$ is not a pole of t.) Now, since $\mathfrak{p}$ is not ramified over $C(t)$ we must have $\operatorname{ord}_{\mathfrak{p}} P(t) = 1$. Further, $dP(t)/dt$ is a polynomial and therefore $\operatorname{ord}_{\mathfrak{p}} dP(t)/dt \geq 0$. Thus, by Corollary B.9.5, for any $x \in K$ integral at $\mathfrak{p}$ we have that $\operatorname{ord}_{\mathfrak{p}} dx/dt \geq \operatorname{ord}_{\mathfrak{p}} x - 1$. Hence the conclusion of the corollary follows. □

B.10 Some calculations

In this section we carry out some calculations necessary for various proofs in the book but not of any independent interest.

Lemma B.10.1. *Let X_l be a sequence of positive real numbers such that*

$$\limsup_{l \to \infty} \frac{X_l}{l^2 \log X_l} \leq c \in \mathbb{R}^+.$$

Then for some constant $\bar{c} \in \mathbb{R}^+$ we have that

$$\limsup_{l \to \infty} \frac{X_l}{l^2 \log l} \leq \bar{c}.$$

Proof. First of all, we observe that our assumptions imply that for some $C \in \mathbb{R}$ and for all $l \in \mathbb{N} \setminus \{0\}$ it is the case that $X_l/\log X_l < Cl^2$. Thus for some constant $\bar{C} \in \mathbb{R}$, we have that $X_l < Cl^2 \log X_l < \bar{C}l^2\sqrt{X_l}$. Therefore, for all $l \in \mathbb{N} \setminus \{0\}$

and some constant $\tilde{C} \in \mathbb{R}$, we have that $X_l < \tilde{C}l^4$. Now observe that as $l \to \infty$ we have that

$$\frac{X_l}{l^2 \log l} \leq \frac{\bar{C}l^2 \log X_l}{l^2 \log l} \leq \frac{\bar{C} \log(\bar{C}l^2 \log X_l)}{\log l} = D + \frac{\hat{C} \log \log X_l}{\log l}$$

for some positive constants D and $\hat{C}$. Since $X_l < \tilde{C}l^4$, we also have as $l \to \infty$ that

$$\frac{\hat{C} \log \log X_l}{\log l} \leq \frac{\hat{C} \log \log \tilde{C}l^4}{\log l} \to 0.$$

Thus the assertion of the lemma is true. □

Proposition B.10.2. *Let $\{\alpha_i, i \in \mathbb{N}\} \subset \mathbb{R}$ be such that*

1. $\sum_{i=1}^{\infty} \alpha_i = \infty$,
2. $\sum_{i=1}^{\infty} \alpha_i^2 < \infty$,
3. $0 < \alpha_i < 1$,
4. *for all $i \in \mathbb{N}$ we have that $\alpha_i > \alpha_{i+1}$* and $\lim_{i \to \infty} \alpha_i = 0$.

Let $S(n) = \sum_{i=1}^{n} \alpha_i$. Let $\mathcal{A}(n) = \{\alpha_1, \ldots, \alpha_n\}$. Then the following statements are true.

1. $G(n) = \prod_{i=1}^{n}(1 - \alpha_i) = O(1)\mathrm{e}^{-S(n)}$.
2. *For all k we have that*

$$S_k(n) = G(n) \sum_{\text{all } k\text{-element subsets of } \mathcal{A}(n)} \frac{\alpha_{i_1} \cdots \alpha_{i_k}}{(1 - \alpha_{i_1}) \cdots (1 - \alpha_{i_k})} \to 0$$

as $n \to \infty$.

Proof.

1. Observe that $\log G(n) = \sum_{i=1}^{n} \log(1 - \alpha_i) = -\sum_{i=1}^{n} \alpha_i + O(1) = -S(n) + O(1)$. Therefore $G(n) = O(1)e^{-S(n)}$.
2. We show that $S_k(n) < O(1)S^k(n)G(n)$. Indeed, using the fact that all the terms in the sum are positive, we see the following:

$$S_k(n) = G(n) \sum_{\text{all } k\text{-element subsets of } \mathcal{A}(n)} \frac{\alpha_{i_1} \cdots \alpha_{i_k}}{(1 - \alpha_{i_1}) \cdots (1 - \alpha_{i_k})}$$

$$< (1 - \alpha_1)^{-k}(S(n))^k G(n) \to 0$$

as $n \to \infty$.

□

Lemma B.10.3. *Let F/G be a finite field extension. Let $\Omega = \{\omega_1, \ldots, \omega_n\}$ be a basis of F over G. Then for $l = 1, \ldots, n$ there exist*

$$P_l(X_1, \ldots, X_n, Y_1, \ldots, Y_n) \in G[X_1, \ldots, X_n, Y_1, \ldots, Y_n],$$

depending only on Ω, such that for all $a_1, \ldots, a_n, b_1, \ldots, b_n \in G$ we have that

$$\sum_{i=1}^{n} a_i\omega_i \sum_{j=1}^{n} b_j\omega_j = \sum_{l=1}^{n} P_l(a_1, \ldots, a_n, b_1, \ldots, b_n)\omega_i.$$

Proof. Let $A_{i,\,j,l} \in G$ be such that $\omega_i\omega_j = \sum_{l=1}^{n} A_{i,\,j,l}\omega_l$. Then

$$\sum_{i=1}^{n} a_i\omega_i \sum_{j=1}^{n} b_j\omega_j = \sum_{i,j} a_ib_j\omega_i\omega_j = \sum_{i,j,l} a_ib_jA_{i,\,j,l}\omega_l = \sum_{l=1}^{n}\left(\sum_{i,j} A_{i,\,j,l}a_ib_j\right)\omega_l.$$

Thus we can set $P_l(X_1, \ldots, X_n, Y_1, \ldots, Y_n) = \sum_{i,j} A_{i,\,j,l}X_iY_j$. □

Lemma B.10.4. *Let R be an integral domain with a quotient field F. Let $A_1, \ldots, A_K \in F$, $a, a_1, \ldots, a_k \in R$ and assume that for $i = 1, \ldots, k$ we have that $aA_i = a_i$. Let*

$$P(X_1, \ldots, X_k) = \sum_{i_1+\cdots+i_k \le d} a_{i_1,\ldots,i_k} X_1^{i_1} \cdots X_k^{i_k}$$

be a polynomial over F of degree d. Let b be a common denominator of all the coefficients of P with respect to R. Let

$$P_R(Y_1, \ldots, Y_k, Z) = \sum_{i_1+\cdots+i_k \le d} ba_{i_1,\ldots,i_k} Y_1^{i_1} \cdots Y_k^{i_k} Z^{d-(i_1+\cdots+i_k)}$$

be a polynomial over R. Then

$$a^d bP(A_1, \ldots, A_k) = P_R(a_1, \ldots, a_k, a) \in R,$$

and $P(X_1, \ldots, X_k) = 0$ has solutions in F if and only if $P_R(Y_1, \ldots, Y_k, Z) = 0$ has solutions in R with $Z \neq 0$.

Proof. Since the second assertion of the lemma is obvious, we will verify the first one only, as follows:

$$\begin{aligned} a^d bP(A_1, \ldots, A_k) &= a^d b \sum_{i_1+\cdots+i_k \le d} a_{i_1,\ldots,i_k} A_1^{i_1} \cdots A_k^{i_k} \\ &= \sum_{i_1+\cdots+i_k \le d} (ba_{i_1,\ldots,i_k})(a^{i_1}A_1^{i_1}) \cdots (a^{i_k}A_k^{i_k})a^{d-i_1-\cdots-i_k} \in R, \end{aligned}$$

since every term in the product is now in R. □

Lemma B.10.5. *Let $F \subseteq G \subseteq L$ be finite extensions of fields. Let $\Omega = \{\omega_1, \dots, \omega_k\}$ be a basis of L over F. Let $\Lambda = \{\lambda_1, \dots, \lambda_m\}$ be a basis of G over F. (Given our assumptions, $m \leq k$.) Assume that Ω is ordered in such a way that the matrix $(c_{i,j})$, $i, j = 1, \dots, m$, where $c_{i,j} \in F$ and $\lambda_i = \sum_{j=1}^{k} c_{i,j}\omega_j$, is non-singular. Then there exist $P_1, \dots, P_m, T_1, \dots, T_{k-m} \in F[x_1, \dots, x_m]$, depending on G, L, Ω, and Λ only, such that for any $a_1, \dots, a_k, b_1, \dots, b_m \in F$ the following two statements are equivalent:*

$$\sum_{i=j}^{k} a_j\omega_j = \sum_{i=1}^{m} b_i\lambda_i, \tag{B.10.1}$$

and

$$\begin{cases} b_i = P_i(a_1, \dots, a_m), & i = 1, \dots, m, \\ a_{m+j} = T_j(a_1, \dots, a_m), & j = 1, \dots, k-m. \end{cases} \tag{B.10.2}$$

Proof. The matrix $(c_{i,j})$ has rank $m \leq k$. By assumption, the first m columns are linearly independent. We can rewrite (B.10.1) as

$$\sum_{i=j}^{k} a_j\omega_j = \sum_{i=1}^{m} b_i \sum_{j=1}^{k} c_{i,j}\omega_j = \sum_{j=1}^{k}\left(\sum_{i=1}^{m} b_i c_{i,j}\right)\omega_j.$$

This equality leads to the system

$$\begin{cases} \sum_{i=1}^{m} b_i c_{i,1} = a_1, \\ \quad\vdots \\ \sum_{i=1}^{m} b_i c_{i,m} = a_m, \\ \quad\vdots \\ \sum_{i=1}^{m} b_i c_{i,k} = a_k, \end{cases} \tag{B.10.3}$$

where we consider $b_1, \dots, b_m$ as variables. Given our assumptions, we have the following information about the system (B.10.3):

1. this system has a solution;
2. the first m equations of the system have a unique solution.

Thus, if we solve the first m equations then the rest of the system will be satisfied automatically. Using Cramer's rule to solve the $m \times m$ system we deduce that $b_i = P_i(a_1, \dots, a_m)$, where for each i we have that P_i is a fixed polynomial over F depending on G, L, Ω, and Λ only. Further, considering the remaining $k - m$ equations we conclude that

$$a_{m+j} = \sum_{i=1}^{m} b_i c_{i,m+j} = \sum_{i=1}^{m} P_i(a_1, \dots, a_m)c_{i,m+j}, \qquad j = 1, \dots, k-m.$$

This proves that (B.10.1) implies (B.10.2). All the steps in the proof above are reversible, however. Thus (B.10.2) implies (B.10.1) also. □

Lemma B.10.6. *Let G/F be a finite field extension and let*

$$\Omega = \{\omega_1, \ldots, \omega_k\}$$

be a basis of G over F. Let $a_1, \ldots, a_k \in F$. Then there exist

$$P_1, \ldots, P_k, Q \in F[x_1, \ldots, x_k]$$

depending on F, G, and Ω only such that $\sum_{i=1}^{k} a_i\omega_i \neq 0$ if and only if

$$Q(a_1, \ldots, a_k) \neq 0$$

and

$$\sum_{i=1}^{k} \frac{P_i(a_1, \ldots, a_k)}{Q(a_1, \ldots, a_k)}\omega_i \left(\sum_{i=1}^{k} a_i\omega_i \right) = 1.$$

Proof. Let $A_1, \ldots, A_k \in F$ be such that $\sum_{i=1}^{k} A_i\omega_i = 1$. For all $i, j = 1, \ldots, k$, let $B_{i,j,1}, \ldots, B_{i,j,k} \in F$ be such that

$$\omega_i\omega_j = \sum_{r=1}^{k} B_{ijr}\omega_r.$$

Note that

$$A_1, \ldots, A_k, B_{1,1,1}, \ldots, B_{k,k,k}$$

depend on Ω only. Next note that $\sum_{i=1}^{k} a_i\omega_i \neq 0$ if and only if there exist $c_1, \ldots, c_k \in F$ such that

$$\sum_{i=1}^{k} a_i\omega_i \left(\sum_{i=1}^{k} c_i\omega_i \right) = 1.$$

The last equation holds if and only if

$$\sum_{i=1}^{k} A_i\omega_i = 1 = \sum_{i=1}^{k} a_i\omega_i \sum_{i=1}^{k} c_i\omega_i$$

$$= \sum_{i,j=1}^{k} a_i c_j \omega_i\omega_j = \sum_{i,j,r=1}^{k} a_i c_j B_{i,j,r}\omega_r.$$

Thus, $\sum_{i=1}^{k} a_i\omega_i \neq 0$ if and only if there exist $c_1, \ldots, c_k$ such that the following system is satisfied:

$$\sum_{i,\,j=1}^{k} a_i c_j B_{i,\,j,r} = A_r, \qquad r = 1, \ldots, k,$$

$$\sum_{j=1}^{k} \left(\sum_{i=1}^{k} a_i B_{i,\,j,r} \right) c_j = A_r.$$

By Cramer's rule, we can conclude that $\sum_{i=1}^{k} a_i \omega_i \neq 0$ if and only if $c_1, \ldots, c_k$ have the required form. (The polynomial in the denominator is the determinant of the system, which is non-zero if and only if the system has a unique solution. The last condition is true if and only if $\sum_{i=1}^{k} a_i \omega_i \neq 0$.) □

Let $\bar{\delta} = \{\delta_1, \ldots, \delta_p\}$, $\bar{\tau} = \{\tau_1, \ldots, \tau_q\}$ be two sets of complex numbers such that $\bar{\delta} \cap \bar{\tau} = \emptyset$. Let

$$C_{\bar{\delta},\bar{\tau}} = \min_{i=1,\ldots,q,\,j=1,\ldots,p} (|\tau_i - \delta_j|).$$

Let $z \in \mathbb{C}$ and let $C_{\bar{\tau},z} = \min_{i=1,\ldots,q}(|\tau_i - z|)$, $C_{\bar{\delta},z} = \min_{j=1,\ldots,p}(|z - \delta_j|)$. Then $\max(C_{\bar{\tau},z}, C_{\bar{\delta},z}) \geq \frac{1}{2} C_{\bar{\delta},\bar{\tau}}$.

Proof. The proof of this lemma is a simple consequence of the triangular inequality. Indeed, suppose that $C_{\bar{\tau},z} < \frac{1}{2} C_{\bar{\delta},\bar{\tau}}$. Note that for all i, j we have that

$$|\tau_i - \delta_j| \leq |z - \delta_j| + |z - \tau_i|.$$

Thus for all i, j it is the case that $|z - \delta_j| \geq |\tau_i - \delta_j| - |z - \tau_i| \geq C_{\bar{\delta},\bar{\tau}} - |z - \tau_i|$. Let i_0 be such that $\frac{1}{2} C_{\bar{\delta},\bar{\tau}} > C_{\bar{\tau},z} = |z - \tau_{i_0}|$. Then for all j we have that $|z - \delta_j| \geq \frac{1}{2} C_{\bar{\delta},\bar{\tau}}$. Hence, $C_{\bar{\delta},z} \geq \frac{1}{2} C_{\bar{\delta},\bar{\tau}}$. □

Lemma B.10.8. *Let $n \in \mathbb{Z}_{>0}$ and let $\bar{\tau}_1 = \{\tau_{1,1}, \ldots, \tau_{1,q_1}\}, \ldots, \bar{\tau}_{n+1} = \{\tau_{n+1,1}, \ldots, \tau_{n+1,q_{n+1}}\}$ be a collection of $n + 1$ pairwise-disjoint sets of complex numbers. Let*

$$C = \min_{u \neq j,\, l_u = 1,\ldots,q_u,\, l_j = 1,\ldots,q_j} (|\tau_{u,l_u} - \tau_{j,l_j}|).$$

Let $\{z_1, \ldots, z_n\}$ be a set of complex numbers. Let

$$C_u = \min_{j=1,\ldots,q_u,\, l=1,\ldots,n} (|z_l - \tau_{u,j}|).$$

Then for some u we have that $C_u \geq \frac{1}{2} C$.

Proof. For $l = 1, \ldots, n$ and $u = 1, \ldots, n + 1$, call z_l close to $\bar{\tau}_u$ if

$$C_{u,l} = \min_{j=1,\ldots,q_u} (|z_l - \tau_{u,j}|) < \tfrac{1}{2} C.$$

By Lemma B.10.6, each z_l can be close to at most one $\bar{\tau}_u$. Thus, there is at least one $\bar{\tau}_u$ such that there is no z_l close to it. □

Lemma B.10.9. *Let K be a field. Let x be an element of the algebraic closure of K. Let $\{F_i(T) = a_{i,0} + a_{i,1}T + \cdots + a_{i,n}T^n,\ i = 0, \ldots, n\}$ be a finite collection of polynomials with $a_{i,\,j} \in K$ and such that the matrix $(a_{i,\,j})$ is non-singular. Suppose that for all $i = 0, \ldots, n$ we have that $F_i(x) \in K$. Then $x \in K$.*

Proof. By assumption, for $i = 0, \ldots, n$ it is the case that $F_i(x) = c_i \in K$. Therefore we have the following linear system:

$$(a_{i,\,j}) \begin{pmatrix} 1 \\ x \\ \vdots \\ x^n \end{pmatrix} = \begin{pmatrix} c_0 \\ c_1 \\ \vdots \\ c_n \end{pmatrix}, \qquad c_i \in K.$$

Let C_j be the matrix obtained by replacing the jth column of $A = (a_{i,\,j})$ by the column vector $\begin{pmatrix} c_0 \\ c_1 \\ \vdots \\ c_n \end{pmatrix}$. Since the system is non-singular, we can solve for x using Cramer's rule to obtain

$$x = \frac{\det C_1}{\det A},$$

where the numerator and the denominator of the fraction are clearly in K. Therefore $x \in K$. □

Corollary B.10.10. *Let K be a field of characteristic 0. Let x be an element of the algebraic closure of K. Let $\{F_i(T) = (T + i + 1)^n, i = 0, \ldots, n\}$. Assume that $F_i(x) \in K$. Then $x \in K$.*

Proof. By assumption, $F_i(T) = \sum_{j=0}^{n} \binom{n}{j} (i+1)^j T^{n-j}$. Therefore, in the notation of Lemma B.10.9, $a_{i,\,j} = \binom{n}{j} (i+1)^j$. Thus

$$\det a_{i,\,j} = \prod_{j=0}^{n} \binom{n}{j} \det (i+1)^j,$$

where $\det (i+1)^j$ is a Vandermonde determinant not equal to 0. Consequently, the corollary holds by Lemma B.10.9. □

References

[1] Emil Artin. *Algebraic Numbers and Algebraic Functions*. New York, Gordon and Breach, 1986.

[2] David C. Cantor and Peter Roquette. On Diophantine equations over the ring of all algebraic integers. *J. Number Theory*, **18**(1): 1–26, 1984.

[3] Claude Chevalley. *Introduction to the Theory of Algebraic Functions of One Variable*, volume 6 of *Mathematical Surveys*. Providence RI, American Mathematical Society, 1951.

[4] Jean-Louis Colliot-Thélène, Alexei Skorobogatov, and Peter Swinnerton-Dyer. Double fibres and double covers: paucity of rational points. *Acta Arithmetica*, **79**: 113–135, 1997.

[5] Gunther Cornelissen. Stockage diophantien et hypothèse *abc* généralisée. *C. R. Acad. Sci. Paris Sér. I Math.*, **328**(1): 3–8, 1999.

[6] Gunther Cornelissen and Karim Zahidi. Topology of diophantine sets: remarks on Mazur's conjectures. In Jan Denef, Leonard Lipshitz, Thanases Pheidas, and Jan Van Geel, editors, *Hilbert's Tenth Problem: Relations with Arithmetic and Algebraic Geometry*, volume 270 of *Contemporary Mathematics*, pp. 253–260. Providence RI, American Mathematical Society, 2000.

[7] Gunther Cornelissen and Karim Zahidi. Complexity of undecidable formulae in rationals and inertial Zygmondy theorems for elliptic curves. Preprint.

[8] Gunther Cornelissen, Thanases Pheidas, and Karim Zahidi. Division-ample sets and diophantine problem for rings of integers. *J. Théorie des Nombres Bordeaux*, **17**: 727–735, 2005.

[9] Luck Darnière. Decidability and local–global principles. In Jan Denef, Leonard Lipshitz, Thanases Pheidas, and Jan Van Geel, editors, *Hilbert's Tenth Problem: Relations with Arithmetic and Algebraic Geometry*, volume 270 of *Contemporary Mathematics*, pp. 145–167. Providence RI, American Mathematical Society, 2000.

[10] Luck Darnière. Nonsingular Hasse principle for rings. *J. Reine Angew. Math.*, **529**: 75–100, 2000.

[11] Luck Darnière. Pseudo-algebraically closed rings. *Manuscripta Math.*, **105**(1): 13–46, 2001.

[12] Martin Davis. Hilbert's tenth problem is unsolvable. *Amer. Mathematical Monthly*, **80**: 233–269, 1973.

[13] Martin Davis and Elaine J. Weyuker. *Computability, Complexity and Languages, Fundamentals of Theoretical Computer Science*. New York, Academic Press, 1983.

[14] Martin Davis, Yurii Matijasevich, and Julia Robinson. Hilbert's tenth problem. Diophantine equations: positive aspects of a negative solution. In *Proc. Symp. Pure Math.*, volume 28, pp. 323– 378. American Mathematical Society, 1976.

[15] Jan Denef. Hilbert's tenth problem for quadratic rings. *Proc. Amer. Math. Soc.*, **48**: 214–220, 1975.

[16] Jan Denef. The diophantine problem for polynomial rings and fields of rational functions. *Trans. Amer. Math. Soc.*, **242**: 391–399, 1978.

[17] Jan Denef. The diophantine problem for polynomial rings of positive characteristic. In M. Boffa, D. van Dalen, and K. MacAloon, editors, *Proc. Logic Colloquium 78*, pp. 131–145. Amsterdam, Netherlands, 1979.

[18] Jan Denef. Diophantine sets of algebraic integers, II. *Trans. Amer. Math. Soc.*, **257**(1): 227–236, 1980.

[19] Jan Denef and Leonard Lipshitz. Diophantine sets over some rings of algebraic integers. *J. London Math. Soc.*, **18**(2): 385–391, 1978.

[20] Jan Denef, Leonard Lipshitz, and Thanases Pheidas, editors. *Hilbert's Tenth Problem: Relations with Arithmetic and Algebraic Geometry*, volume 270 of *Contemporary Mathematics*. Providence RI, American Mathematical Society, 2000. Papers from the workshop held at Ghent University, November 2–5, 1999.

[21] Martin Eichler. *Introduction to the Theory of Algebraic Functions and Algebraic Numbers*. New York, Academic Press, 1966.

[22] Kirsten Eisenträger. Hilbert's tenth problem for algebraic function fields of characteristic 2. *Pacific J. Math.*, **210**(2): 261–281, 2003.

[23] Kirsten Eisenträger. Hilbert's tenth problem for function fields of varieties over $\mathfrak{C}$. *Int. Math. Res. Not.*, **59**: 3191–3205, 2004.

[24] Kirsten Eisenträger. Integrality at a prime for global fields and the perfect closure of global fields of characteristic $p > 2$. *J. Number Theory*, **114**: 170–181, 2005.

[25] Kirsten Eisenträger. Hilbert's tenth problem for function fields of varieties over number fields and p-adic fields. Preprint.

[26] Ju. L. Ershov. The undecidability of certain fields. *Dokl. Akad. Nauk SSSR*, **161**: 27–29, 1965.

[27] Yu. L. Ershov. Nice locally global fields. I. *Algebra i Logika*, **35**(4): 411–423, 497, 1996.

[28] Yu. L. Ershov. Nice locally global fields. II. *Algebra i Logika*, **35**(5): 503–528, 624, 1996.

[29] Yu. L. Ershov. Nice locally global fields. III. *Sibirsk. Mat. Zh.*, **38**(3): 526–532, 1997.

[30] Yu. L. Ershov. Nice locally global fields. IV. *Sibirsk. Mat. Zh.*, **43**(3): 526–538, 2002.

[31] Graham Everest, Alf van der Poorten, Igor Shparlinski, and Thomas Ward. *Recurrence Sequences*, volume 104 of *Mathematical Surveys and Monographs*. Providence RI, American Mathematical Society, 2003.

[32] Michael D. Fried, Dan Haran, and Helmut Völklein. Real Hilbertianity and the field of totally real numbers. In *Arithmetic Geometry (Tempe AZ, 1993)*, volume 174 of *Contemp. Mathematics*, pp. 1–34. Providence RI, American Mathematical Society, 1994.

[33] Michael D. Fried and Moshe Jarden. *Field Arithmetic*, volume 11 of *Ergebnisse der Mathematik und ihrer Grenzgebiete. 3. Folge.* A Series of Modern Surveys in Mathematics (*Results in Mathematics and Related Areas. 3rd Series*. A Series of Modern Surveys in Mathematics). Berlin, Springer-Verlag, second edition, 2005.

[34] Wulf-Dieter Geyer and Moshe Jarden. Bounded realization of l-groups over global fields. The method of Scholz and Reichardt. *Nagoya Math. J.*, **150**: 13–62, 1998.

[35] Barry Green, Florian Pop, and Peter Roquette. On Rumely's local–global principle. *Jahresber. Deutsch. Math.-Verein.*, **97**(2): 43–74, 1996.

[36] G. H. Hardy and E. M. Wright. *An Introduction to the Theory of Numbers*. Oxford, Oxford Science Publications, fifth edition, 1989.

[37] Gerald Janusz. *Algebraic Number Fields*. New York, Academic Press, 1973.

[38] Moshe Jarden and Aharon Razon. Rumely's local–global principle for algebraic psc fields over rings. *Trans. Amer. Math. Soc.*, **350**(1): 55–85, 1998.

[39] Moshe Jarden and Aharon Razon. Skolem density problems over large Galois extensions of global fields. In Jan Denef, Leonard Lipshitz, Thanases Pheidas, and Jan Van Geel, editors, *Hilbert's Tenth Problem: Relations with Arithmetic and Algebraic Geometry*, volume 270 of *Contemporary Mathematics*, pp. 213–235. Providence RI, American Mathematical Society, 2000. With an appendix by Wulf-Dieter Geyer.

[40] Carl Jockusch and Alexandra Shlapentokh. Weak presentations of computable fields. *J. Symbolic Logic*, **60**: 199–208, 1995.

[41] H. K. Kim and F. W. Roush. Diophantine undecidability of $\mathbb{C}(t_1, t_2)$. *J. Algebra*, **150**(1): 35–44, 1992.

[42] H. K. Kim and F. W. Roush. Diophantine unsolvability for function fields over certain infinite fields of characteristic p. *J. Algebra*, **152**(1): 230–239, 1992.

[43] H. K. Kim and F. W. Roush. An approach to rational Diophantine undecidability. In *Proc. Asian Mathematical Conf. (Hong Kong, 1990)*, pp. 242–248. River Edge NJ, World Science, 1992.

[44] H. K. Kim and F. W. Roush. Diophantine unsolvability over p-adic function fields. *J. Algebra*, **176**: 83–110, 1995.

[45] Helmut Koch. *Number Theory. Algebraic Numbers and Functions*. Providence RI, American Mathematical Society, 2000.

[46] Serge Lang. *Algebraic Number Theory*. Reading MA, Addison Wesley, 1970.

[47] Serge Lang. *Algebra*. Reading MA, Addison Wesley, 1971.

[48] Leonard Lipshitz. Undecidable existential problems for addition and divisibility in algebraic number rings, II. *Proc. Amer. Math. Soc.*, **64**(1): 122–128, 1977.

[49] Leonard Lipshitz. The diophantine problems for addition and divisibility. *Trans. Amer. Math. Soc.*, **235**: 271–283, 1978.

[50] Leonard Lipshitz. Undecidable existential problems for addition and divisibility in algebraic number rings. *Trans. Ameri. Math. Soc.*, **241**: 121–128, 1978.

[51] R. C. Mason. *Diophantine Equations over Function Fields*, volume 96 of *London Mathematical Society Lecture Notes*. Cambridge UK, Cambridge University Press, 1996.

[52] Yu. V. Matiyasevich. *Десятая проблема Гилберта*, volume 26 of *Математическая Логика и Основания Математики* (*Monographs in Mathematical Logic and Foundations of Mathematics*). Moscow, Nauka, 1993.

[53] Yuri V. Matiyasevich. *Hilbert's Tenth Problem*. Foundations of Computing Series. Cambridge MA, MIT Press, 1993. Translated from the 1993 Russian original by Yuri Matiyaseuch, with a foreword by Martin Davis.

[54] Barry Mazur. Rational points of abelian varieties with values in towers of number fields. *Invent. Math.*, **18**: 183–266, 1972.

[55] Barry Mazur. The topology of rational points. *Experimental Mathematics*, **1**(1): 35–45, 1992.

[56] Barry Mazur. Questions of decidability and undecidability in number theory. *J. Symbolic Logic*, **59**(2): 353–371, 1994.

[57] Barry Mazur. Speculation about the topology of rational points: an update. *Asterisque*, **228**: 165–181, 1995.

[58] Barry Mazur. Open problems regarding rational points on curves and varieties. In A. J. Scholl and R. L. Taylor, editors, *Galois Representations in Arithmetic Algebraic Geometry*. Cambridge UK, Cambridge University Press, 1998.

[59] Barry Mazur and Karl Rubin. Elliptic curves and class field theory. In *Proc. Inte. Congress of Mathematicians, (Beijing, 2002)*, volume 2, pp. 185–195. Beijing, Higher Education Press, 2002.

[60] Laurent Moret-Bailly. Elliptic curves and Hilbert's Tenth Problem for algebraic function fields over real and p-adic fields. *J. Reine und Angewandte Math.*, **587**: 77–143, 2005.

[61] Laurent Moret-Bailly. Groupes de Picard et problèmes de Skolem. I, II. *Ann. Sci. École Norm. Sup. (4)*, **22**(2): 161–179, 181–194, 1989.

[62] Laurent Moret-Bailly. Applications of local-global principles to arithmetic and geometry. In *Hilbert's Tenth Problem: Relations with Arithmetic and Algebraic Geometry (Ghent, 1999)*, pp. 169–186. Providence RI, American Mathematical Society, 2000.

[63] Laurent Moret-Bailly. Problèmes de Skolem sur les champs algébriques. *Compositio Math.*, **125**(1): 1–30, 2001.

[64] O. T. O'Meara. *Introduction to Quadratic Forms*. Berlin, Springer Verlag, 1973.

[65] Ju. G. Penzin. Undecidability of fields of rational functions over fields of characteristic 2. *Algebra i Logika*, **12**: 205–210, 244, 1973.

[66] Thanases Pheidas. An undecidability result for power series rings of positive characteristic. *Proc. Amer. Math. Soc.*, **99**(2): 364–366, 1987.

[67] Thanases Pheidas. An undecidability result for power series rings of positive characteristic. II. *Proc. Amer. Math. Soc.*, **100**(3): 526–530, 1987.

[68] Thanases Pheidas. Hilbert's tenth problem for a class of rings of algebraic integers. *Proc. Amer. Math. Soc.*, **104**(2): 611–620, 1988.

[69] Thanases Pheidas. Hilbert's tenth problem for fields of rational functions over finite fields. *Inventiones Mathematicae*, **103**: 1–8, 1991.

[70] Thanases Pheidas. An effort to prove that the existential theory of $\mathbb{Q}$ is undecidable. In Jan Denef, Leonard Lipshitz, Thanases Pheidas, and Jan Van Geel, editors, *Hilbert's Tenth Problem: Relations with Arithmetic and Algebraic*

Geometry, volume 270 of *Contemporary Mathematics*, pp. 237–252. Providence RI, American Mathematical Society, 2000.

[71] Thanases Pheidas and Karim Zahidi. Undecidability of existential theories of rings and fields: a survey. In Jan Denef, Leonard Lipshitz, Thanases Pheidas, and Jan Van Geel, editors, *Hilbert's Tenth Problem: Relations with Arithmetic and Algebraic Geometry*, volume 270 of *Contemporary Mathematics*, pp. 49–105. Providence RI, American Mathematical Society, 2000.

[72] Bjorn Poonen. Elliptic curves whose rank does not grow and Hilbert's Tenth Problem over the rings of integers. Private communication.

[73] Bjorn Poonen. Using elliptic curves of rank one towards the undecidability of Hilbert's Tenth Problem over rings of algebraic integers. In C. Fieker and D. Kohel, editors, *Algorithmic Number Theory*, volume 2369 of *Lecture Notes in Computer Science*, pp. 33–42. New York, Springer-Verlag, 2002.

[74] Bjorn Poonen. Hilbert's Tenth Problem and Mazur's conjecture for large subrings of $\mathbb{Q}$. *J. Amer. Math. Soc.*, **16**(4): 981–990, 2003.

[75] A. Prestel and J. Schmid. Existentially closed domains with radical relations: an axiomatization of the ring of algebraic integers. *J. Reine Angew. Math.*, **407**: 178–201, 1990.

[76] A. Prestel and J. Schmid. Decidability of the rings of real algebraic and p-adic algebraic integers. *J. Reine Angew. Math.*, **414**: 141–148, 1991.

[77] Mihai Prunescu. Defining constant polynomials. In Jan Denef, Leonard Lipshitz, Thanases Pheidas, and Jan Van Geel, editors, *Hilbert's Tenth Problem: Relations with Arithmetic and Algebraic Geometry*, volume 270 of *Contemporary Mathematics*, pp. 139–143. Providence RI, American Mathematical Society, 2000.

[78] M. Rabin. Computable algebra. *Trans. Amer. Math. Soc.*, **95**: 341–360, 1960.

[79] I. Reiner. *Maximal Orders*, volume 28 of *London Mathematical Society Monographs. New Series*. Oxford, Clarendon Press, 2003. Corrected reprint of the 1975 original, with a foreword by M. J. Taylor.

[80] Paulo Ribenboim. *The Theory of Classical Valuations*. Springer Monographs in Mathematics. New York, Springer-Verlag, 1999.

[81] Julia Robinson. Definability and decision problems in arithmetic. *J. Symbolic Logic*, **14**: 98–114, 1949.

[82] Julia Robinson. The undecidability of algebraic fields and rings. *Proc. Amer. Math. Soc.*, **10**: 950–957, 1959.

[83] Julia Robinson. On the decision problem for algebraic rings. In *Studies in Mathematical Analysis and Related topics*, pp. 297–304. Stanford CA, Stanford University Press, 1962.

[84] Hartley Rogers. *Theory of Recursive Functions and Effective Computability*. McGraw-Hill, New York, 1967.

[85] Robert Rumely. Undecidability and definability for the theory of global fields. *Trans. Amer. Math. Soc.*, **262**(1): 195–217, 1980.

[86] Robert S. Rumely. Arithmetic over the ring of all algebraic integers. *J. Reine Angew. Math.*, **368**: 127–133, 1986.

[87] Jean-Pierre Serre. Propriétés galoisiennes des points d'ordre fini des courbes elliptiques. *Invent. Math.*, **15**(4): 259–331, 1972.

[88] Jean-Pierre Serre. Quelques applications du théorème de densité de Chebotarev. *Inst. Hautes Etudes Sci. Publ. Math.*, **54**: 323–401, 1981.

[89] Jean-Pierre Serre. *Lectures on Mordell–Weil Theorem*. Braunschweig Wiesbaden, Vieweg, third edition, 1997.

[90] H. N. Shapiro. *Introduction to the Theory of Numbers*. New York, John Wiley and Sons, 1983.

[91] Alexandra Shlapentokh. Extension of Hilbert's tenth problem to some algebraic number fields. *Comm. Pure and Appl. Math.*, **XLII**: 939–962, 1989.

[92] Alexandra Shlapentokh. Hilbert's tenth problem for rings of algebraic functions of characteristic 0. *J. Number Theory*, **40**(2): 218–236, 1992.

[93] Alexandra Shlapentokh. Diophantine relations between rings of S-integers of fields of algebraic functions in one variable over constant fields of positive characteristic. *J. Symbolic Logic*, **58**(1): 158–192, 1993.

[94] Alexandra Shlapentokh. Diophantine classes of holomorphy rings of global fields. *J. Algebra*, **169**(1): 139–175, 1994.

[95] Alexandra Shlapentokh. Diophantine equivalence and countable rings. *J. Symbolic Logic*, **59**: 1068–1095, 1994.

[96] Alexandra Shlapentokh. Diophantine undecidability for some holomorphy rings of algebraic functions of characteristic 0. *Communi. Algebra*, **22**(11): 4379–4404, 1994.

[97] Alexandra Shlapentokh. Algebraic and Turing separability of rings. *J. Algebra*, **185**: 229–257, 1996.

[98] Alexandra Shlapentokh. Diophantine undecidability of algebraic function fields over finite fields of constants. *J. Number Theory*, **58**(2): 317–342, 1996.

[99] Alexandra Shlapentokh. Diophantine definability over some rings of algebraic numbers with infinite number of primes allowed in the denominator. *Invent. Math.*, **129**: 489–507, 1997.

[100] Alexandra Shlapentokh. Diophantine definability over holomorphy rings of algebraic function fields with infinite number of primes allowed as poles. *Int. J. Math.*, **9**(8): 1041–1066, 1998.

[101] Alexandra Shlapentokh. Defining integrality at prime sets of high density in number fields. *Duke Math. J.*, **101**(1): 117–134, 2000.

[102] Alexandra Shlapentokh. Hilbert's tenth problem for algebraic function fields over infinite fields of constants of positive characteristic. *Pacific J. Math.*, **193**(2): 463–500, 2000.

[103] Alexandra Shlapentokh. Defining integrality at prime sets of high density over function fields. *Monatshefte für Mathematik*, **135**: 59–67, 2002.

[104] Alexandra Shlapentokh. Diophantine undecidability of function fields of characteristic greater than 2 finitely generated over a field algebraic over a finite field. *Compos. Math.*, **132**(1): 99–120, 2002.

[105] Alexandra Shlapentokh. On diophantine decidability and definability in some rings of algebraic functions of characteristic 0. *J. Symbolic Logic*, **67**(2): 759–786, 2002.

[106] Alexandra Shlapentokh. On diophantine definability and decidability in large subrings of totally real number fields and their totally complex extensions of degree 2. *J. Number Theory*, **95**: 227–252, 2002.

[107] Alexandra Shlapentokh. Existential definability with bounds on archimedean valuations. *J. Symbolic Logic*, **68**(3): 860–878, 2003.

[108] Alexandra Shlapentokh. A ring version of Mazur's conjecture on topology of rational points. *Int. Math. Res. Not.*, **2003**(7): 411–423, 2003.

[109] Alexandra Shlapentokh. Diophantine undecidability for some function fields of infinite transcendence degree and positive characteristic. *Zapiski Seminarov POMI*, **304**: 141–167, 2003.

[110] Alexandra Shlapentokh. On diophantine definability and decidability in some infinite totally real extensions of $\mathbb{Q}$. *Trans. Amer. Math. Soc.*, **356**(8): 3189–3207, 2004.

[111] Alexandra Shlapentokh. Elliptic curves retaining their rank in finite extensions and Hilbert's tenth problem. To appear in *Trans. Amer. Math. Soc.*

[112] Alexandra Shlapentokh. Rational separability of the integral closure. Preprint.

[113] Joseph Silverman. *The Arithmetic of Elliptic Curves*. New York, Springer-Verlag, 1986.

[114] Joseph Silverman. *Advanced Topics in the Arithmetic of Elliptic Curves*. New York, Springer-Verlag, 1994.

[115] Lou van den Dries. Elimination theory for the ring of algebraic integers. *J. Reine Angew. Math.*, **388**: 189–205, 1988.

[116] Lou van den Dries and Angus Macintyre. The logic of Rumely's local–global principle. *J. Reine Angew. Math.*, **407**: 33–56, 1990.

[117] Carlos Videla. Hilbert's tenth problem for rational function fields in characteristic 2. *Proc. Amer. Math. Soc.*, **120**(1): 249–253, 1994.

[118] I. M. Vinogradov. *The Method of Trigonometrical Sums in the Theory of Numbers*. London and New York, Interscience, 1954. Translated, revised and annotated by K. F. Roth and Anne Davenport.

[119] André Weil. *Basic Number Theory*. New York, Springer-Verlag, 1974.

[120] Karim Zahidi. The existential theory of real hyperelliptic fields. *J. Algebra*, **233**(1): 65–86, 2000.

[121] Karim Zahidi. Hilbert's tenth problem for rings of rational functions. *Notre Dame J. Formal Logic*, **43**: 181–192, 2003.

Index